Global Energy
Interconnection

Global Energy Interconnection

Zhenya Liu

AMSTERDAM • BOSTON • HEIDELBERG • LONDON
NEW YORK • OXFORD • PARIS • SAN DIEGO
SAN FRANCISCO • SINGAPORE • SYDNEY • TOKYO

Academic Press is an Imprint of Elsevier

Academic Press is an imprint of Elsevier
125, London Wall, EC2Y 5AS, UK
525 B Street, Suite 1800, San Diego, CA 92101-4495, USA
225 Wyman Street, Waltham, MA 02451, USA
The Boulevard, Langford Lane, Kidlington, Oxford OX5 1GB, UK

Notices
Knowledge and best practice in this field are constantly changing. As new research and experience broaden our understanding, changes in research methods, professional practices, or medical treatment may become necessary.

Practitioners and researchers must always rely on their own experience and knowledge in evaluating and using any information, methods, compounds, or experiments described herein. In using such information or methods they should be mindful of their own safety and the safety of others, including parties for whom they have a professional responsibility.

To the fullest extent of the law, neither the Publisher nor the authors, contributors, or editors, assume any liability for any injury and/or damage to persons or property as a matter of products liability, negligence or otherwise, or from any use or operation of any methods, products, instructions, or ideas contained in the material herein.

British Library Cataloguing-in-Publication Data
A catalogue record for this book is available from the British Library

Library of Congress Cataloging-in-Publication Data
A catalog record for this book is available from the Library of Congress

ISBN: 978-0-12-804405-6

For information on all Academic Press publications
visit our website at http://store.elsevier.com/

Typeset by Thomson Digital

Printed and bound in the United States and Singapore

www.elsevier.com • www.bookaid.org

Publisher: Jonathan Simpson
Acquisition Editor: Simon Tian
Editorial Project Manager: Naomi Robertson
Production Project Manager: Jason Mitchell
Designer: Victoria Pearson Esser

Contents

Foreword

Energy provides an important physical foundation of socioeconomic development. Mankind's exploitation of energy has evolved over time, from firewood to fossil fuels, such as coal, oil, and natural gas, and then to clean energy resources derived from hydropower, wind power, and solar power. Each evolution has seen tremendous progress in productivity and human civilization. As a motive power for modernization, energy is closely associated with the national economy as well as people's livelihood and human well-being. Countries all over the world now share a strategic goal to expedite the development of a safe, reliable, economical, efficient, clean, and environment-friendly modern energy supply system by capitalizing on a new round of energy evolution.

Over the past three centuries since the advent of industrialization, the global energy industry has witnessed rapid growth, vigorously bolstering the global economy and social development. At the same time, overdevelopment of conventional fossil energy resulted in a host of problems, such as insufficient resources, environmental pollution, and climate change, which are threatening human existence and sustainable development. Global fossil fuel resources are insufficient and their distribution and consumption imbalanced. As a result, energy development is increasingly controlled by a minority of countries and regions. Some resource-scarce countries are depending more and more on imported energy supply and facing highly pressing energy security concerns. Besides, the use of fossil energy has caused severe pollution and damage to air, water, and soil during the production, transportation, and utilization processes. Carbon dioxide emissions from fossil fuel combustion have become a significant factor contributing to global warming, glacial melting, and rising sea levels. In the foreseeable future, energy demand will continue to grow and the conventional energy development model based on fossil fuels is difficult to sustain amid the expanding global economy and increasing world population.

Seeking a solution to these energy and environmental concerns with the objective of removing the bottleneck in socioeconomic development is now a matter of utmost urgency. In a world abundant with clean energy resources, hydropower, onshore wind energy and solar power amount to 10,000, 1,000,000, and 100,000,000 GW respectively, far more than enough to meet global energy needs. Since the twenty-first century, the development of an energy structure centered on electricity and characterized by clean energy has gathered momentum. Massive development and utilization of clean energy such as wind and solar power has become a common option among the world's major nations. Supported by the advancement of technology and the application of new materials, the development of wind energy, solar power, marine energy, and other clean energy sources has become more efficient, resulting in stronger technical economies and market competitiveness. These alternative energy sources will probably become the world's dominant form of energy. Most of the clean energy sources can be used in a highly efficient way only after transformation into electric power, which, as a quality, clean, and efficient form of secondary energy, can meet the great majority of energy demand and may therefore become the most important energy for end-users in the future. From the perspective of global energy development trends and resource endowment, the pursuit of a "two-replacement" strategy that encompasses clean energy and electricity as two energy substitutes, is an important direction of development for global energy sustainability.

Marking a fundamental reform in the conventional models and concepts of energy production and consumption, the "two-replacement" strategy requires us to develop a global energy view to address energy concerns with a global, historical, forward-looking, and systematic approach. We are encouraged to step up the construction of a global energy interconnection for larger-scale development, allocation, and more efficient utilization of clean energy. The ultimate objective is to coordinate energy development in line with progress on the political, economic, social, and environmental fronts, and secure a safe, clean, efficient, and sustainable energy supply. With UHV grids as its backbone, such a global energy interconnection transmits primarily clean energy, connects large clean energy bases with various distributed power generation facilities, and delivers clean energy to different types of end-users. It is a widespread, highly deployable, safe, reliable, green, and low-carbon global energy distribution platform. In energy development, the substitution of clean energy for fossil fuels based on a global energy interconnection is becoming a viable low-carbon and green solution with the focus being shifted from fossil fuels to clean energy. In energy consumption, the substitution of electricity for coal and oil and the growing popularity of electric boilers, electric heaters, electric coolers, electric cookers, and electrified transport should be encouraged to increase the share of electricity in the energy end-use sector and solve the pollution and GHG emission problems caused by fossil fuels.

At the heart of the "two-replacement" strategy is the global energy interconnection, which works as an important platform on which to develop clean energy efficiently and ensure a reliable energy supply to the world. It will bring about comprehensive adjustments to different areas of the energy sector, including development strategy, development roadmap, structural layout, means of production and consumption, and energy technology. From a global energy perspective, the author of the book analyzes the status quo and challenges of global energy development and the inevitability of what he describes as "two replacements". By building on a projection of future electricity supply and demand, the concept of developing a global energy interconnection using UHV AC/UHV DC grid technology and smart grid technology is put forward and the relevant strategic thinking, overall objective, basic planning, structuring method and development roadmap are dealt with. A brand new solution to promote safe, clean, efficient, and sustainable energy development worldwide is offered.

The book consists of eight chapters:

Chapter 1 analyses the current distribution and development of global energy resources, together with a summary of the major challenges now facing global energy development.

Chapter 2 introduces the development trends of global wind power and solar power generation, and demonstrates the inevitability of the "two-replacement" strategy and its importance to an energy revolution.

Chapter 3 deals with the trends in world energy development and suggests a global view of energy development. It points to the need to explore and address global energy concerns from a global, historical, differentiated, and open perspective.

Chapter 4 analyzes the main factors that influence electricity supply and demand and predicts total global energy demand and the energy structure involved. The chapter also describes the focus and pattern of global energy development in the future and presents an outlook on global power demand and electricity flows.

Chapter 5 explains the correlation between the global energy interconnection and the robust smart grid. In conceptualizing the development of a global energy interconnection and its framework, the chapter studies plans for the intercontinental UHV grid, the intracontinental UHV grid, and the ubiquitous national smart grid. It also explores the establishment of a mechanism for supporting cooperation in global energy interconnection. In addition, a preliminary assessment of the overall benefits of global energy interconnection has been conducted.

Chapter 6 expounds on the significance and the important fields of technological innovation in global energy interconnection. It highlights the newest breakthroughs, direction of development, and prospects of application in such important technological fields as power sources, grids, energy storage, and information communication.

Chapter 7 touches on the foundation of global energy interconnection research and application. The focus is on the progress, locally and internationally, in such fields as UHV grids, smart grids, clean energy and megagrid networking in terms of technical research, standard formulation, planning development, and project construction.

Chapter 8 presents an outlook on the significant impact of the global energy interconnection on the future world, from the four aspects of energy, economy, society, and civilization.

As a major infrastructure system with both short-term and long-term interests, the global energy interconnection plays a global and strategic leading role in the world's energy sustainability. It is in line with the common interests of all mankind. In a few decades to come, the global energy interconnection will enter a growth period of critical importance and may require proactive involvement of all governments, international organizations, social groups, and energy developers. Efforts should be stepped up in theoretical research, communication of ideas, research and development, and international cooperation with the ultimate aim of developing a global energy interconnection as a bulwark for socioeconomic development.

In this book, the author draws on his years of reflections on China's and the world's energy development strategies to elaborate on the concepts of and thinking behind the global energy interconnection, integrating in particular China's successful application of UHV grid technology and with reference to the research findings of local and foreign academics and organizations. By doing so, the author hopes to contribute to global energy strategy research and policy making.

Author
January 2015

About the Book

This book discusses the strategic issues regarding sustainable development of the world's energy resources from a global perspective and offers solutions. It analyzes the current distribution and development of the world's fossil fuels and clean energy resources at the same time as it reveals the critical challenges the world is facing in relation to energy development. The book also dissects the trends in the development of clean energy and electricity as fuel substitutes. The author postulates the study and resolution of the world's energy problems from a perspective that is global, historical, respectful of differences, and open-minded. He goes on to describe the concept of building up a global network of energy interconnections to achieve the sustainable development of energy. Based on his analyses of the world's electricity demand and supply, and his examination of the development of major energy bases including those in the Arctic and equatorial regions, the writer systematically describes a plan for constructing a global energy interconnection, the roadmap for its implementation, the innovative technologies involved, and the engineering practices required. This book presents an outlook on the development of a global energy interconnection that the whole world can tap into.

My heartfelt thanks go to the readers of this book, for your encouragement and support.

GLOBAL ENERGY DEVELOPMENT: THE REALITY AND CHALLENGES

1

CHAPTER OUTLINE

1 GLOBAL ENERGY DEVELOPMENT: THE REALITY

Global energy development has gone through a course of evolution from firewood to coal and further to oil, gas, and electricity. Currently, world energy supply is dominated by fossil fuels as a gigantic motive force for economic development. Meanwhile, hydro, wind, solar power, and other clean energy alternatives are being developed and applied at increasingly high speeds to accommodate future energy demands, thereby playing an increasingly significant part in ensuring security of global energy supply and promoting clean energy.

1.1 BACKGROUND

Total world energy consumption has maintained a long-standing growing trend with constant adjustments to the energy structure. In the mid-nineteenth century, firewood was the primary source of energy consumed by humans, compared to coal that accounted for a less than 20% share of total energy consumption. With the progress of the Industrial Revolution, the proportion of coal consumption soared significantly to more than 70% till the beginning of the twentieth century. In the twentieth century, the share of coal consumption plummeted along with the growing popularity of oil and natural gas. In the 1960s, oil surpassed coal as the most widely used energy source. The proportion of oil consumption peaked in 1973 before gradually falling after the two global oil crises from the 1970s to the 1980s. In the meantime, natural gas consumption rose constantly, while coal consumption rebounded slightly.

Global Energy Interconnection. http://dx.doi.org/10.1016/B978-0-12-804405-6.00001-4

It is most notable that profound changes in the global energy structure in the recent two decades have brought about a new pattern marked by equal predominance of coal, oil, and natural gas as well as rapid development of clean energy. See Fig. 1.1 for the changes in the composition of world energy consumption since 1850.

1.1.1 Energy Resources

Global energy resources include primarily fossil fuels (e.g., coal, oil, and natural gas) and clean energy (e.g., hydro,[1] wind, solar, and marine energy). *Despite its massive fossil energy resources, the world is facing many practical problems, such as serious resource depletion and waste emissions, which are the legacy of large-scale exploration since the Industrial Revolution a couple of centuries ago. By contrast, clean energy is abundant, low-carbon, environment-friendly and renewable, with huge potential for future development.*

By 2013, the world's remaining proven recoverable reserves of coal, oil, and natural gas were estimated at 891.5, 238.2 billion tons, and 186 trillion m^3, respectively, accounting for 52.0, 27.8, and 20.2%, respectively, of a total of 1.2 trillion tons of standard coal.[2] Based on the current average mining intensity, the global reserves of coal, oil, and natural gas can sustain 113, 53, and 55 years, respectively. Distribution of these fossil fuels is extremely unbalanced on a global basis. Ninety-five percent of coal is distributed in Europe, the Eurasian continent, Asia-Pacific, and North America (Fig. 1.2); 80% of oil is distributed in the Middle East, North, South, and Central America; and Europe, the Eurasian

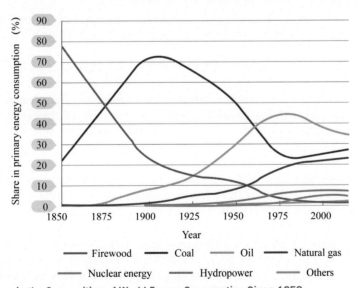

FIGURE 1.1 Changes in the Composition of World Energy Consumption Since 1850

Source: Ref. [23].

---

[1]Broadly speaking, hydropower includes river energy and marine energy. In this publication, hydropower refers to river energy.
[2]One ton of raw coal is equivalent to 0.714 tons of standard coal, one ton of crude oil is equivalent to 1.43 tons of standard coal, and 1000 m^3 of natural gas is equivalent to 1.33 tons of standard coal.

FIGURE 1.2 Global Distribution of Coal, Oil, and Natural Gas Resources

Note: Regional classification in the diagram is based on British Petroleum (BP) statistical standards. Data in the subsequent text follow the same regional classification as long as they are sourced from BP.

continent, and the Middle East are home to over 70% of natural gas reserves. See Table 1.1 for the distribution of coal, oil, and natural gas resources in the world. Coal dominates China's fossil energy structure,[3] while oil and natural gas resources are comparatively scarce. The remaining proven recoverable reserves of fossil fuel in China amount to about 89.6 billion tons of standard coal, including coal (91.2), oil (3.9), and natural gas (4.9%). The reserves-to-production ratio (R/P ratio) of the three fuel types is 31, 12, and 28 years, respectively.[4]

The earth is endowed with various forms of clean energy, such as hydro, wind, and solar power. Based on World Energy Council (WEC) estimates, the theoretical developable potential of clean energy worldwide surpasses 150,000,000 TWh a year, amounting to 45 trillion tons of standard coal (coal consumption rate: 300 g/kWh) or 38 times the remaining proven recoverable reserves of fossil energy on earth. Clean energy resources are distributed very unevenly. Water resources are distributed primarily in the drainage basins of Asia, South America, North America, and Central Africa. Wind resources are distributed mainly in the Arctic, Central and Northern Asia, Northern Europe, Central North America, and East Africa. To a lesser extent, quality wind resources are also found in the near-shore regions of each continent. Solar energy resources are distributed primarily in North Africa, East Africa, the Middle East, Oceania, Central and South America, and other regions near the Equator. Besides, other areas of arid climate, like the Gobi and other deserts, are also endowed with quality solar

---

[3]Unless otherwise specified herein, the term "China" in this publication does not include Hong Kong, Macau, and Taiwan.
[4]The R/P ratio refers to the ratio of the remaining proven reserves in any year to production in that year, namely, the number of years over which the reserves could be recovered at the current production rate.

Table 1.1 Global Distribution of Coal, Oil, and Natural Gas Resources

Regions	Coal			Oil			Natural Gas		
	Remaining Proven Recoverable Reserves (billion tons)	Percentage	R/P ratio (year)	Remaining Proven Recoverable Reserves (billion tons)	Percentage	R/P ratio (year)	Remaining Proven Recoverable Reserves (trillion m^3)	Percentage	R/P ratio (year)
North America	245.1	27.5	250	35.0	13.6	37	12	6.3	13
Central and South America	14.6	1.6	149	51.1	19.5	>100	8	4.1	44
Europe and the Eurasian Continent	310.5	34.8	254	19.8	8.8	23	57	30.6	55
Middle East	1.1	0.1	>500	109.4	47.9	78	80	43.2	>100
Africa	31.8	3.6	122	17.3	7.7	41	14	7.6	70
Asia-Pacific	288.4	32.4	54	5.6	2.5	14	15	8.2	31
Total	891.5	100	113	238.2	100	53	186	100	55

Source: Ref. [65].

Table 1.2 Global Distribution of Hydro, Wind, and Solar Energy Resources

Regions	Hydropower		Wind Energy		Solar Energy	
	Theoretical Reserves (TWh/year)	Percentage	Theoretical Reserves (TWh/year)	Percentage	Theoretical Reserves (TWh/year)	Percentage
Asia	18,000	46	500,000	25	37,500,000	25
Europe	2,000	5	150,000	8	3,000,000	2
North America	6,000	15	400,000	20	16,500,000	11
South America	8,000	21	200,000	10	10,500,000	7
Africa	4,000	10	650,000	32	60,000,000	40
Oceania	1,000	3	100,000	5	22,500,000	15
Total	39,000	100	2000,000	100	150,000,000	100

Source: Refs. [77] and [84].

resources. Mostly concentrated in sparsely populated areas several hundred to thousands of kilometers away from population and production centers, clean energy cannot be explored and utilized without the capability of allocation over vast areas. See Table 1.2 for the global distribution of hydro, wind, and solar energy resources.

1.1.2 Energy Consumption

Global energy consumption has maintained a growth momentum on an aggregate and per capita basis. From 1965 to 2013, world population growth, industrialization, urbanization, and numerous other factors led to a sharp rise in annual primary energy consumption from 5.38 billion tons of standard coal to 18.19 billion tons of standard coal (or approximately 19.5 billion tons of standard coal if noncommercial energy is included), registering a 2.4-fold increase in about half a century or average annual growth of 2.6%. At the same time, per capita annual energy use rose from 2.1 tons of standard coal to 2.6 tons of standard coal, representing a 23.8% increase or an average annual growth rate of 0.4%.

The Asia-Pacific region is gradually becoming the world's largest energy consumer in terms of total consumption and demand growth. Due to the shift of industries and the changing demographic structure, the world's developed countries registered a decreasing trend in primary energy demand, whereas developing countries registered an increasing trend. From 1965 to 2013, the share of the Asia-Pacific in global energy consumption rose from 11.7% to 40.5%, representing an annual growth rate of 5.2% and making this region, the fastest-growing energy consumer in the recent five decades. Of the increase of 12.81 billion tons of standard coal in global primary energy consumption, 52.6% came from Asia-Pacific. Since 2003, Asia-Pacific has surpassed North America and Europe in terms of total energy use, ranking as the largest energy consumer in the world. See Fig. 1.3 for changes in the composition of world primary energy consumption during 1965–2013.

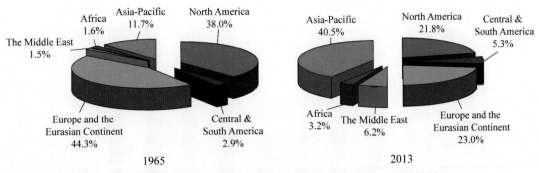

FIGURE 1.3 Changes in the Composition of Global Primary Energy Consumption, 1965–2013

Source: Ref. [65].

Since the implementation of its reform and policies of opening up, China has maintained continued progress in economic growth and national quality of life. This drives energy consumption year by year such that China has now replaced the US as the largest energy consumer in the world. China's total annual energy consumption soared from 600 million tons of standard coal to 3.75 billion tons of standard coal, registering average annual growth of 5.5% or 2.8 times the world's average annual growth in the same period. Average per capita energy consumption climbed from 0.6 tons of standard coal to 2.8 tons of standard coal, representing an increase from 26% of the global average to 104% of the global average.

Although the world's energy consumption has been dominated by fossil energy, the share of fossil energy in global consumption is declining gradually. During 1965–2013, global annual fossil energy consumption soared from 5.05 billion tons of standard coal to 15.75 billion tons of standard coal, registering a 2.1-fold increase and annual growth of 2.3%. The share of fossil fuels in primary energy consumption fell by approximately 7.6 percentage points from 94.3% to 86.7%. See Fig. 1.4 for the world's total primary energy consumption and the share of fossil energy in total consumption during 1965–2013.

The share of electric power in terminal energy consumption is gradually increasing. With higher levels of electrification, the share of fossil power in the world's terminal energy consumption is falling as ever-larger quantities of fossil fuels (e.g., coal and gas) are being transformed into electricity. From 1973 to 2012, the shares of coal and oil in the world's terminal energy consumption fell by 3.6 and 7.5 percentage points, respectively, while the share of electricity increased from 9.4% to 18.1%, second only to oil. In 2012, the share of electricity exceeded 20% to reach 22.6% of China's total terminal energy consumption, higher than the world average but still lower than Japan and some other highly electrified countries. See Fig. 1.5 for the change in the world's energy end-use structure during 1973–2012.

1.1.3 Energy Production

Currently, world energy production is rising steadily; fossil energy production is going up gradually, and clean energy is developing leaps and bounds. Since the age of industrialization, fossil energy has been fuelling the growth of the global economy. In fossil energy production, oil plays the most important role, followed by coal and natural gas. From 1980 to 2013, world annual oil production increased by 33.7% from 3.09 billion tons to 4.13 billion tons, registering annual growth of 0.9%. While the Middle

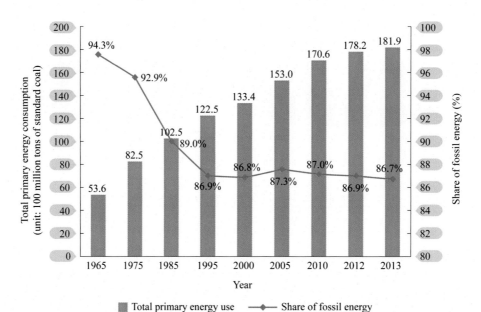

FIGURE 1.4 The World's Total Primary Energy Consumption and the Share of Fossil Energy, 1965–2013

Source: Ref. [65].

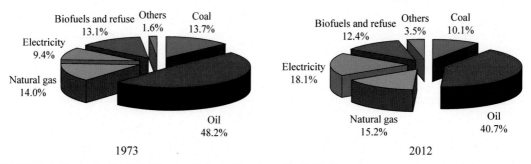

FIGURE 1.5 Change in the Global Energy End-use Structure, 1973–2012

Source: IEA, Key World Energy Statistics 2014.

East and Africa were playing a more prominent role in global oil production, North America showed an opposite trend. Annual gas production increased from 1.4 trillion m^3 to 3.4 trillion m^3, registering a 1.3-fold increase or average annual growth of 2.6%, as Europe and the Eurasian continent became the most important gas producer. Annual coal production jumped from 3.84 billion tons to 7.90 billion tons, registering a 1.1-fold increase or average annual growth of 2.3%, as the Asia-Pacific became the world's largest producer of coal. Development of wind, solar and other clean energy alternatives has experienced rapid growth in the twenty-first century. From 2000 to 2013, the installed capacity of wind farms and solar farms in the world went up from 17400 MW to 320 GW and from 1640 MW to 140 GW, respectively, registering a 17-fold increase (annual growth rate: 24.8%) and a 111-fold increase (annual growth rate: 43.7%), respectively. However, starting from a small base, wind, solar, and other types of nonhydro renewables accounted for a still relatively limited share at just 2.2% of the world's total primary energy production. See Fig. 1.6 for the world's primary energy production structure in 2013.

1.1.4 Energy Trading

Taking place primarily in the fossil fuels sector, global energy trading is rising steadily on a total volume basis. The distribution of fossil energy production and consumption is highly imbalanced, requiring the capability to optimize allocations of energy resources across the world. Transnational and intercontinental energy trade flows have been expanding increasingly along with the development and improvement of energy transport networks, including ocean transport, railway, and oil/gas transmission networks. In 2013, transcontinental fossil energy trade flows globally amounted to 6.3 billion tons of standard coal, with oil, gas, and coal accounting for 63, 22, and 15%, respectively. Fig. 1.7 shows the change in the trade volume of different fossil energies as a percentage of global consumption between

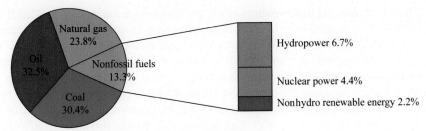

FIGURE 1.6 World's Primary Energy Production Structure, 2013

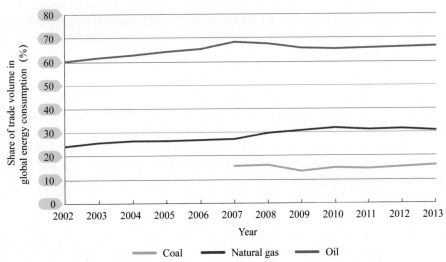

FIGURE 1.7 Change in Trade Volume of Different Fossil Energies as a Percentage of Total Global Consumption, 2002–2013

Note: Statistics of global coal trading volume are not available for the period 2002–2007.

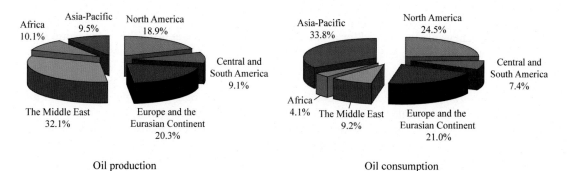

FIGURE 1.8 Distribution of Global Oil Production and Consumption, 2013

Source: Ref. [65].

2002 and 2013. As shown, the trade volumes of oil, gas, and coal accounted for 66.4, 31.9, and 17.1% of global consumption, respectively. Due to grid transmission capacity constraints, electric power is geared mainly toward achieving a balance at the local and regional levels, while transnational and transcontinental trade operates on a small scale. In terms of calorific value equivalents, transnational and transcontinental electricity trade accounted for only 1.3% of global fossil energy trade.

Currently, oil trade accounts for the largest share of global energy trade volume. In 2013, the Middle East and Russia produced 45% of the world's oil[5], but accounted for only 12.9% of global consumption. By comparison, North America, Europe, and Asia-Pacific produced 35.8% of the world's oil, while accounting for 75.6% of global consumption. See Fig. 1.8 for global oil production and consumption in 2013.

Global energy production and consumption still depends primarily on fossil energy, while the share of clean energy and electricity is increasing relatively rapidly. Imbalanced energy distribution leads to a widening gap between demand and supply, although global energy trade keeps growing.

1.2 FOSSIL ENERGY

Fossil energy refers primarily to coal, oil, natural gas, and other nonrenewable energy alternatives formed from organic remains over hundreds of millions of years ago. Since the first Industrial Revolution, fossil energy has been a pillar of modern and contemporary industrial development. Global fossil energy consumption is growing and exhibiting a trend marked by structural optimization and expansion of long-distance distribution. Coal, oil, natural gas, and other forms of fossil energy account for over 80% of global primary energy consumption.

---

[5]Given that the territory of Russia spans across Asia and Europe, this publication categorizes Russia under "Europe and the Eurasian Continent" as per BP's classification method. As Russia's hydropower and wind energy resources are concentrated primarily in Asia, it is categorized under "Asia" for clean energy discussion purposes. As around 90% of Russia's electric energy resources are distributed in Europe, Russia is categorized under "Europe" for the purpose of discussing future electricity demand. As Russia has a vast Asian territory, it is categorized under "Asia" for the purpose of discussing the current situation of power networks.

1.2.1 Coal

As the first fossil energy developed on a large scale by humans, coal has been used as a fuel for more than three millenniums till date. In the late eleventh century, coal began to be utilized as a building material and a metallurgy-purpose fuel. In the 1780s, James Watt invented an improved steam engine that contributed to large-scale exploration and consumption of coal and brought about the first Industrial Revolution, with the rapid growth of the textiles, steel, machinery, and railway industries driven by mechanization to propel the world into the Steam Age. By the end of the nineteenth century, coal became the world's predominant energy option but its share in global consumption had since fallen somewhat. Until the mid-twentieth century, coal had accounted for the largest share of the world energy structure. In recent years, the share of coal in the world energy structure has decreased slightly, although coal exploration and utilization on a global level keeps growing all the time.

Coal is the most abundant fossil energy resource on earth. In 2013, global remaining proven recoverable reserves of coal amounted to 891.5 billion tons and, in terms of calorific value, represented 1.8 and 2.5 times the remaining proven recoverable reserves of oil and natural gas, respectively. Europe and the Eurasian Continent are richest in coal deposits, totaling 310.5 billion tons or approximately 34.8% of the world's total reserves. The second on the list is the Asia-Pacific, totaling 288.4 billion tons or 32.4% of the world's total reserves. North America is also rich in coal resources, totaling 245.1 billion tons or approximately 27.5% of the world's total. By contrast, Central and South America, the Middle East, and Africa have only very limited coal resources, totaling merely 5.3% of the world's total. See Fig. 1.9 for the regional distribution of the world's remaining proven recoverable reserves of coal in 2013.

Coal production is continuing to grow as the Asia-Pacific has become the world's most important coal producer. Global coal production in 2013 was 7.9 billion tons, up 100% from 1980 or representing annual growth of 2.3%. Currently, coal production is concentrated primarily in the Asia-Pacific, North America, Europe, and the Eurasian continent. From 1980 to 2013, the share of the Asia-Pacific in global coal production rose from 26.7% to 68.8%. By contrast, other coal production areas experienced different levels of decrease. As a share of world total, production in North America dropped from 26.3% to 14.1%, and production in Europe and the Eurasian Continent declined from 42.4% to 11.6%. Since 1985, China has replaced the United States as the largest coal producer in the world, turning out 3.68 billion tons in 2013 alone or approximately half of the world's total production. See Fig. 1.10 for the change in the regional distribution of the world's coal production during 1980–2013.

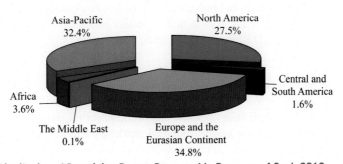

FIGURE 1.9 Global Distribution of Remaining Proven Recoverable Reserves of Coal, 2013

Source: Ref. [65].

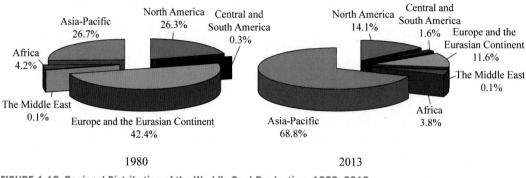

FIGURE 1.10 Regional Distribution of the World's Coal Production, 1980–2013

Source: Ref. [65].

Coal consumption is on an increase in overall terms, albeit falling as a share of total energy consumption. World coal consumption climbed from 2.58 billion tons of standard coal in 1980, to 5.47 billion tons of standard coal in 2013, representing annual growth of about 2.3% and basically in proportion to the growth in the world's total energy consumption. However, from the second half of the twentieth century, coal, as a percentage of world primary energy consumption, began to fall from 38.1% in 1965 to 30.1% in 2013, representing an 8 percentage point decrease. From the 1980s, the world's new economies, represented by China and India, entered a stage of fast economic growth, supporting a rapid increase in coal consumption and stalling the otherwise falling trend of coal as a percentage of energy consumption.

International coal trade is predominantly sea based. Due to the high delivery costs involved and concerns about environmental pollution, coal is consumed primarily in producing countries and within a limited geographical scope, while long-distance trade takes place on a relatively small scale. Typically, international coal trade among neighboring countries is conducted primarily through sea transportation except for a few interconnected inland countries and regions where coal is transported by rail or highway. Global coal trade in 2013 came to around 1.33 billion tons or 17% of global coal production, with sea transportation accounting for 90% of the trade volume.[6] Today's international coal market is divided roughly into two regional markets – Asia-Pacific and the Atlantic across the United States and Europe. In the Asia-Pacific, exporters include primarily Australia and Indonesia, and importers include primarily China, Japan, South Korea, and India. In the United States/European Atlantic market, exporters include primarily South Africa and Russia and importers include primarily the United Kingdom, France, and Germany. See Fig. 1.11 for world coal trade flows.

1.2.2 Oil

Oil is the predominant energy source supporting the modern industrial system. Humans began to develop and utilize oil in the nineteenth century. After the world's first oil well was drilled and operated in the state of Pennsylvania in 1859, the United States became one of the major oil producers and consumers in the early years. Later, the Soviet Union also began drilling oil, marking the nascent development of the modern oil industry. With the wide application of internal combustion engines, the demand for fuel oil soared and

---

[6]Source: World Coal Association (WCA), World Coal Association Statistics 2014.

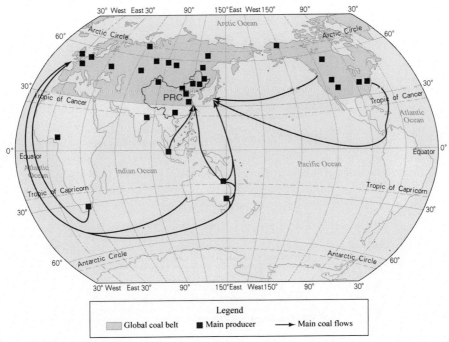

FIGURE 1.11 World Coal Trade Flows

some countries commenced oil development and refining on a large scale, leading to skyrocketing production. After the 1920s, oil began to be widely used and after the 1940s, major developed nations began to shift the focus of energy consumption from coal to oil. In the 1960s, oil surpassed coal as a share of energy consumption to become the world's predominant energy option. In the 1990s, oil accounted for more than 40% of world primary energy consumption. It can be said that world energy development entered into the Oil Age after the mid-twentieth century. Thanks to the development of the oil industry and the discovery and application of electricity, the second Industrial Revolution ensued and drove significant growth in the transportation, chemical engineering, electrical engineering, automobiles, and electric appliance industries.

Global distribution of oil resource is highly unbalanced. The Middle East, Central and South America, and North America are richest in oil resources. As at the end of 2013, the world's remaining proven recoverable oil reserves were estimated at 238.2 billion tons, concentrating mainly on the Middle East (approximately 109.4 billion tons, or 47.9% of the world total). Central and South America had 51.1 billion tons of remaining proven recoverable oil reserves (19.5%), compared to North America's 35 billion tons (13.6%). Together, Europe, the Eurasian continent, Africa, and Asia-Pacific accounted for a relatively small share (only 19.0%). See Fig. 1.12 for the global distribution of remaining proven recoverable oil reserves as at the end of 2013.

On the whole, global oil production maintains steady growth as the world economy develops quickly and oil demand keeps rising. For nearly half a century, global oil production has been rising steadily, reversed only in the two oil crisis periods (i.e., 1973–1974 and 1979–1980). From 1965 to 1980, the world's annual oil production rose from 1.57 billion tons to 3.09 billion tons, representing a onefold increase or average annual growth of 4.6%. Since the mid-1980s, global oil production has experienced

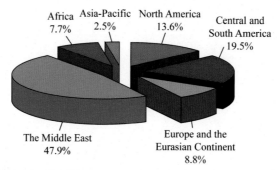

FIGURE 1.12 Global Distribution of Remaining Proven Recoverable Oil Reserves as at End of 2013

Source: Ref. [65].

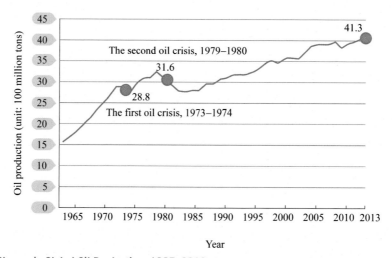

FIGURE 1.13 Change in Global Oil Production, 1965–2013

a marked slowdown. Between 1980 and 2013, world production registered average annual growth of a mere 0.9%, amounting to 4.13 billion tons in 2013. See Fig. 1.13 for the change in global oil production from 1965 to 2013.

The Middle East, together with Central and South America, is playing an increasingly important role in global oil production, while production in Europe and North America tends to decline. During 1980–2013, as a share of global total, production in the Middle East increased from 30.2% to 32.1%, Central and South America increased from 6.4% to 9.1%, and North America decreased from 21.7% to 18.9%. The three largest oil producers were Saudi Arabia, Russia, and the United States, which produced 13.1, 12.9, and 10.8%, respectively, of the world total in 2013. As the fourth largest oil producer in the world, China has maintained steady growth in oil production. China's production in 2013 totaled to 208 million tons, approximately 5% of world oil production. However, due to resource constraints, annual oil production in China is close to peak levels, with limited growth potential in the future. See Fig. 1.14 for the change in oil production of different producing regions during 1980–2013.

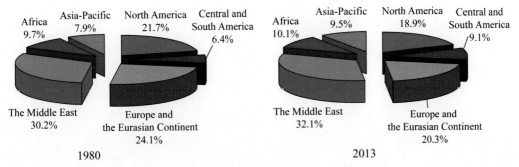

FIGURE 1.14 Change in Oil Production of Different Producing Regions, 1980–2013

Source: Ref. [65].

Asia-Pacific is becoming the center of consumption, with year-on-year growth of global oil consumption. During 1965–2013, global oil consumption leaped from 1.53 billion tons to 4.19 billion tons, representing a 1.7-fold crisis or an average annual growth rate of approximately 2.1%. Oil consumption in the Asia-Pacific jumped from 10.8% to 33.8% of global consumption; at the same time, oil consumption in North America declined from 40.5% to 24.5%, and oil consumption in Europe fell from 38.6% to 21.0%.

Oil is the most heavily traded fossil energy resource worldwide. Global oil trade volume in 2013 totaled 2.78 billion tons, accounting for 63% of fossil energy trade globally. Global oil trade volume has been increasing year by year. Between 2003 and 2013, global annual oil trade volume rose from 2.26 billion tons to 2.78 billion tons, representing an annual growth rate of 2.1%. Oil trade as a percentage of global oil consumption rose from 62.1% to 66.4%. The Middle East and Russia are the most important exporters in world oil trade. Africa and Central and South America are also seeing higher exports year by year. Although export flows still center on developed countries (e.g., those in North America and Europe), exports to developing countries are also on the rise year by year. Currently, more than 60% of the world's oil trade is conducted by ocean-going tankers and the remaining less than two-thirds transported by pipeline. See Fig. 1.15 for the world's oil trade flows in 2013.

1.2.3 Natural Gas

Natural gas is a relatively clean fossil fuel. The commercialization of natural gas first commenced in the state of Pennsylvania in the United States in 1821, after which a large number of natural gas fields were gradually found. However, the safety risks involved in gas pipeline delivery led to the natural gas industry falling seriously behind the oil industry. Between 1945 and 1970, worldwide efforts were stepped up to invest more in oil and gas drilling, with a sharp increase in natural gas reserves and production levels around the world. Production in the Soviet Union, the United States, and the Netherlands grew the fastest of all. In recent years, the share of natural gas in global primary energy consumption has increased, narrowing the gap between oil and coal.

Global distribution of natural gas resources is highly unbalanced. At the end of 2013, the world's remaining proven recoverable reserves of natural gas amounted to 186 trillion m^3, distributed primarily in the Middle East and Europe and the Eurasian Continent. The Middle East had 80 trillion m^3 of remaining proven recoverable reserves of natural gas, accounting for 43.2% of the world total, while Europe and the Eurasian Continent had 57 trillion m^3 of remaining proven recoverable reserves of

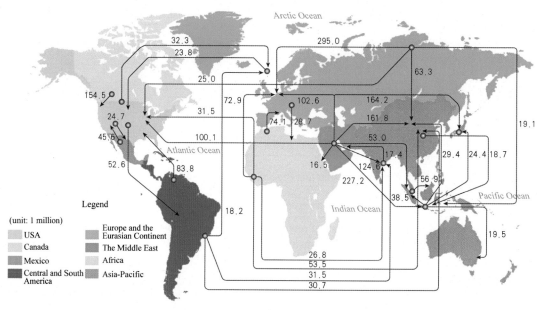

FIGURE 1.15 World Oil Trade Flows in 2013

Source: Ref. [65].

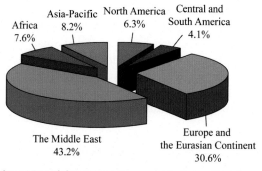

Asia-Pacific 8.2%
Africa 7.6%
North America 6.3%
Central and South America 4.1%
The Middle East 43.2%
Europe and the Eurasian Continent 30.6%

FIGURE 1.16 Global Distribution of Remaining Proven Recoverable Natural Gas Reserves

Source: Ref. [65].

natural gas, accounting for 30.5% of the world total. Together the two regions accounted for 73.7% of the world total. The rest of the world's remaining proven recoverable reserves of natural gas were divided roughly equally among Asia-Pacific (8.2%), Africa (7.6%), and North America (6.3%). Progress in gas exploration technology continues to push up the remaining proven recoverable gas reserves. Between 1980 and 2013, the world's remaining proven recoverable gas reserves rose from 72 trillion m^3 to 186 trillion m^3, representing an annual growth rate of 2.9%. On the other hand, increasing exploration activities have driven the R/P ratio from 57 years down to 55 years. See Fig. 1.16 for the global distribution of remaining proven recoverable reserves of natural gas.

Global natural gas production has continued to grow, with Europe and the Eurasian continent, and North America as the major producing regions. Global natural gas production in 2013 totaled 3.4 trillion m^3, representing 2.3 times the level in 1980 or an annual growth rate of 2.6%. World gas production is concentrated primarily in Europe and the Eurasian continent and North America, which accounted for 88.6% of the world total in 1980, falling to 57.5% in 2013. With gradually rising gas production in recent years, the Middle East, Asia-Pacific, and Africa as a share of global production rose 14.4, 9.6, and 4.3 percentage points, respectively, in 2013 from 1980. As the world's most productive countries/regions in natural gas production currently, the United States, Russia, and the Middle East account for 20.6, 17.9, and 16.8% of global production, respectively in 2013 or collectively over one half of the world total. In recent years, China has also entered into a period of fast-growing natural gas production. In 2013, as the world's sixth largest gas producer, China turned out 117.05 billion m^3 of natural gas, 8.2 times the level in 1980 or an approximately 3.5% share of global production. See Fig. 1.17 for the distribution of the world's natural gas producing areas during 1980–2013.

Total natural gas consumption continues to grow, accounting for a steadily increasing share of global energy consumption. From 1965 to 2013, global annual natural gas consumption shot up from 644.5 billion to 3.3476 trillion m^3, representing an approximately fourfold increase. In 1971, global natural gas consumption topped 1 trillion m^3 for the first time, rising to 2 trillion m^3 in 1991 and further to 3 trillion m^3 in 2008. In 1965, natural gas consumption accounted only for 15.6% of global primary energy consumption, soaring to 23.7% in 2013, representing an approximately 8 percentage point increase within nearly five decades.

World natural gas trade registers a relatively high growth rate. World natural gas trade is conducted primarily through pipeline and in the form of liquefied natural gas (LNG). In regard of pipeline transmission, gas is delivered from Russia to Europe and from Canada to the United States. With regard to LNG, natural gas is delivered primarily from the Middle East and North Africa to East Asia, Europe, and North America. Between 2003 and 2013, global annual natural gas trade volume increased from 623.7 billion m^3 to 1.0360 trillion m^3, representing annual growth of 5.2%. International gas trade as a percentage of total natural gas consumption had also climbed from 24.7% in 2003 to 31.9% in 2013. In 2013, piped natural gas and LNG accounted for 68.6 and 31.4% of total natural gas trade, respectively. See Fig. 1.18 for the world's natural gas trade flows in 2013.

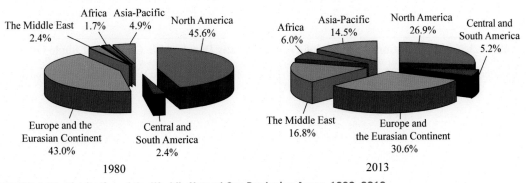

FIGURE 1.17 Distribution of the World's Natural Gas Producing Areas, 1980–2013

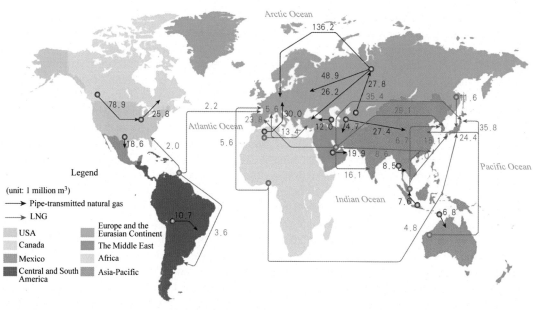

FIGURE 1.18 World Natural Gas Trade Flows, 2013

Source: Ref. [65].

1.2.4 Unconventional Oil and Gas

The world has large, albeit unevenly distributed, reserves of unconventional oil and gas. Unconventional oil mainly consists of heavy oil, oil sand, shale oil, etc. Heavy oil is distributed mainly in South America, Middle Asia, Russia, and the Middle East while most oil sand is in North America, Africa, Middle Asia, and Russia. The world has recoverable shale oil reserves of 47.1 billion tons, mainly in Russia and the United States. See Table 1.3 for the world's top five nations in terms of recoverable shale oil reserves in 2013.

Unconventional natural gas includes combustible ice, shale gas, coal seam gas, tight sandstone gas, shallow biogas, water-soluble gas, abiogenetic gas, etc. With the advantages of abundant reserves, high

Table 1.3 Top Five Nations in Terms of Technologically Developable Capacity of Shale Oil

Rankings	Nations	Technologically Developable Capacities (million tons)
1	Russia	10,200
2	USA	7,900
3	China	4,400
4	Argentina	3,700
5	Libya	3,500

Source: Ref. [92].

energy density and low pollution from burning, combustible ice (also known as natural gas hydrate) is generally found in the slopes off continental shelves, deep seas, deep lakes, and permanent tundras. It is estimated that the world has combustible ice reserves of 20,000 trillion m^3. The world's shale gas resource is mainly distributed in Asia, North America, and other regions, with technologically developable capacity of 207 trillion m^3. See Table 1.4 for the world's top five nations in terms of technologically developable capacity of shale gas in 2013. The world's coal seam gas is mainly distributed in North America, Middle Asia, Russia, and other Asia-Pacific regions, with total reserves estimated at 225 trillion m^3.

China has large reserves of unconventional natural gas. Combustible ice has been found in the South China Sea, the East China Sea, and Qinghai–Tibet Plateau tundra. The combustible ice reserves in the north of the South China Sea alone are equivalent to half of China's onshore oil reserves. China's prospective onshore combustible ice reserves are estimated at more than 50 billion tons of standard coal. The potential shale gas reserves are estimated at 134 trillion m^3,[7] with 25 trillion m^3 (excluding the reserves in the Qinghai–Tibet Plateau) mainly distributed in offshore shale formations in the south and also in onshore sedimentary basins, such as the Songliao Basin in the northeast, the Erdos Basin in Inner Mongolia, the Turpan–Hami Basin, and the Dzungaria Basin in Xinjiang. Coal seam gas reserves with embedment depth of less than 2000 m are estimated at approximately 36.8 trillion m^3, ranking third in the world. The reserves are mainly distributed in the Erdos Basin in Inner Mongolia, the Qinshui Basin in Shanxi, the Turpan–Hami Basin, and the Dzungaria Basin in Xinjiang.

Due to cost and technology constraints, most nations have not achieved large-scale development of unconventional oil and gas. Shale is characterized by high hardness, low porosity, and low permeability. Development of unconventional oil and gas requires higher processing and refining technologies, such as horizontal drilling, hydraulic fracturing, logging while drilling, geosteering drilling, and microseismic detection, which entail much higher requirements compared to conventional oil and gas development. Unconventional oil and gas wells generally have a pressure that is more than three times that of conventional oil and gas wells. The cost of a fracturing truck for unconventional oil and

Table 1.4 2013 Top Five Nations in Terms of Technologically Developable Capacity of Shale Gas

Rankings	Nations	Technologically Developable Capacities (Trillion m^3)
1	China	32
2	Argentina	23
3	Algeria	20
4	USA	19
5	Canada	16

Source: Ref. [92].

[7]Source: National Energy Strategy Report for 2030 Scientific Development from Development & Planning Department of National Energy Administration, 2012; Bulletin of National Oil & Gas Resource Reserve from Ministry of Land and Resources of the People's Republic of China, 2014.

gas wells is more than 10 times that of a fracturing truck used for conventional oil and gas wells. As shale formations in the United States are marine sediments distributed in relatively stable geotectonic elements, shale oil and gas development in the United States is relatively more economically viable, based on several decades of development experiences. Currently, just the United States and a few other nations have achieved larger scale development of unconventional oil and gas (the development cost of unconventional oil and gas in China is two to three times the cost in the United States). In addition, development of shale oil and gas is water-intensive, and the chemicals for fracture processing pollute the environment and underground water. Such impacts on the ecogeological environment warrant further research and evaluation. Though an abundant source of energy, combustible ice is still in the stage of resource investigation and technology R&D, far from the level of commercialized development and utilization on a large scale.

1.3 CLEAN ENERGY

Clean energy refers to hydro, wind, solar, nuclear, marine, and biomass energy, among other renewable energy alternatives. Clean energy is an abundant resource with great development potential. Following breakthroughs in development technologies, the economics of clean energy have greatly improved and the substitution of clean energy for fossil fuels will become an important trend in global energy development. With more than 10,000 GW of hydropower resources, more than 1,000 TWh of onshore wind resources and more than 10,000 GW of solar energy resources, the global developable capacity of clean energy is far more than enough to meet all mankind's energy needs.

1.3.1 Hydropower

Hydropower is currently the most technologically sophisticated, the most economically viable and the most extensively developed type of clean energy. According to the WEC, the theoretical reserves of the world's hydropower resources are estimated at 39,000 TWh per year, being mainly concentrated in Asia (18,000 TWh or 46% of the world total), South America (8,000 TWh or 21% of the world total), and North America (6,000 TWh or 15% of the world total).

The developable capacity of the world's water resources is estimated at 16,000 TWh or 41% of the global theoretical reserves. Of this total capacity, Asia accounts for 7,200 TWh (46% of the global total), South America for 2870 TWh (18% of the global total), North America for 2420 TWh (16% of the global total), and Europe for 1040 TWh (7% of the global total). See Table 1.5 for the global distribution of hydropower resources.

The world's top five nations in terms of theoretical hydropower reserves are China, Brazil, India, Russia, and Indonesia, amounting to 6080, 3040, 2640, 2300, and 2150 TWh, respectively. The world's top five nations in terms of developable hydropower capacity are China, Russia, the United States, Brazil, and Canada, amounting to 2470, 1670, 1340, 1250, and 830 TWh, respectively on a yearly basis. See Table 1.6 for the top three nations on each continent in terms of water resources.

The future development of large-scale hydropower bases is focused on Asia, Africa, and South America. Most of Asia's hydropower resources are concentrated along the Yangtze River, the Yarlung Zangbo River, and the Ganges River with theoretical installed capacity of more than 1100 GW. Africa has the second largest theoretical installed capacity at approximately 580 GW, concentrated along the Congo River and the Zambezi River. The current low level of development in Africa indicates huge growth potential in the future. Most of South America's hydropower

Table 1.5 World's Hydropower Resources by Continent

Regions	Theoretical Reserves	Technologically Developable Capacities (TWh/year)
Asia	18,310	7,200
Europe	2,410	1,040
North America	5,510	2,420
South America	7,770	2,870
Africa	3,920	1,840
Oceania	650	230

Source: Ref. [77].

Table 1.6 Top Three Nations on Each Continent by Water Resources

Region	Nation	Theoretical Reserves	Technologically Developable Capacity (TWh/year)
Asia	China	6080	2470
	Russia	2300	1670
	India	2640	660
Europe	Norway	600	240
	Turkey[1]	430	220
	Sweden	200	130
North America	USA	2040	1340
	Canada	2070	830
	Mexico	430	140
South America	Brazil	3040	1250
	Venezuela	730	260
	Columbia	1000	200
Africa	Democratic Republic of Congo	1450	780
	Ethiopia	65	260
	Cameroon	29	120
Oceania	Australia	270	100
	New Zealand	210	80
	Papua New Guinea	180	50

[1]Turkey's territory stretches across Europe and Asia. It is grouped under "Europe" for the research purpose of this book.
Source: Ref. [77]; 2005 Investigation Results for Hydropower in the People's Republic of China from National Leading Group for Nationwide Hydropower Review.

resources are found along the Amazon River and the Orinoco River, amounting to approximately 463 GW. The Amazon River, with a theoretical installed capacity of 279 GW, holds great potential for large-scale development. See Table 1.7 for the theoretical installed capacities of the world's major hydropower bases.

China ranks first in the world in terms of water resources. It has more than 3,800 rivers each with a theoretical reserve of 10,000 kW or above, producing theoretical energy output of 6,080 TWh per year. Technologically developable installed capacity amounts to 570 GW with annual energy output of 2470 TWh. China's water resources are distributed mainly along the Yangtze River, the Yarlung Zangbo

Table 1.7 Theoretical Installed Capacities of World's Major Hydropower Bases		
Regions	**Hydropower Bases**	**Theoretical Installed Capacity (GW)**
Asia	The Yangtze River	268
	The Yarlung Zangbo River	160
	The Ganges River	153
	The Yenisei River	149
	The Lena River	121
	The Indus River	88
	The Irrawaddy River	79
	The Mekong River	75
	The Yellow River	43
	River Ob	42
	Total	1178
Europe	The Danube	40
North America	The Columbia River	54
	The Mississippi River	49
	Total	103
South America	The Amazon River	279
	The Orinoco River	95
	The Parana River	89
	Total	463
Africa	The Congo River	390
	The Zambezi River	137
	The Nile	50
	Total	577

Note: Figures for the Yenisei River, the Lena River, and River Ob refer to technologically developable installed capacity.
Source: see Ref. [34] [102] [105] [106] [107].

River, and the Yellow River, respectively accounting for 47, 13, and 7% of the country's technologically developable capacity. By the end of 2013, China's installed hydropower capacity was 280 GW or about half of its technologically developable capacity, indicating huge potential for future expansion.

Despite the continued growth in the world's installed hydropower capacity in recent years, the share of hydropower in total installed capacity has been declining. During 1990–2013, the world's hydropower capacity increased 57.6%, from 640 GW to 1010 GW, representing an average annual growth rate of 2.0%. In 2013, hydropower capacity accounted for 17.7% of the world's total installed generation capacity, came down to 5.6 percentage points compared to 1990. See Fig. 1.19 for details.

In 2013, Asia had the world's largest installed hydropower capacity at 370 GW, accounting for 36.7% of the global total. It is followed by Europe, North America, South America, Africa, and Oceania, with installed hydropower capacities of 250 GW (24.8% of the global total), 200 GW (19.8% of the global total), 140 GW (13.7% of the global total), 30 GW (3.0% of the global total), and 20 GW (2.0% of the global total), respectively.

During 1990–2013, the world's hydropower output increased from 2210 TWh to 3780 TWh, representing an average annual growth rate of 1.6%. See Table 1.8 for the world's hydropower installed capacity and electricity generation in 2013.

The world's top five nations in terms of installed hydropower capacity are China, the United States, Brazil, Canada, and Russia, boasting installed capacities of 280, 101, 81, 76, and 49 GW, respectively. The top three in Asia are China, Russia, and Japan. Placed among the top three in North America are the United States, Canada, and Mexico, while the top three positions in South America are held by Brazil, Venezuela, and Columbia. The top three in Europe are Norway, France, and Italy. The top three

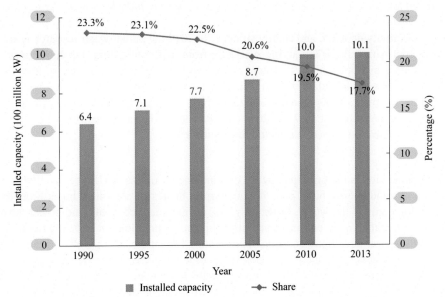

FIGURE 1.19 The World's Installed Hydropower Capacity and Share of Hydropower in Global Installed Capacity, 1990–2013

Table 1.8 World's Hydropower Installed Capacity and Electricity Generation in 2013

Regions	Installed Capacities		Electricity Generation	
	Installed Hydropower Capacities (GW)	Installed Hydropower Capacity as a Percentage of World's Total (%)	Hydropower Generation (TWh)	Hydropower Generation as a Percentage of World's Total (%)
World's Total	1010	100	3780	100
Asia	370	36.7	1400	37.0
Europe	250	24.8	830	22.0
North America	200	19.8	690	18.3
South America	140	13.7	710	18.8
Africa	30	3.0	110	2.9
Oceania	20	2.0	40	1.0

Source: Ref. [77] and [65].

in Africa are Egypt, Democratic Republic of Congo, and Mozambique. The top three in Oceania are Australia, New Zealand, and Papua New Guinea. See Table 1.9 for the top three nations on each continent in terms of water resources.

The current low level of the world's hydropower development indicates relatively significant scope for future expansion. A nation's water resource utilization rate is determined mainly by its hydropower endowment, electricity demand, and energy strategy. In 2013 the global water resource utilization rate amounted to approximately 20%, varying markedly from nation to nation. Japan, Norway, Sweden, and Canada started hydropower development earlier, with a high utilization rate of above 40%. The United States, despite its abundant water resources, has a relatively low utilization rate of 20.3% in hydropower development. Water resource utilization rates in China, Brazil, and Venezuela are between 30% and 40%, compared to just 10.9 and 20.0% for Russia and India, respectively. See Table 1.10 for the hydropower utilization rates of a selected few countries.

1.3.2 Wind Energy

Wind power generation is the principal means of wind energy utilization. Since the 1990s, the development costs of wind energy have gone down sharply, thanks to continued breakthroughs in wind power technology around the world. In recent years, development costs of wind power have become increasingly comparable to the costs of conventional power generation, on basically the same scale of expansion with nuclear development. Though accounting for a mere 3% of global generating capacity at present, wind power is being incorporated as a part of the national energy development strategy and plan by more and more countries. In the future, with the growing economic benefits and market competitiveness of wind power technology, wind power will become one of the world's most important energy option.

Table 1.9 Top Three Nations on Each Continent by Water Resources (year 2013)

Regions	Nations	Installed Hydropower Capacities (GW)	Hydropower Generations (GWh)
Asia	China	280	892,100
	Russia	49	181,200
	Japan	49	82,200
Europe	Norway	30	129,000
	France	25	68,400
	Italy	22	51,500
North America	USA	101	271,900
	Canada	76	391,600
	Mexico	12	27,400
South America	Brazil	81	385,400
	Venezuela	15	83,800
	Columbia	15	44,400
Africa	Egypt	3	12,900
	Democratic Republic of Congo	2	7,800
	Mozambique	2	16,800
Oceania	Australia	9	20,100
	New Zealand	5	23,000
	Papua New Guinea	0.25	1,000

Source: United Nations World Energy Statistics Yearbook 2013; Ref. [65].

Table 1.10 Hydropower Utilization Rates of Selected Countries in 2013

Rankings	Nations	Hydropower Utilization Rates (%)	Rankings	Nations	Hydropower Utilization Rates (%)
1	Japan	60.2	6	Venezuela	32.2
2	Norway	53.8	7	Brazil	30.8
3	Sweden	47.2	8	USA	20.3
4	Canada	47.2	9	India	20.0
5	China	36.1	10	Russia	10.9

Note: Hydropower utilization rate (%) = annual generation capacity/technologically developable capacity/year $\times$ 100%.

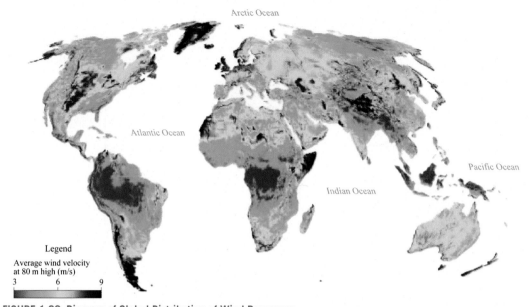

FIGURE 1.20 Diagram of Global Distribution of Wind Resources

Source: American 3TIER Resources Evaluation Company for Wind Energy and Solar Energy.

Global wind resources are very abundant. See Fig. 1.20 for the global distribution of wind resources. The theoretical reserves of global wind resources are estimated at 2,000,000 TWh/year, distributed unevenly around the world as determined by factors such as atmospheric circulation, terrain, land–sea topography and water bodies. By continent, Africa, Asia, North America, South America, Europe, and Oceania account for 32, 25, 20, 10, 8, and 5%, respectively, of the world's theoretical reserves of wind energy. See Table 1.11 for the wind resources of each continent in the world.

The theoretical reserves of Asia's wind energy are estimated at 500,000 TWh/year, located mainly in Russia, China, and Kazakhstan. Wind resources in Russia are concentrated on the coast of the Arctic Rim in Siberia, especially the Kara Sea and the Bering Strait, where the average wind speed reaches

Table 1.11 Wind Resources of Each Continent		
Areas	**Theoretical Reserves (TWh/Year)**	**Shares of Global Total (%)**
Asia	500,000	25
Europe	150,000	8
North America	400,000	20
South America	200,000	10
Africa	650,000	32
Oceania	100,000	5

7–9 m/s. Wind resources in China are mainly found in the so-called "Three North" region (north-west, north-east, and northern China), the south-eastern coastal region and the islands in its vicinity. Among them, the average annual wind speed of "Three North" region reaches 6–9 m/s. Kazakhstan is the richest country in Central Asia in terms of wind resources, with half of its territory providing average wind speeds over 4 m/s. Quality wind resources are concentrated in the central and southern parts and Caspian Sea region of the country.

The theoretical reserves of Europe's wind resources are estimated at 150,000 TWh/year, found mainly in the Nordic countries such as Denmark (including Greenland[8]) and Norway. The coastal regions of these three countries boast the richest wind energy resources with average wind speeds of 9 m/s. On the European continent, wind speeds in most areas basically exceed 6–7 m/s, with the exception of central Iberian Peninsula, northern Italy, Romania, Bulgaria, and Turkey.

The theoretical reserves of North America's wind resources are estimated at 400,000 TWh/year, located mainly in the United States, Canada, and Mexico. Wind resources in the United States are mainly in the midwest, eastern, and western coastal regions, and the Caribbean Coast, especially the central region with the expansive North American prairies where the open, flat terrain supports annual average wind speeds more than 7 m/s, and even up to 9 m/s across the eastern and western coasts. Quality wind resources in Canada are located mainly in the central and north-eastern regions of the country where 40% of power demand nationwide can be met by the wind resources of northern Quebec alone. The wind resources in Mexico are also abundant, concentrated principally in the Yucatan Peninsula, Campeche, and Oaxaca.

The theoretical reserves of South America's wind resources are estimated at 200,000 TWh/year, located mainly in Brazil, Argentina, and Chile. Wind resources in Brazil are concentrated in the south-eastern plateau with wind speeds above 7 m/s. Argentina occupies wind resource-rich areas where wind speeds are over 6 m/s and can even reach 8–9 m/s in the south with a flat terrain and low altitudes. Wind speeds in southern Chile are between 7 m/s and 9 m/s.

The theoretical reserves of Africa's wind resources are estimated at 650,000 TWh/year, located mainly in Sudan, Somalia, and Egypt, among other nations. Sudan is the richest country in terms of wind resources in Africa, with half of its territory yielding annual wind speeds above 5 m/s and over 5% of its land area boasting quality wind resources with wind speeds between 7 m/s and 9 m/s. Annual average wind speeds covering 90% of Somalia's land area are above 5 m/s and can reach between 7 m/s and 10 m/s in 70% of its land area. Average wind speeds in the Gulf of Suez in Egypt reach 10.5 m/s, with abundant wind resources in the desert on both sides of the Nile and some parts of the Sinai Peninsula.

The theoretical reserves of Oceania's wind resources are estimated at 100,000 TWh/year, found mainly in Australia and New Zealand. Wind speeds in Australia's land territory exceed 7 m/s, going up between 8 m/s and 9 m/s in all coastal regions along its coastline. New Zealand is also abundant in wind resources, which is concentrated mainly in the coastal regions along its coastline having wind speeds between 8 m/s and 9 m/s.

Regions with average wind speeds below 4.5 m/s are generally considered not suitable for wind power development on a large scale, given the current technical and economic conditions. By eliminating land area where low wind speeds and the presence of natural reserves render wind power development unsuitable, we can work out the technologically developable capacity of wind power for each

---

[8]Greenland is grouped under "Europe" for the purpose of discussing clean energy in this publication.

continent based on the installed capacity per unit area and the number of operating hours at full load. See Table 1.12 for the countries with abundant onshore wind resources on each continent.

For the development of wind power bases, consideration must be given not just to wind resources, but also engineering geology, site coverage, natural disaster, potential land or nearshore development, and utilization. At present, wind power development is mainly restricted by space and other economic constraints. In the future, wind power development for Asia will be concentrated in China's "Three North" region, northern Russia, and Kazakhstan in central Asia due to a number of factors. In North America, development will be focused on the central and eastern and western coastal regions of the United States with great potential. Wind power development in South America will center on the south-western plateau of Brazil, southern Argentina, and southern Chile. In Europe, the development focus will be on the North Sea and the coastal regions of the Atlantic Ocean. In Africa, development will be focused on countries in the eastern and northern parts of the continent, such as Sudan, Somalia, Egypt, and others. In Oceania, wind power development will be concentrated in the coastal regions of Australia and New Zealand. See Table 1.13 for the major wind resource–rich regions around the world.

Currently, wind power, as one of the fastest growing options for power generation, has become the third most popular source of clean energy after hydropower and nuclear power. In 2013, global installed capacity of wind power totaled 320 GW. Of the 24 countries each having an installed wind

Table 1.12 Countries with Abundant Onshore Wind Resources on Each Continent		
Regions	**Nations**	**Technologically Developable Capacity (TWh)**
Asia	Russia	68,000
	China	20,000
	Kazakhstan	3,000
Europe	Denmark (including Greenland)	26,000
	Norway	2,000
	Spain	1,000
North America	USA	33,000
	Canada	25,000
	Mexico	4,000
South America	Brazil	10,000
	Argentina	6,000
	Chile	3,000
Africa	Sudan	46,000
	Somalia	44,000
	Egypt	37,000
Oceania	Australia	21,000
	New Zealand	1,000

Table 1.13 Major Wind Resource-Rich Regions

Regions	Countries	Areas of Abundance	Average Wind Speeds (m/s)	Technologically Developable Capacities (TWh)
Asia	Russia	Siberia	6–9	45,000
		Kara Sea and Bering Strait and surrounding regions	7–9	10,000
	China	"Three North" region	6–9	14,000
	Kazakhstan	Central and southern parts and Caspian Sea region	>6	2,000
Europe	Denmark	Greenland	5–14	95,000
		North Sea	5–12	30,000
North America	USA	Central region of USA	7–10	11,000
South America	Brazil	Brazil	>6, southern parts 8–9	4,000
	Argentina	Argentina	Southeast parts >7	2,000
Africa	Sudan, Ethiopia, Somalia	East Africa	8–9	60,000
Oceania	Australia	Northwest, southeast	Land >7	10,000

power capacity of over 1 GW, 16 are in Europe, including Germany, Spain, and Britain; 3 are in Asia, including China, India, and Japan; 3 are in North America, including the United States, Canada, and Mexico; in addition to Australia in Oceania and Brazil in South America. In June 2012, China overtook the United States as the world's largest nation in terms of installed wind power capacity. Globally, developed capacity is found mainly in the regions blessed with quality wind resources, close proximity to load centers, and excellent access to power grids. So far global wind resources have only been developed to a very limited extent. With the development and application of long-distance power transmission technology, pockets of quality wind resources far away from load centers can also be developed efficiently.

1.3.3 Solar Energy

Originating from solar radiation, solar energy is the most abundant and widely dispersed form of clean energy. Solar power generation is the most important means of solar energy development and utilization. Since the twenty-first century, global solar power generation has demonstrated strong growth momentum to surpass wind power as the fastest-growing source of renewable energy generation. Solar power generation in Germany, the United States, Japan, and other countries or regions has developed relatively early and quickly to the large scale we see today. Despite a late start, solar power generation in China has seen a rapid growth. It is now on a scale second only to Germany in the world. Due to technology and cost constraints, the current capacity of global solar power generation is not significant, with an installed capacity less than half than that of wind power. However, given the abundance of solar resources, solar power generation holds great growth potential to become the most important source of energy in the future, provided that these costs can be brought down significantly through technological breakthroughs.

Solar energy holds great growth potential. With the exception of nuclear energy, tidal energy, and geothermal energy, all other energies on Earth come from the sun directly or indirectly. The amount of solar radiation falling on the Earth's surface is equivalent to 116 trillion tons of standard coal per year or 6500 times the world's total consumption of primary energy in 2013 (18.19 billion tons of standard coal). See Fig. 1.21 for the global distribution of solar resources.

The amount of solar resources in a given location is determined by two major factors. *One is the angle of sunlight.* The energy from direct sunlight on the Earth's surface on a per unit-area basis definitely exceeds that from slanting rays of sunlight. Therefore, solar resources are richest between the Tropic of Cancer and the Tropic of Capricorn with the equator at the center. *The other one is atmospheric scattering.* The more particles there are in the atmosphere, the stronger scattering is and the less sunlight reaches the Earth's surface. In a plateau region where the air is thin, the atmosphere's diffusion effect on the sun's rays is limited, which explains why solar resources in high terrain are more abundant compared to regions at the same latitude but lower sea levels. China's Qinghai–Tibet Plateau, for instance, has much more solar resources than many regions at low latitudes. With a higher water content in the atmosphere and more scattering of sunlight, regions covered by tropical rainforests close to the Equator have a lower amount of solar radiation than the arid and semiarid zones near the Equator. See Table 1.14 for the theoretical reserves of solar energy by continent.

The theoretical reserves of Asia's solar energy are estimated at 37,500,000 TWh, found mainly in China, Saudi Arabia, and Kazakhstan, among other nations. China's solar resources are concentrated in the Qinghai–Tibet Plateau, northern Gansu, northern Ningxia, and southern Xinjiang, with an annual irradiation intensity that exceeds 1600 kWh/m^2, compared to over 2300 kWh/m^2 in some parts of the

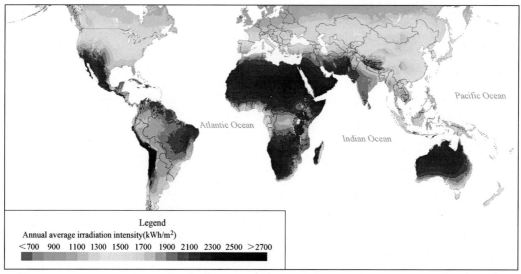

FIGURE 1.21 Diagram of the Global Distribution of Solar Resources

Source: SolarGIS, http://solargis.info.

Table 1.14 Theoretical Reserves of Solar Energy by Continent		
Areas	**Theoretical Reserves (TWh/Year)**	**Shares of Global Total (%)**
Asia	37,500,000	25
Europe	3,000,000	2
North America	16,500,000	11
South America	10,500,000	7
Africa	60,000,000	40
Oceania	22,500,000	15

Qinghai–Tibet Plateau. Half of Saudi Arabia is desert land with an annual irradiation intensity of over 2200 kWh/m^2. The comparative figure for Kazakhstan is 1300–1800 kWh/m^2.

The theoretical reserves of Europe's solar energy are estimated at 3,000,000 TWh, found mainly in Spain, Italy, Portugal, and other southern European countries. An annual irradiation intensity of 1600–1800 kWh/m^2 is recorded in over 60% of Spain, 50% of Italy, and 70% of Portugal. As for the other regions, the comparative figure is 1400–1600 kWh/m^2.

The theoretical reserves of North America's solar energy are estimated at 16,500,000 TWh, concentrated in the United States and Mexico, among other countries. Solar resources in the United States are found mainly in the south-west and one-third of the country's land territory has an annual irradiation intensity of 1700 kWh/m^2 to 2200 kWh/m^2. The comparative figure for most regions of Mexico is comparable to that of the American southwest. In the northern region of Mexico, 20% of the land has an annual irradiation intensity of over 2200 kWh/m^2.

The theoretical reserves of South America's solar energy are estimated at 10,500,000 TWh, concentrated in Chile, Peru, and Brazil, among other nations. The solar resources in Peru and Chile are found mainly in the Atacama Desert. The annual irradiation intensity of southern Peru and northern Chile exceeds 2100 kWh/m^2. Solar resources in Brazil are mainly found in the country's plateau regions. Regions with an annual irradiation intensity of above 1500 kWh/m^2 account for 20% of Brazil's total land area.

The theoretical reserves of Africa's solar energy are estimated at 60,000,000 TWh, distributed mainly in Sudan, South Africa, Tanzania, etc. The regions in Sudan, South Africa, and Tanzania with an annual irradiation intensity of 1500–2000 kWh/m^2 account for 20, 25, and 18% of the total land area of the respective countries. The comparative figures for regions with an annual irradiation intensity of 2000–2500 kWh/m^2 are 56, 8, and 5%, respectively.

The theoretical reserves of Oceania's solar energy are estimated at 22,500,000 TWh, distributed mainly in Australia and New Zealand. Australia has abundant solar resources, where regions with annual irradiation intensities of over 2200, 1900–2200, and 1600–1900 kWh/m^2 accounting for 54, 35, and 8%, respectively, of the country's land territory. Most regions of New Zealand also have an annual irradiation intensity of about 2000 kWh/m^2.

The technologically developable capacity of solar energy is determined mainly by factors such as solar irradiation intensity, conversion efficiency, and space availability. Generally, only regions with an annual irradiation intensity of over 1500 kWh/m^2 are suitable for solar power development. At present, the conversion efficiencies of photovoltaic modules and photothermal power stations are 16.5 and 14%, respectively. Space availability refers to the area suitable for installation of solar power generation systems after exclusion of natural reserves, especially disadvantageous terrain, etc. See Table 1.15 for countries with abundant solar energy resources on each continent around the world.

A mix of centralized and distributed systems is the way forward for solar power generation. Large-scale centralized solar power bases that deliver power to load centers through extra-high voltage and ultrahigh voltage transmission corridors are well attuned to regions with high irradiation intensities, sparse populations, and expansive desert areas, such as North Africa, East Africa, the Middle East, Australia, western China, American south-west, and Chile. The development of distributed photovoltaic power generation is suitable for cities and villages with high population and building densities where local demand load is satisfied partly by access to local grids. See Table 1.16 for the world's major solar resource-rich regions.

Solar power generation is one of the most important ways to use solar energy efficiently. As the basic principle of power generation, there are two major types of solar power namely photovoltaic and photothermal. Recently, photovoltaic power generation has entered the stage of large-scale commercial development. By the end of 2013, the total installed capacity of global photovoltaic power generation reached approximately 140 GW. In terms of annual installed capacity additions, solar power is at par with hydropower and has exceeded wind power for the first time. Countries with an installed capacity of photovoltaic power generation of over 10 GW include Germany, China, Italy, Japan, and

Table 1.15 Countries with Abundant Solar Resources on Each Continent		
Region	**Nation**	**Technologically Developable Capacity (TWh)**
Asia	China	110,000
	Saudi Arabia	98,000
	Kazakhstan	74,000
Europe	Turkey	13,000
	Spain	5,000
	Italy	3,000
North America	USA	254,000
	Mexico	78,000
South America	Chile	35,000
	Peru	25,000
	Brazil	20,000
Africa	Sudan	66,000
	South Africa	43,000
	Tanzania	40,000
Oceania	Australia	251,000
	New Zealand	7,000

Table 1.16 Major Solar Resource-Rich Regions

Region		Areas of Abundance	Annual Irradiation Intensity (kWh/m²)	Technologically Developable Capacity Per Year (TWh/h)
Asia	The Middle East	Israel, Jordan, Saudi Arabia, United Arab Emirates	2000–2700	120,000
	Western China	Five provinces (municipalities) in the west and northwest: Xinjiang, Inner Mongolia, Tibet, Gansu, Qinghai	1500–2150	14,000
Europe	South Europe	Portugal, Spain, Italy, Greece, Turkey	1600–2100	3000
North America	The Southwest	California, Kansas, Colorado, Oklahoma, Texas, Utah, New Mexico, Nevada, Arizona	2100–2500	80,000
South America	Peru, Chile	Atacama Desert	2000–2500	15,000
Africa	North Africa	The Sahara and its north	2000–2700	141,000
	East Africa	Ethiopia, Sudan, Kenya, etc.	1900–2800	187,000
Oceania	Australia	Northern region	1800–2500	65,000

Source: Huangxiang, International Solar Energy Resource and Solar Power Generation Trend, Huadian Technology; 2009(12); Refs. [89] and [85].

the United States, compared to 1 000 MW for 17 countries. By project type, the share of photovoltaic power stations in global newly installed capacity additions, had increased gradually from 23% in 2009 to 45% in 2013, whereas the share of building-integrated photovoltaics (including applications in residential projects and industrial and commercial buildings) had decreased from 77% to 55%, in the same period.

Since 2008, photothermal power generation has experienced rapid growth. Accumulative installed capacity posted annual average growth of 47.6% from 2008 to 2013. But compared to photovoltaic power generation, the scale of photothermal power generation has remained small. By the end of 2013, photothermal power generation projects had been built and in operation in nine countries, including Spain, the United States, and India, with a total installed capacity of 3630 MW. In addition, projects under construction and projects approved for development are expected to provide installed capacities of 2000 MW and 10 GW, respectively. In Spain, 2206 MW has come from completed photothermal power generation projects, with a further 50 MW to come from projects under construction and 185 MW from projects approved for development. For the United States, the comparative figures are 1073 MW (completed), 615 MW (under development), and 3615 MW (approved for development), compared to 156, 425, and 551 MW, respectively, for India. Based on project developments and approvals announced, photothermal power generation in China, India, South Africa, Morocco, the United Arab Emirates, Chile, and other nations is expected to grow substantially in the coming years. See Table 1.17 for the development of photothermal power generation by country as at end of 2013.

Table 1.17 Photothermal Power Generation Development by Country as at End of 2013 (MW)

Nation	Completed Capacities	Capacities Under Construction	Approved Capacities
Spain	2,206	50	185
USA	1,073	615	3,615
India	156	425	551
UAE	100	0	0
China	21	170	1,670
Egypt	20	0	350
Morocco	20	160	300
Algeria	20	0	150
Australia	10	44	31
South Africa	0	300	350
Chile	0	110	765
Others	37	18	2,951
Total	3,663	1,892	10,918

Source: China New Energy Chamber of Commerce, Global New Energy Development Report 2014.

1.3.4 Nuclear Energy

Natural uranium resources are relatively rich and intensively distributed in the world. In 2013, the world's total proven uranium resources minable at a cost of less than 260 USD/ton amounted to 7.6352 million tons,[9] compared to less than 130 USD/ton for 5.9029 million tons, less than 80 USD/ton for 1.9567 million tons, and less than 40 USD/ton for 682,900 tons (Table 1.18). According to preliminary estimates, global nuclear fuel resources are ten times of all fossil energy, with proven uranium resource mainly distributed in Australia, Kazakhstan, Russia, Canada, Niger, Namibia, South Africa, Brazil, United States, China, and other countries. These 10 countries combined boast proven uranium resources of 5.1936 million tons minable at a cost of less than 130 USD/ton, representing approximately 88.0% of the global total.

In 2013, there were 21 countries in the world with uranium mining operations, producing total annual output of approximately 59,500 tons. Kazakhstan is the world's largest producer of uranium with an output of about 22,500 tons, representing 37.8% of the world total. It is followed by Canada, with a uranium output of about 9000 tons or 15.1% of the world total. Australia ranks third, with a uranium output of about 6700 tons or 11.3% of the world total. In recent years, China's uranium output has basically remained at 1300–1500 tons.

The share of nuclear power in the world's total installed capacity has been declining. In the 1970s to 1980s, nuclear leaking accidents in Three Mile Island in the United States and Chernobyl in the Soviet Union plunged the global nuclear power industry into a "trough" of slow development. In 2011,

---

[9]A 1000 MW-class nuclear power station typically consumes 25 tons of uranium.

Table 1.18 Global Uranium Resources in 2013

Item		Resources Classification (USD/t)	Quantity of Resources (10,000 tons)
Proven resource		<260	763.52
		<130	590.29
		<80	195.67
		<40	68.29
Including	Reliable resources	<260	458.72
		<130	369.89
		<80	121.16
		<40	50.74
	Inferred resources	<260	304.80
		<130	220.40
		<80	74.51
		<40	17.55

Source: Ref. [80].

nuclear safety sparked global concerns after the nuclear leakage accident in Fukushima. Comprehensive safety inspections and assessments of nuclear power stations in operation were then conducted in different parts of the world. Switzerland, Germany, Italy, and other countries announced plans to abandon nuclear power development. The United States, France, United Kingdom, Russia, Vietnam, the United Arab Emirates, Turkey, and many other countries expressed intentions to continue with nuclear power development subject to stringent safety standards. Currently, the major obstacles that stand in the way of nuclear power development are safety, nuclear waste disposal and other issues. As at the end of 2013, 434 nuclear power generating units were in operation in 30 countries or regions around the world, with an installed capacity of about 370 GW distributed mainly in the United States, France, Japan, and other developed countries. See Fig. 1.22 for the installed capacity of global nuclear power and its share in total electricity generation in 1990–2013.

Nuclear fission technology is a common means of utilizing nuclear power, and nuclear fission reactors are adopted for global nuclear power stations. *Nuclear fusion is the way forward for nuclear power development, despite an uncertain future given the current technology constraints.* Nuclear fusion has certain advantages namely: power from nuclear fusion on Earth is richer than energy generated by nuclear fission, and deuterium, abundantly present in seawater, is the main fuel for controllable fusion power stations. Safe and clean, nuclear fusion does not emit radioactive substances that pollute the environment. However, commercial application of power generated by nuclear fusion poses grave challenges in the coming 30 years as technology in this area is far from being mature.

1.3.5 Other Clean Energies

Marine energy refers to seawater-based renewable energy, including tidal energy, wave energy, ocean current energy, and energy created from temperature and salinity differences. The energy density of

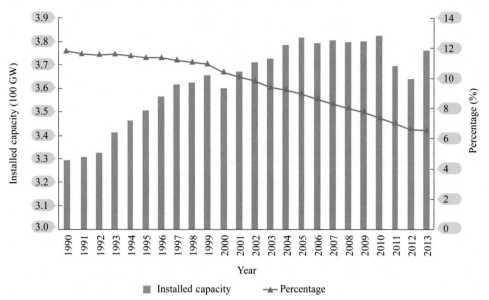

FIGURE 1.22 Installed Capacity of World Nuclear Power and its Share in Total Electricity Generation During 1990–2013

Source: Ref. [66]; World Nuclear Association, http://world-nuclear.org/nucleardatabase/advanced.aspx.

various ocean energies is relatively low in general. The maximum tidal range of tidal energy in the world is about 17 m, and the maximum range in China is 9.3 m. The average wave height of wave energy for the largest single station in the world is 2 m above, and that for the largest single station in China is 1.6 m. The maximum flow rate of ocean current is 2.5 m/s, and the maximum in China is 1.5 m/s. For energy created from temperature differences, the maximum temperature difference between surface seawater and deep seawater in the world is 24°C, comparable to the value in China. Energy created from salinity differences has the greatest energy density among all ocean energies, with an osmotic pressure of 24 atm, equivalent to a head of 240 m and also comparable to the value in China.[10] Statistics show that the theoretically developable capacity of ocean energies is 76,600 GW, including a technologically developable capacity of 6,400 GW (Table 1.19). Tidal power generation is a relatively well-developed ocean power generation technology. As at the end of 2013, the installed capacity of ocean power generation in the world had reached approximately 530 MW, and the world's largest ocean power station is the 254 MW tidal power generation station in South Korea.

Biomass energy is a clean energy with biomass as its carrier. It comes mainly from agricultural waste, forestry waste, domestic waste, and industrial waste as well as potential artificial biomass energy, energy crops, and energy forests. Currently, the theoretical productive potential of biomass energy in the world is 37.6–51.2 billion tons of standard coal each year. Given environmental and other constraints, a more practical productive potential could reach 6.8–17 billion tons of standard coal.[11]

---

[10]Data source: Ref. [35].

[11]Data source: New Opinion of Sweden Scientists on Global Biomass Resources, www.most.gov.cn/gnwkjdt/201008/t20100805_78751.htm.

Table 1.19 World Marine Energy Resources

Type	Theoretical Reserves	Technologically Developable Capacity (GW)
Tidal energy	3,000	100
Wave energy	3,000	1,000
Ocean current energy	600	300
Energy from temperature differences	40,000	2,000
Energy from ocean salinity	30,000	3,000

Source: UNESCO, ocean energy development.

Globally, biomass energy is mainly distributed in South America, South Africa, Eastern Europe, Oceania, and East Asia. Biomass is mainly used for supply of heat, power generation and production of liquid biofuels, but is not used to any significant extent for generating power. As at the end of 2013, the installed capacity of world biomass power generation reached about 76.4 GW, and annual energy output was 257.6 TWh. The European Union ranks first in biomass power generation in terms of scale.

Geothermal energy is a general term for thermal energies in the Earth's crust. It is categorized mainly into categories such as hot water type, steam type, ground pressure type, hot dry rock type, and lava type. It is estimated that the workable reserves of global geothermal energy is equivalent to 5 billion tons of standard coal, mainly distributed in the Pacific Rim geothermal belt, the Mediterranean–Himalaya geothermal belt, the Atlantic midocean ridge geothermal belt, and the Red Sea–Aden–East African Rift Valley geothermal belt. Heat utilization and geothermal power generation are two major means of utilizing geothermal energy. As at the end of 2013, the installed capacity of global geothermal power generation reached approximately 11.71 GW.

1.4 ENERGY DEVELOPMENT IN THE ARCTIC AND EQUATORIAL REGIONS

In terms of the distribution of world clean energy resources, the Arctic Circle and its surrounding areas (the Arctic region[12]) are found to have rich wind energy resources, while solar energy resources are also abundant at the Equator and its vicinity (the equatorial region[13]). Centralized development of wind resource in the Arctic and solar resources at the Equator will be an important direction of development for world resources. Resources will be delivered to load centers across the continents via ultra-high voltage and other electric power transmission techniques in mutual support for large energy bases and distributed generation to ensure a safer and more reliable supply of clean energy.

---

[12]The North Pole refers to the area to the north of 66°N34", which is located at the north of Europe, Asia, and North America. In this book, "the Arctic Region" refers to the area to the north of 66°N (Arctic for short), including the coastal and offshore areas of the Arctic Ocean, Northern Europe, Siberia of Russia, Alaska of the US, Northern Canada, and Greenland, which involves in relevant regions in eight countries including Canada, Denmark, Finland, Iceland, Norway, Russia, and USA.

[13]In this book, the equatorial region mainly refers to the area between 30°S and 30°N (equatorial region for short), which is a heartland connecting the south and north and mainly involves the following countries and regions: North Africa, East Africa, the Middle East, Australia, south-west of North America, north of South America, etc.

1.4.1 Energy Development in "The Arctic Region"

1.4.1.1 Overview of Resources

Wind energy resources are rich and widely distributed in the Arctic region, with a technologically developable capacity of 100 TW or 20% of global onshore wind resources. Wind resources are most abundant in the Kara Sea, the Barents Sea, the Bering Strait, and Greenland along the Arctic Ocean. The Arctic region schematic diagram is shown in Fig. 1.23.

Wind energy resources in the Arctic Ocean and offshore areas are mainly distributed in Greenland, the Norwegian Sea, the Barents Sea, the Kara Sea, and the Bering Strait, with relatively high annual mean wind speeds. The annual mean wind speeds within the Arctic Circle are up to 10–11 m/s, being present in Greenland and north of Iceland. The second highest wind speed is 9–10 m/s, present in Norwegian Sea. Wind force on one side of the Atlantic is relatively strong, with an average speed of higher than 7 m/s. The wind speed across most land in Greenland is 5–8 m/s. The sea surface wind speed (Kara Sea) to the east of Novaya Zemlya on the north side of Eurasia is 6–8 m/s. Wind speed at the Bering Strait is 8–9 m/s. The average wind speed at the eastern islands of North America and its sea surface

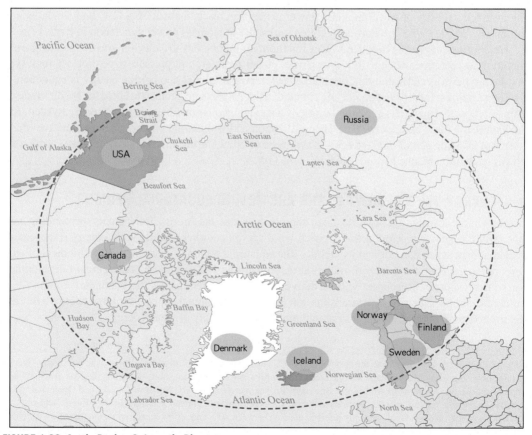

FIGURE 1.23 Arctic Region Schematic Diagram

is mainly below 5 m/s, going up to 5– 9 m/s in a few zones. Future development of wind power in the Arctic is focused on Greenland, the Norwegian Sea, the Barents Sea, the Kara Sea, the Bering Strait, and other regions rich in wind energy.

With a total area of approximately 2 million km^2, the Scandinavian Peninsula in Northern Europe and the North Sea region are blessed with favorable wind energy resources along the coasts where the annual mean wind speed is 8–9 m/s. Half of the Russian territory of Siberia is located to the north of 66°N, especially the areas along the Arctic Ocean with the most abundant wind energy resources where annual mean wind speeds reach over 8 m/s and the capacity coefficient of wind power exceeds 40%. Wind energy resources in Alaska are mainly distributed at lower sea levels in the west as well as the Bering Strait and the areas along the Arctic Ocean. The annual mean wind speed in more than half of Canada is higher than 7 m/s, and it is up to 10 m/s or above in some north-eastern parts of the country.

1.4.1.2 Development Status

Currently, the development and utilization of wind power is on a relatively small scale though the Arctic is rich in wind energy resources. As one of the Arctic Rim nations, Russia holds great potential in wind energy; however, wind power development is relatively slow. At the end of 2013, the installed gross capacity of wind power in Russia amounted to 105 MW, consisting mainly of small wind turbines of approximately 30k W. The largest wind power plant in Russia is Kulikovo Wind power plant located in Kaliningrad, with an installed capacity of 20 MW through decades of expansion. Grid-connected wind power plants have been built in Russia, including 2200 kW Qiujinli wind power plant in the Republic of Bashkortostan, 1000 kW Kalmyk wind power plant, and 200 kW Malaboza wind power plant in the Republic of Chuvashia. In addition, 2500 kW Anadyr wind power plant in Chukotka Autonomous Okrug, 1500 kW Polar wind power plant in the Republic of Komi, 1200 kW San Nicolas wind power plant in Kamchatka of the Bering Strait, and 300 kW wind power plant in Rostov have been developed but not connected to the grid yet. These projects were built mainly at the end of 1990s or the beginning of the twenty-first century.

With the exception of Russia, large-scale development of wind power has been realized in many other Arctic Rim nations, such as Denmark, Sweden, Canada, and the United States. However, as the built wind power projects are mainly located to the south of the Arctic Circle, the Arctic region with untapped wind energy resources will be the focal point of future wind power development. Table 1.20 shows the installed capacity of wind power in selected Arctic Rim nations from 2000 to 2013.

Table 1.20 Installed Wind Power Capacity of Selected Arctic Rim Nations from 2000 to 2013				
Countries	**2000 (10 MW)**	**2005 (10 MW)**	**2010 (10 MW)**	**2013 (10 MW)**
Denmark	249	313	380	477
Sweden	30	53	216	447
Canada	20	68	401	780
USA	428	915	4030	6109
Source: See Ref. [108].				

1.4.2 Energy Development in "The Equatorial Region"
1.4.2.1 Overview of Resources

Areas near the Equator are located at low latitudes and directly exposed to solar radiation. Arid, semi-arid, or desertous, with little scattering, some of these regions are extremely rich in solar energy resources, making them key areas for large-scale development and utilization of solar energy in the future. There are bountiful wind and hydropower resources in areas near the Equator, examples being the Congo River in Africa and the Amazon River in South America where there are abundant water resources.

Global solar energy resources are mainly distributed in North Africa, East Africa, the Middle East, Australia, and other regions where the development potential accounts for more than 30% of the world total. North Africa and East Africa have an annual irradiation intensity of solar energy of mostly 2000–2800 kWh/m^2 where solar energy resources are concentrated in Algeria, Morocco, Libya, and Sudan. The comparative intensity in the Middle East is 2200–2400 kWh/m^2, where solar energy resources are concentrated in Iran, Saudi Arabia, the UAE, and other countries. In Australia, the irradiation intensity is 2000 kWh/m^2 for Class I and II resource areas. Apart from areas near the Equator, regions rich in solar energy resources also include southern Europe, with an annual irradiation intensity of 1600–2100 kWh/m^2; American south-west, with an annual irradiation intensity of 2100–2500 kWh/m^2; the west coast of South America, with an annual irradiation intensity of 2000–2500 kWh/m^2. With improvements in conversion efficiency, the potential of solar power development will be greater. See Fig. 1.24 for the distribution of solar energy resources in the equatorial region.

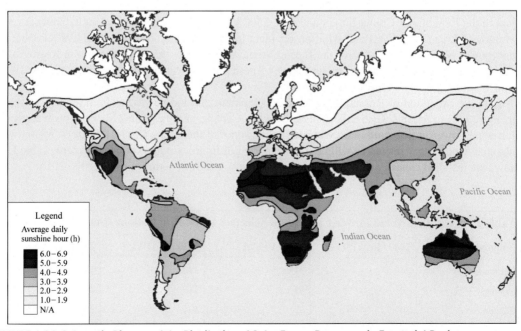

FIGURE 1.24 Schematic Diagram of the Distribution of Solar Energy Resources in Equatorial Region

Source: Australian Government Department of Industry and Science and others, Australian Energy Resource Assessment.

1.4.2.2 Development Status

While Europe and the United States are undoubtedly the world center of solar power development, countries and regions near the equatorial region have paid more attention in recent years to the development and utilization of solar power generation resources with the selection of different technological pathways and development models in line with local daylight resources and development conditions. Some photothermal power stations have been built in Morocco, Tunisia, and Algeria in North Africa as well as Saudi Arabia, the United Arab Emirates, and other countries in the Middle East, with an installed capacity of 50–100 MW for individual projects. Photovoltaic power generation is mainly adopted by Australia in Oceania, Italy, Spain, and Portugal in Europe and the United States in North America, with the development of some photothermal power stations as well. Brazil, Chile, and Peru in South America have pursued solar power generation on a small scale. Table 1.21 shows the development of solar energy resources in the equatorial region as at the end of 2013.

In general, the "Arctic and equatorial regions" are rich in clean energy with great development potential. In the future, with the full development of renewable energy sources in the major countries on each continent and the growing maturity of technology for large-scale clean energy development and long-distance power transmission, the "Arctic and equatorial region" will become important

Table 1.21 Development of Solar Energy Resources in the Equatorial Region toward the end of 2013				
Regions	**Countries**	**PV Power Generation (MW)**	**Photothermal Power Generation (MW)**	**Typical Solar Power Generation Projects**
North Africa	Morocco	15	20	160 MW Ouarzazate slot-type photothermal power station phase I (under construction)
	Tunisia	7	–	–
	Algeria	7	20	–
Middle East	Saudi Arabia	19	–	–
	UAE	33	100	"Solar I" 100 MW photothermal power station
Oceania	Australia	3300	10	20 MW tower-type photothermal power station (under planning)
South Europe	Italy	17930	–	–
	Spain	5340	2206	Gemasolar photothermal power station
	Portugal	280	–	Serpa PV power station
North America	USA	13518	1073	Ivanpah photothermal power station
South America	Brazil	20	–	–
	Chile	180	–	–
	Peru	100	–	44 MW Arequipa PV Power Station

strategic base for global resources development and provide continued energy security for the world's economic and social development. Centralized development of wind power in the Arctic region to deliver power southward to load centers in East Asia, North America, and Europe will form a "north-to-south power transmission" system. Large-scale development of solar energy resources in the equatorial region integrated with local water and wind energy resources can not only meet local power requirements, but also provide a supply of cleaner energy for Europe, Asia, North America, and the southern parts of South America.

1.5 ELECTRIC POWER DEVELOPMENT

Electric power is a clean and efficient form of secondary energy. The invention and utilization of electrical power marked a revolution in the global energy industry, ushering in the "Age of Electrification." From its beginnings in the 1880s, the global electric power industry has gone through more than a century of growth. Since the 1970s, the global power industry has undergone profound changes in terms of power generation, capacity expansion, as well as power source and grid technologies. The industry has gradually entered into a new era characterized by coordinated development of large energy bases with distributed power sources, large power grids and micro grids.

1.5.1 Power Source Development

The world's installed power capacity and electricity generation has continued to grow rapidly. Since the 1990s, with the fast-growing global economy and continued technology breakthroughs, the world's installed power capacity and electricity generation has witnessed enormous growth. From 1990 to 2013, global installed power capacity increased from 2760 GW to 5730 GW, representing an average annual growth of 3.2%, while annual electricity generation grew to 22,500 TWh from 11,770 TWh, representing an average annual growth rate of 3.1%. See Table 1.22 for the world's installed power capacity and electricity generation from 1990 to 2013.

The power supply structure remains dominated by fossil energy generation, including coal and gas, with a growing trend toward clean energy. As at the end of 2013, global installed capacity reached 5730 GW, with fossil fuel-based power capacity accounting for 66.1% of the global total. In recent years, the installed generating capacity of clean energy has increased rapidly. As at the end of 2013, the installed gross capacity of nuclear, water, wind, solar and other green energy resources were estimated at approximately 1940 GW or 33.9% of the global total. Table 1.22 shows the change in the structure of installed power capacity from 1990 to 2013 (Fig. 1.25).

The world's installed power capacity is mainly distributed in Asia, North America, Europe, and other regions. In 2013, the installed power capacity in Asia, North America, and Europe accounted for

Table 1.22 World's Installed Power Capacity and Electricity Generation in 1990–2013						
Items	**1990**	**1995**	**2000**	**2005**	**2010**	**2013**
Installed capacity (100 GW)	27.6	30.6	34.4	42.1	50.9	57.3
Electricity generation (1000 TWh)	11.8	13.1	15.5	18.4	21.4	22.5
Source: Ref. [66]; China New Energy Chamber of Commerce, Global Energy Development Report in 2014.						

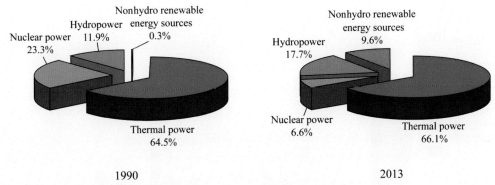

FIGURE 1.25 Change in the Structure of Installed Power Capacity During 1990–2013

42.5, 24.5, and 24.3%, respectively, of the global total. From the level in 1990, Asia's capacity as a percentage of the global total was up 17.2 percentage points, compared to a decrease of 12.1 percentage points for North America and a decrease of 4.1 percentage points for Europe. By energy type, the world's installed coal power capacity is mainly distributed in Asia and North America with rich coal resources, together accounting for 73% of the global total. Installed hydropower capacity is mainly distributed in Asia and Europe, together accounting for 62% of the global total. Installed nuclear power capacity is mainly distributed in Europe and North America, together accounting for 75% of the global total. Installed capacity of gas-fired generation is mainly distributed in Europe and North America, together representing about 80% of the global total. Installed wind power capacity is mainly distributed in Asia and Europe, which together account for 75% of the global total. Installed solar power capacity is mainly distributed in Asia, which accounts for 60% of the global total. Fig. 1.26 shows the changing share of installed power capacity by continent from 1990 to 2013.

In the recent years, Asia has maintained a leadership position among all continents in terms of annual average growth of installed power capacity, while growth in Europe and North America has slowed down. See Fig. 1.27. Now the world's largest power producer, China's installed power capacity reached 1258 GW with annual electricity generation of 5370 TWh as at the end of 2013.

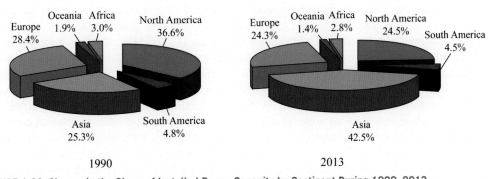

FIGURE 1.26 Change in the Share of Installed Power Capacity by Continent During 1990–2013

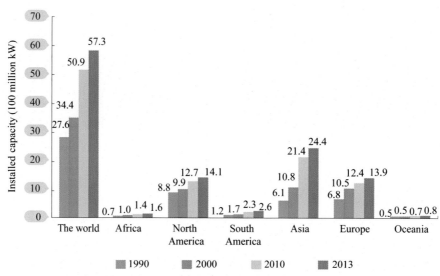

FIGURE 1.27 World's Installed Power Capacity by Continent During 1990–2013

Power development varies significantly from country to country on each continent, reflecting different levels of economic capacity and power consumption. In terms of installed power capacity and electricity generation, the top three countries, in descending order, are China, Japan, and Russia in Asia; Germany, France, and Italy in Europe; the United States, Canada, and Mexico in North America; Brazil, Argentina, and Venezuela in South America; South Africa, Egypt, and Algeria in Africa; and Australia, New Zealand, and Papua New Guinea in Oceania. Table 1.23 shows the installed power capacity and annual electricity generation of the major countries on each continent as at the end of 2013.

Power source technology has been developing rapidly around the world, with fast expansion of single unit capacity. In recent years, the technological level of thermal power equipment has improved continually around the world, along with the promotion and application of ultra-supercritical thermal power generating units with high parameters and large capacity. Currently, the maximum single unit capacity of thermal power in the world is 1300 MW. There have been significant breakthroughs in hydropower technology with respect to operational control techniques, closure and cofferdam techniques, design, and manufacture techniques of water-turbine generator sets, etc. At present, the maximum single unit capacity of hydropower generation is 1000 MW. Nuclear power is moving into the fourth generation of technological development marked by improved economics and safety with production of less waste and larger unit capacity. The world's largest nuclear power plant is in Kariwa, Kashiwazaki, Japan, with seven boiling-water reactor sets and total capacity of approximately 8210 MW. Commercialized development of PV power generation technology has been realized, with improved production efficiency of polycrystalline silicon and amorphous silicon cells and continued breakthroughs in PV system accumulators. Wind power equipment technologies are moving in the direction of larger single unit capacity and variable pitch rather than fixed pitch, as well as direct drive, combination drive, etc. On January 28, 2014, the first 8 MW offshore wind turbine generator in the world was put into operation, with mass production scheduled for 2015. On October 29, 2014,

Table 1.23 Installed Power Capacity and Annual Electricity Generation of Major Countries on Each Continent in 2013

Continents	Nations	Installed Capacities (MW)	Annual Electricity Generation (TWh)
Asia	China	1,257,680	5372.1
	Japan	295,230	1052.3
	Russia	243,100	1049.9
	Germany	184,620	596.4
Europe	France	128,060	550.8
	Italy	124,230	276.0
North America	USA	1,067,900	4274.5
	Canada	134,200	652.0
	Mexico	62,140	298.2
	Brazil	117,130	552.5
South America	Argentina	33,810	134.8
	Venezuela	25,710	122.1
	South Africa	44,170	255.1
Africa	Egypt	30,050	164.4
	Algeria	11,550	51.2
	Australia	63,220	247.0
Oceania	New Zealand	9,490	42.5
	Papua New Guinea	690	3.5

Source: Refs. [21], [69], and [97].

installation was completed of an onshore permanent magnetic direct-drive fan, with a maximum single unit capacity of 5 MW, as part of China's National Wind/Photovoltaic/Energy Storage and Transmission Demonstration Project.

1.5.2 Development of Power Grids

After decades of growth between the late nineteenth century and the mid twentieth century, the electricity industry has seen the emergence of power grids dominated by alternating current (AC) generation and transmission and distribution technology, with a voltage class below 220 kV and grid capacity focused on urban grids, standalone grids and small grids. Since the mid twentieth century, the capacity of power grids has been increasing continuously. Many transnational interconnected mega grids have been developed, like those seen in North America, Europe, and the Russia–Baltic Sea region, with 330 kV-plus DC–AC transmission systems. As at the end of 2013, transmission lines above 220 kV measured a total length of 2.5 million km, with transforming capacity totaling around 12,000 GVA.

With growing capacity, the voltage levels of power grids in the world have also been increasing. Studies on UHV electricity transmission have been conducted since the 1960s and stepping into the twenty-first century, China has spared no efforts in promoting UHV DC–AC transmission technologies and projects in order to secure the optimal allocation of energy resources across vast areas. China has now completed and commissioned three 1000 kV UHV AC transmission and transformation projects, namely: Southeast Shanxi-Nanyang-Jinmen, Huainan–North Zhejiang-Shanghai, and North Zhejiang–Fuzhou. Six ±800 kV UHV DC transmission projects have also been completed to supply power to Xiangjiaba–Shanghai, Jinping–South Jiangsu, South Hami–Zhengzhou, Xiluodu–Jinhua, Chuxiong–Zengcheng, and Puer–Jiangmen.

It has become a global trend to step up the construction of transnational interconnected power grids, expand coverage of power grids, and achieve optimal allocation of energy resources over more expansive areas. With the establishment of transnational power grids and the interconnection among mega grids, the level of electric power exchange among countries has expanded dramatically. In 2013, electricity exports and exports reached 441.7 and 444.4 TWh, respectively, among countries of the Organization for Economic Cooperation and Development, or OECD. Between 2000 and 2013, OECD countries saw a rising trend in imports and exports, up 26.7 and 27.7%, respectively. An overview of power grid development in each continent has been described further.

1.5.2.1 Asia's Power Grids

Asia's power grids include regionally interconnected grids in China, the Russia–Baltic Sea region and the Gulf region as well as national power grids in Japan, South Korea, India, and south-east Asian countries. No integrated continent-wide power grid has been built in Asia yet. Asia's power grids cover a population of 4 billion in 48 countries and regions, with a total installed capacity of 2,400 GW and a total electricity consumption of 10,000 TWh. The power grids have a maximum voltage of 1000 kV, including approximately 1.5 million km of transmission lines with a voltage above 220 kV. Among these power grids, the interconnected grid in the Russia–Baltic Sea has an installed capacity of 300 GW, covering an area of 22.54 km^2 and providing power to 280 million people. The Gulf region's interconnected power grid features an installed capacity of 94.81 GW, covering 2.67 million km^2 of area and providing power to 41.98 million people.

Compared to other Asian countries, the power grids in China, Japan, and Russia are relatively large, with installed capacities of 1258, 295, and 243 GW, respectively, and maximum demand loads of 830, 156,and 130 GW, respectively. The highest operating AC voltage classes are 1,000, 500, and 765 kV, respectively, with 540,000, 40,000, and 130,000 km of transmission lines at or above 220 kV, respectively.

In recent years, China has continued to quicken the pace of power grid development, resulting in vastly improved capability to optimize resource allocation. After the completion and commissioning of Tibet's ±400 kV DC interconnected power grid in December 2011, China has achieved nationwide interconnections covering all its territories other than Taiwan. In April 2012, the back-to-back ±500 kV DC interconnected power grid between China and Russia was put into commercial operation, marking China's largest transmission and transformation project, in terms of voltage and capacity, for import of electricity. China also delivers power to Vietnam through three 220 and four 110 kV lines, and to northern Laos through a 115 kV line. Ranking first in the world in terms of installed capacity, length of transmission lines and voltage levels, China boasted 543,000 km of transmission lines at 220 kV or above and transformer capacity of 2720 GVA as at the end of 2013. Table 1.24 shows the length of China's transmission lines at 220 kV or above in 1995–2013.

Table 1.24 Length of China's transmission lines at 220 kV or above from 1995 to 2013

Years	UHV	750 kV	±660 kV	500 kV	330 kV	220 kV	Total (km)
1995	–	–	–	13,052	5,609	96,913	115,574
2000	–	–	–	26,837	8,669	128,114	163,620
2005	–	141	–	62,866	13,059	177,617	253,683
2010	3,972	6,685	1,400	135,180	20,338	277,988	445,563
2011	3,973	10,005	1,400	140,263	22,267	295,978	473,886
2012	6,105	10,088	1,400	146,250	22,701	318,217	504,761
2013	8,840	12,666	1,400	156,818	24,065	339,075	542,864

Source: the China Electricity Council, Compilation of Electric Power Industry Data (2013).

Japan has also realized nationwide power interconnections for all its territories except Okinawa. The Japanese power grid is comprised of 50 and 60 Hz system. The 50 Hz system has been adopted for the three power grids in Hokkaido, Northeast Japan, and Tokyo, with the grids of Tokyo and Northeast Japan linked up through a 500 kV transmission line, and the grids of Northeast Japan and Hokkaido linked up through a ±250 kV undersea DC transmission line. The 60 Hz system has been adopted for Chubu, Hokuriku, Kansai, Chugoku, Shikoku, and Kyushu, with interconnections achieved through a 500 kV transmission line. These two different systems are connected back to back by three DC lines in Sakuma (300 MW), East Shimizu (100 MW), and Shin Shinano (600 MW).

The Russian power grid consists of seven transregional grids – East Power Grid, Siberia Power Grid, Ural Power Grid, Middle Volga River Power Grid, South Power Grid, Middle Power Grid, and West Power Grid, covering 79 Oblasts of Russia. Six of these regional grids, except for the East Power Grid, which operates independently, interconnect and synchronize with each other.

1.5.2.2 Europe's Power Grids

The European power grid is the most extensively interconnected continental power grid, including five synchronous grids in continental Europe, Northern Europe, the Baltic Sea, the United Kingdom, and Ireland, together with two independent power grids in Iceland and Cyprus. Power grid operators in these countries and regions have come together to form a European alliance of transmission network operators. As at the end of 2013, the alliance was comprised of 34 member countries, with a combined installed capacity of 1,007 GW, electricity generation of 3,350 TWh, 300,000 km of transmission lines at 220 kV or above, a total area of 4.5 million km^2, and a customer base of 700 million people.

Compared to other European countries, the power grids in Germany, France, and Italy are relatively large, with installed capacities of 184.62, 128.06, and 124.23 GW, respectively, and maximum demand loads of 83.1, 92.9, and 54 GW, respectively. The power grids of Germany, France, and Italy feature highest voltage levels of 380, 400, and 400 kV, respectively, while transmission lines at or above 220 kV measure 35,000, 48,000, and 22,000 km, respectively. With the progress of grid interconnection in Europe, the level of electricity exchange between different countries has increased dramatically. Table 1.25 shows the length of AC power grids of major European countries in 2013.

Table 1.25 Length of Transmission Lines of AC Power Grids of Major European Nations in 2013

Countries/Regions	220/285 kV	330 kV	380/400 kV	750 kV	Total (km)
France	26,640	–	21,752	–	48,392
Germany		35,147			35,147
Italy	11,149	–	10,746	–	21,895
UK	6,264	–	11,829	–	18,093
Europe	141,359	9,141	151,272	471	302,243

Source: Based on statistics of the electric power industries of different countries; Ref. [97].

1.5.2.3 North America's Power Grids

North America's power grids include the national power grids of the United States, Canada, and Mexico, together with Central America's interconnected power grid. The North American interconnected grid is comprised of four synchronous power grids in eastern North America, western North America, the United States state of Texas and the Canadian province of Quebec, covering all United States territories, most parts of Canada, and the Baja California region of Mexico. The eastern North American power grid is the largest of the four synchronous grids, covering extensive areas from central Canada eastward to the Atlantic coast (except Quebec), southward to Florida, and westward to the Rocky Mountains (except Texas). It provides power to the eastern and central states of the United States and five Canadian provinces. Second only to the grid of eastern North America by capacity, the western North American power grid covers broad areas from West Canada to the Baja California peninsula, stretching eastward across the Rocky Mountains to the Eastern Plains. The western North American power grid provides power to 14 states of the United States, two Canadian provinces, and parts of one Mexican state.

At the end of 2013, the interconnected power grid of North America featured a total installed capacity of approximately 1200 GW, with the highest voltage class at 765 kV. The grid is supported by 760,000 km of transmission lines at 100 kV or above, covering an area of 11.39 million km^2 and providing electric power to some 500 million people. Table 1.26 shows the length of transmission lines at 100 kV or above of North America's interconnected power grid in 2008–2011.

Table 1.26 Length of Transmission Lines at 100 kV or Above of North America's Interconnected Power Grid from 2008 to 2011

Countries/Regions	2008 (km)	2009 (km)	2010 (km)	2011 (km)
USA	587,378	599,095	622,590	627,502
Canada	126,591	127,041	128,157	130,834
Mexico (Baja California)	2,113	2,256	2,293	2,346
Total	716,082	728,393	753,039	760,682

Based on the statistics of North American Electric Reliability Corporation (NERC).

Central America's interconnected power grid boasts an installed capacity of 11.48 GW, covering an area of 500,000 km^2 and providing electric power to 39 million people. The grid consists of six national power grids in Panama, Costa Rica, Honduras, El Salvador, Guatemala, and Nicaragua.

Compared to other North American countries, the power grids in the United States, Canada, and Mexico are relatively large, with installed capacities of 1068, 134, and 62 GW, respectively, and maximum demand loads of 782, 92, and 50 GW, respectively. For the United States, Canadian, and Mexican grids, the highest voltage classes are 765, 735, and 400 kV, respectively, with 630,000, 130,000, and 50,000 km of transmission lines at or above 100 kV, respectively. There is a high level of electricity exchange between the United States and Canada, while the electricity exchange with Mexico is relatively small. The United States is a net importer of electric power, while Canada and Mexico are net exporters. In 2013, the power imports and exports of the United States totaled 74.9 TWh.

1.5.2.4 South America's Power Grids

No interconnected power grid on a continental level has yet been developed in South America. There are two major transnational interconnected grids in the north and south of South America, covering 14 countries, with a combined installed capacity of 240 GW and electricity consumption totaling 1000 TWh. The two grids provide power to about 400 million people and cover vast areas measuring 15.5 million km^2. The highest voltage of the South American power grids is 750 kV, with the length of transmission lines at 220 kV or above estimated at approximately 250,000 km.

Compared to other nations in South America, Brazil, and Argentina maintain relatively large power grids, with installed capacities of 117.13 and 33.81 GW, respectively, and the highest voltages at 750 and 500 kV, respectively, based on 2013 figures. The transmission lines at 220 kV or above in Brazil and Argentina measured 100,000 and 10,000 km, respectively. Table 1.27 shows the length of transmission lines at 220 kV or above in the two countries in 2013.

1.5.2.5 Africa's Power Grids

Africa's power grids cover more than 50 countries and regions, with a total installed capacity of 150 GW and electricity consumption of 700 TWh. The grids provide electric power to approximately 1 billion people. Connections between the national grids of African countries are relatively weak, with efforts focused mainly on self-balancing of supply and demand. Other than the one in southern Africa, no regionally interconnected grids are present.

The southern African interconnected power grid covers nine countries, namely – Botswana, Mozambique, South Africa, Lesotho, Namibia, Congo, Swaziland, Zambia, and Zimbabwe. At the end of 2013, the total installed capacity of the southern African interconnected grids stood at 57.18 GW, covering an area of 6.96 million km^2 and supplying power to a population of 176 million people, with

Countries	220/230 kV	345 kV	440 kV	500 kV	600 kV	750 kV	Total (km)
Table 1.27 Length of Transmission Lines at 220 kV or Above in Brazil and Argentina in 2013							
Brazil	45,709	10,062	6,681	35,003	3,224	2,683	103,362
Argentina	11,113	1,116	0	1,884	0	0	14,113

Source: See Ref. [109].

Table 1.28 Length of Transmission Lines at 220 kV or Above in South Africa and Egypt in 2013

Countries	220 kV	275 kV	400 kV	500 kV	±533 kV	765 kV	Total (km)
South Africa	1,217	7,360	16,899	–	1,035	1,667	28,178
Egypt	17,001	–	33	2,863	–	–	19,897

Source: Japan Electric Power Information Center, 2013 Statistics of Overseas Electric Industries.

a maximum load of 53.83 GW, an electricity shortfall of 7.71 GW, the highest voltage level of 765 kV, and 30,000 km of transmission lines at 220 kV or more.

South Africa and Egypt have relatively large power grids, with installed capacities of 44.17 and 30.05 GW, the highest voltages at 765 and 500 kV and transmission lines at 220 kV or above measuring 30,000 and 20,000 km, respectively. Table 1.28 shows the length of transmission lines at 220 kv or above in the two countries in 2013.

1.5.2.6 Oceania's Power Grids

The Oceanian power grids cover 14 countries and provide 30 million people with electricity, with a total installed capacity of 75 GW and electricity consumption totaling 300 billion kWh. Like Australia, Oceanian countries such as New Zealand and Papua New Guinea are island states where the independent operation of grids allows little room for grid interconnections. The highest voltage level of the Oceanian power grids is 500 kV, with an estimated 30,000 km of transmission lines at 220 kV or above.

Australia and New Zealand have relatively large power grids, with installed capacities of 63220 and 9490 MW, respectively and the highest voltage level at 500 kV. Transmission lines at 220 kV or above in Australia and New Zealand are estimated at 30,000 and 20,000 km, respectively. Divided by administrative region, Australia has nine regional power grids in Victoria, New South Wales, Queensland, South Australia, Australian Capital Territory, Snow Mountains, Tasmania, Northern Territory, and Western Australia. See Table 1.29 for the length of transmission lines at 220 kV or above in Australia from 1995 to 2011.

Table 1.29 Length of Transmission Lines at 220 kV or Above in Australia during 1995–2011

Years	500 kV	330 kV	275 kV	220 kV	Total (km)
1995	2,574	6,261	7,304	7,100	23,239
2000	1,611	6,853	8,547	7,133	24,144
2005	2,574	7,700	9,368	7,245	26,887
2010	2,588	8,028	11,137	7,228	28,981
2011	2,588	8,028	11,137	7,229	28,982

Source: Japan Electric Power Information Center, 2013 Statistics of Overseas Electric Industries.

1.5.3 Electricity Consumption

Global electricity consumption has continued to go up rapidly at a rate faster than energy consumption. Between 1980 and 2013, the world's annual electricity consumption rose from 7300 TWh to 22,100 TWh. Since the twenty first century, global electricity consumption has seen even faster growth, as evidenced by an average annual increase of 3.4%, 1.2 percentage points higher than average annual growth of energy consumption. Fig. 1.28 shows global electricity consumption during 1980–2013.

Emerging economies in Asia and Central and South America have witnessed far more significant growth of electric power consumption compared to the developed regions. In 2013, the levels of electricity consumption in Asia, Central, and South America, and Africa were 9820, 1050, and 710 TWh, respectively, up 5.8, 2.4, and 2.3 times, respectively, from 1980. During the same period, North America and Europe experienced growth of electricity power consumption by 87 and 148%, respectively, far lower compared to growth in the emerging and developing economies. Fig. 1.29 shows the change in the share of each continent's electricity consumption in the world total during 1980–2013.

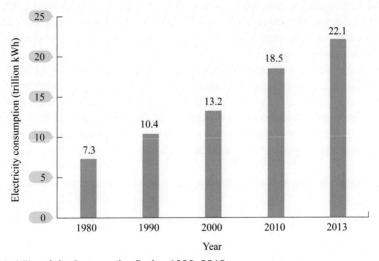

FIGURE 1.28 Global Electricity Consumption During 1980–2013

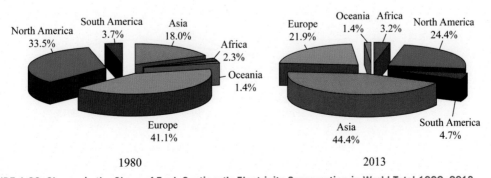

FIGURE 1.29 Change in the Share of Each Continent's Electricity Consumption in World Total 1980–2013

Annual electricity consumption per capita serves as an important measure of a country's electric power development. Generally speaking, electricity consumption grows faster when the industrialization process develops quickly and goes down rapidly when industrialization is completed or near completion. The same can be said about annual electricity consumption per capita. Between the 1950s and the early 1970s, during the fastest growth of the United States economy in history, a threefold increase in annual electricity consumption per capita was recorded from 1990 kWh in 1950 to 7870 kWh in 1973. However, as the oil crisis and the economic meltdown took their toll, the United States went into a period of stagflation, with slower growth in annual electricity consumption per capita. The Japanese economy also grew strongly from the mid-1950s to the late-1960s, triggering a surge of electricity demand and significant growth in electricity consumption per capita. However, the two oil crises in the 1970s left the Japanese economy badly battered, driving down both economic growth and electricity demand growth. The United Kingdom saw rising electricity consumption per capita between World War II and the mid-1970s, driven by faster economic growth. In the 1970s and 1980s when the economy slowed, per capita electricity consumption growth went down as well. After reaching 6115 kWh in 2000, the United Kingdom's electricity consumption per capita has dropped slightly in recent years. After industrialization has run its course, a developed country typically registers annual electricity consumption per capita of 4500–5000 kWh. Fig. 1.30 shows changes in annual electricity consumption per capita of major countries during 1960–2013.

In 2013, the world's annual electricity consumption per capita reached 3084 kWh, up 42.3% from 1990. By continent, Asia recorded annual electricity consumption per capita of 2,355 kWh, with Bahrain, South Korea, and the United Arab Emirates among the top three consumers each exceeding 10,000 kWh. Europe's annual electricity consumption per capita was 6543 kWh, with Iceland, Norway, and Finland ranking among the top three, each consuming over 15,000 kWh. In North America, annual electricity consumption per capita amounted to 10,226 kWh, with Canada and the United States accounting for 16,000 and 13,000 kWh, respectively. South America's annual electricity consumption per capita was 2242 kWh, with Chile, Venezuela, and Argentina among the top three recording annual

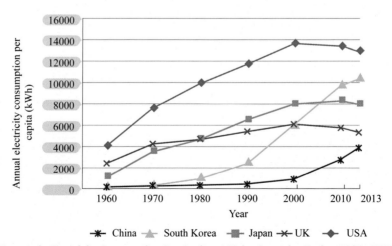

FIGURE 1.30 Changes in Electricity Consumption Per Capita of Major Countries During 1960–2013

Source: Ref. [68]; Ref. [21].

Table 1.30 Top Three Countries/Regions of each Continent by Annual Electricity Consumption Per Capita in 2013

Region	Country/Region	Annual Electricity Consumption Per Capita (kWh)
Asia	Bahrain	17,601
	South Korea	10,382
	UAE	10,175
	Iceland	54,414
Europe	Norway	23,215
	Finland	15,392
	Canada	15,765
North America	USA	12,871
	Mexico	2,099
	Chili	3,807
South America	Venezuela	3,401
	Argentina	3,027
	Gibraltar	5,344
Africa	South Africa	4,410
	Botswana	1,568
	Australia	10,010
Oceania	New Zealand	8,794
	Papua New Guinea	500

Source: IEA, Energy Balances of OECO Countries 2014; IEA, 2014 Key World Energy Statistics.

electricity consumption per capita of 3000–3800 kWh. Africa's annual electricity consumption per capita was 663 kWh, with Gibraltar, South Africa, and Botswana among the top three. In Oceania, annual electricity consumption per capita reached 9,500 kWh, with Australia accounting for 10,000 kWh. Fig. 1.30 shows the top three countries/regions of each continent by annual electricity consumption per capita (Table 1.30).

With the economic restructuring now underway, the composition of the world's electricity consumption is also changing. Reflecting the falling share of the industrial sector, especially energy-intensive industries, in total electricity consumption, industrial use of power, currently accounting for 30% of total consumption, in developed countries like the United States, Japan, and the United Kingdom, is declining continuously, in contrast with the increasing share of commercial service and residential use in total consumption. In developed countries, the composition of electricity consumption shows balanced proportions of the industrial, commercial services, and residential sectors in total consumption, compared to the relatively low shares of transportation and agriculture. Table 1.31 shows the composition of electricity consumption of the world's major countries in 2012.

Table 1.31 Percentage Composition of Electricity Consumption of World's Major Countries in 2012

Countries	Industries (%)	Transport (%)	Agriculture (%)	Commercial Services (%)	Residential Sectors (%)	Others (%)
China	73.7	1.8	2.3	3.1	12.1	7.0
USA	24.9	0.2	0.0	34.2	37.1	3.6
Japan	34.3	1.9	0.1	33.4	30.0	0.4
Russia	57.1	9.5	1.7	18.7	13.0	0.0
Germany	44.2	3.1	1.7	26.1	25.0	0.0
Canada	38.9	0.8	1.9	28.9	29.6	0.0
France	30.8	2.7	0.8	30.8	34.4	0.6
United Kingdom	33.6	1.2	1.2	28.9	35.2	0.0
Italy	44.7	3.4	1.8	27.6	22.4	0.0
South Korea	51.7	0.5	2.2	32.2	13.4	0.0

Source: Ref. [69]; IEA, Energy Statistics of Non-OECD Countries 2014.

2 CHALLENGES TO GLOBAL ENERGY DEVELOPMENT

Fossil fuels have long supported the development of industrial civilization. But in view of the problems created by fossil energy that threaten human existence, such as environmental pollution and climate change, it is time to change energy production and consumption based on fossil fuels. A case in point is the rapid growth of the world's clean energies like wind and solar power, despite the major challenges still confronting technology innovation, equipment R&D, engineering application, system safety, and economics.

2.1 CHALLENGES TO ENERGY SUPPLY

2.1.1 Total Volume Growth

Total energy consumption worldwide will continue to show a rising trend for a relatively long period. Fuelled by global economic growth, world energy consumption rose 2.4 times, from 5.38 billion tons of standard coal in 1965 to 18.19 billion tons of standard coal in 2013.[14] Energy consumption is expected to remain high, as it is difficult to reverse the long-established pattern of intensive energy consumption in developed countries. Due to the continued shift of the heavy chemical industry from developed nations into developing countries,[15] developing nations are seeing a trend of rapid growth in energy consumption. From 1990 to 2013, China's annual energy consumption registered average growth of 6%, accounting for 47.4% of the world's total new energy consumption growth. In particular,

---

[14]Ref. [65].

[15]Heavy chemical industry generally refers to the production of materials, including the energy, machinery manufacturing, chemicals, metallurgical, and building materials industries.

Table 1.32 Forecast average annual growth in energy consumption of the world and major regions from 2012 to 2040

Country/Region	Average Annual Growth 2012–2040 (%)
World	1.5
OECD Countries	0.4
Non-OECD Countries	2.0
China	3.3

Source: Ref. [70].

since 2000, China has recorded new energy consumption growth of 180 million tons of standard coal annually, equivalent to the total annual consumption in Spain.

The world's energy consumption will maintain a growing trend in the future. The International Energy Agency (IEA) has forecasted that 60% or more of the world's primary energy consumption growth will come from developing countries between 2000 and 2030, and the share of developing countries in the world's total energy demand will rise from 30% in 2000 to 43% in 2030, a trend that will continue up to 2050, when average annual growth of more than 1% is still expected. Meeting such huge demands will pose challenges to all aspects of energy development, allocation, and application. Table 1.32 shows the forecasted average annual growth in energy consumption of the world and major regions during 2012–2040.

2.1.2 Resource Constraints

Resource constraints can be perceived from a whole as well as structural perspective. *On the whole, as reserves of fossil fuels are limited and nonrenewable, large-scale exploration will definitely fasten the depletion of these resources.* Oil reserves easily explorable are now diminishing rapidly and concentrated in a very few countries. Coal reserves easily minable can sustain just several decades of exploration. In the foreseeable future, resource constraints will become a major bottleneck inhibiting the sustainable supply of energy. How to resolve this stressful energy resource situation will become an issue with socioeconomic development implications that mankind must be prepared to deal with. The most fundamental solution is to reduce overdependence on fossil fuels and to step up the exploration of renewable clean energies.

Structurally, the world's energy resources and consumption are inversely distributed, with energy development being increasingly concentrated in a very few countries and regions. Some resource-scarce countries are depending more on energy imports, resulting in a fragile supply chain and prominent security issues. In China, for instance, the level of dependence on oil imports has exceeded 60%, compared to over 30% on gas imports. This resource endowment structure has made the supply of resources more challenging and added to the costs of supply. The situation calls for the development of a platform for optimal resource allocation over larger areas to break the bottleneck in the allocation of energy resources.

2.1.3 The Costs of Supply

The costs of energy supply are an important economic factor that affects energy development. Currently, the supply costs of fossil fuels and clean energies show an opposite trend.

The exploration cost of fossil energies continues to rise gradually. As the exploration of coal, oil, and natural gas continues to gain momentum, the marginal development costs of fossil energy will grow gradually, leading directly to higher supply costs of energy resources. According to IEA statistics, the average marginal costs of the world's oil development rose from US$ 30/barrel in 2003 to US$ 80/barrel in 2011. With the growing exploration of oil and gas, and the higher level of resource utilization, the future focus of oil and gas exploration will shift gradually into the deep sea and polar regions where geological and geographical conditions are highly complex, leading to continuously higher exploration costs. The costs of coal have also continued to rise, reflecting increasing difficulty in exploration. Additionally, in view of the scarcity of fossil resources and their impact on the environment, some countries and regions have already begun or are planning to impose taxes on natural resources, carbon, and pollution, which will drive the prices of fossil energy resources even higher.

The exploration costs of clean energies go down gradually but still at a high level. With the rapid growth of clean energy technology, the costs of clean energy development have been declining. From 2009 to 2013, the price of photovoltaic systems around the world dropped 60%, with the average electricity cost going down from 35 US cents/kWh to 23 US cents/kWh. The price of wind turbines in the global market fell 10%, with the costs of onshore wind power down from 13 US cents/kWh to 10 US cents/kWh, and offshore wind power from 28 US cents/kWh to 20 US cents/kWh. Due to limited overall capacity, the average costs per kWh of solar thermal power saw a limited drop and, at close to 30 US cents/kWh, were much higher than the average costs of photovoltaic power and wind power. Fig. 1.31 shows the changes in the costs of wind and solar power from 2009 to 2013. On the whole, the average costs per kWh of renewable power, such as wind and solar energy, are still higher than the average costs, at 6–10 US cents/kWh, of thermal power. The economics of clean energies need to be further enhanced to make them more competitive in the market, so as to truly realize large-scale substitution of clean energies for fossil fuels. The opposite price trends of fossil fuels and clean energies also determine the future direction of clean energies and contribute to their great development potential.

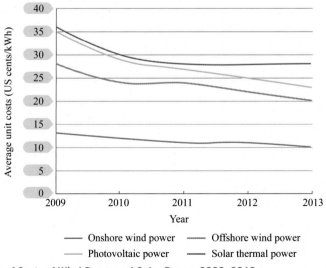

FIGURE 1.31 Changes of Costs of Wind Power and Solar Power, 2009–2013

2.2 CHALLENGES TO THE ENERGY ENVIRONMENT

2.2.1 Global Warming

Burning of fossil fuels is the main source of global greenhouse gas emissions.[16] The world's carbon dioxide from the burning of fossil energy accounts for 56.6% of global greenhouse gas emissions generated by human activity.[17] In the past 160 years, extensive use of fossil fuels have raised the concentration of carbon dioxide in the atmosphere from about 280 ppm to around 400 ppm (Fig. 1.32).[18] Energy production will remain the decisive factor that affects greenhouse gas emissions now and in the distant future. According to the fifth assessment report of the UN's Intergovernmental Panel on Climate Change, global warming is happening due to mankind's extensive use of fossil fuels. Between 1880 and 2012, the global temperature went up by 0.85°C.

Greenhouse gas emissions produce greenhouse effect and pose the subsequent four major threats to human survival and development. *Land area is shrinking.* Global warming leads to the melting of glaciers, permafrosts, as well as rising sea levels. NASA research shows that the melting of Greenland ice and snow will add seven meters to sea level rise such that coastal cities like New York, Shanghai, and London will be swamped by floods. If the Antarctic Ice Sheet melted, sea level would rise by 57 m and some low-lying countries such as Britain, France, and the Netherlands would simply disappear under water. The impact would be disastrous:

- *A large number of species will become extinct.* According to the fourth assessment report of the UN Intergovernmental Panel on Climate Change (IPCC), if the global temperature whould rise by 1.5–2.5°C, 20–30% of all species would face extinction. If the temperature should rise by more

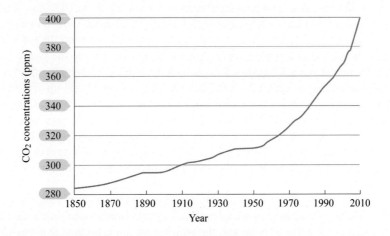

FIGURE 1.32 Atmospheric Concentrations of CO$_2$ During 1850–2010

Source: http://www.globalcarbonproject.org/activities/AcceleratingAtmosphericCO2.htm.

---

[16]The carbon dioxide emission factor of raw coal is approximately 1.902 t/ton, compared with approximately 3.094 t/ton for crude oil and approximately 2.173 kg/m^3 for natural gas.

[17]Source: Intergovernmental Panel on Climate Change (IPCC), Climate Change 2007: Synthesis Report.

[18]ppm refers to part per million concentration.

than 3.5°C, 40–70% of all species would face extinction. If the current mode of fossil energy consumption should go unchecked, the average global temperature would rise by 3–6°C by the year 2100.

- *Global warming will pose a threat to food supplies.* According to China Meteorological Administration research findings, if the current mode of energy development should continue, global warming would cause China's farming production to go down by 5–10% by 2030 and food supplies will decline by 20–30% 50 years from now. By then, food shortage would pose the greatest threat to humanity.[19]
- *Global warming will threaten human health.* Global warming will lead to more frequent extreme weather conditions like floods, droughts, heat waves, and typhoons, which will cause the mortality and disability rates of certain diseases. Hence the incidence of infectious diseases might go up, thus posing a serious threat to human health. According to the fourth assessment report of the UN IPCC, if the average global temperature should increase by 2°C, the proportion of world population at risk of malaria could rise to 60% from the current 45%, and five million to eight million new cases would ensue each year. Global climate change will further threaten human health with its damaging effect on ecosystems like forests, farmlands, and natural wetlands.

2.2.2 Damage to the Ecological Environment

The combustion of fossil fuels produces large quantities of soot and other pollutants, resulting in frequent smogs that seriously undermine human health. Since the Industrial Revolution, compound air pollution, characterized by fine particles of nitrogen oxides, hydrocarbons, and secondary pollutants, has occurred in most developed countries and some developing countries, resulting in lower atmospheric visibility, more hazy days and growing threats to human health. In 1930, the Maas River Valley fog event in Belgium left thousands of people with respiratory diseases, with a death toll that was 10 times higher compared to the deaths from natural causes in the same period. In 1952, the worst smog disaster in history broke out in London, causing over 4000 deaths. Due to the heavy burning of fossil fuels such as coal in recent years, China has also been hit by regional smoggy weather, characterized by extensive affected areas, long duration, high pollution levels, and rapid build-up of pollutant concentrations. In 2013, most areas stretching from the central and southern parts of North China to the northern parts of the southern Yangtze River Basin experienced 50–100 days of foggy and smoggy weather, with some areas affected for more than 100 days, as shown in Fig. 1.33. As indicated by the figures, the smog-inducing air pollution lasted 17 days, causing 74 cities to suffer 677 days of pollution, including 477 days of heavy pollution and 200 days of serious pollution. Such air pollution has become the most deep-seated threat to public health.

Fossil fuel combustion produces large quantities of pollutants such as sulfur dioxide, leading to environmental pollution such as acid rain that seriously affects humans and production activities. Currently, global emissions of sulfur dioxide into the atmosphere are estimated at about 120 million tons every year, 80% of which are caused by fossil fuel combustion. In 2013, China's territories affected by acid rain exceeded 1 million km², concentrating along the Chang Jiang River and areas south of the middle and lower reaches of the river, covering most areas of Jiangxi, Fujian, Hunan, Chongqing, as well as the Chang Jiang River delta, the Pearl River delta, and south-eastern Sichuan province. Severe

---

[19]Source: DONG Chongshan, "Predicament and Breakthroughs – Overall Energy Crisis of Human and Solutions," People's Publishing House, 2006.

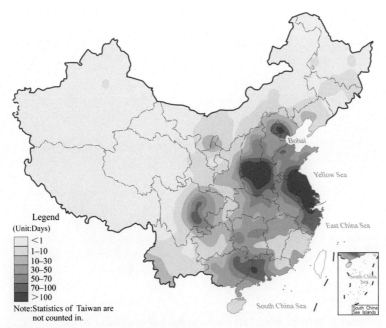

FIGURE 1.33 2013 Distribution of Smoggy Days in China

Source: Environment Bulletin of China 2013, Ministry of Environmental Protection of the People's Republic of China.

acid rain affects an area of about 60,000 km², hitting 44.4% of all cities as shown in Fig. 1.34. Mercury emissions from coal combustion have also caught growing attention in recent years, with 45% of global anthropogenic mercury emissions coming from coal combustion.

Large quantities of fossil fuels cause severe pollution to water, soil, and air, in all the areas of the mining, transportation, and utilization processes. Coal mining brings about ground subsidence, renders land barren, damages vegetation, and seriously undermines the ecological system of the mining areas. Mining operations also lead to lower underground water levels, exposing this body of water to large quantities of pollutants that cause damage to water resources. The area of ground subsidence caused by coal mining in China has reached 7000 km², while approximately 2.2 billion m³ of underground water gets polluted every year. Coal production accounts for about 25% of China's total waste/polluted water effluents every year.[20]

Storage and transport of coal can also affect the environment. Coal gangue piles not only take up land space, but also cause severe pollution to rivers and underground waters if the poisonous and harmful substances and salts contained in the coal gangue go into the soil and water. The spontaneous combustion of coal will release a large quantity of harmful gas, causing air pollution. The transportation of coal will also bring serious dust pollution to the areas along the railway. Within a 100 m radius of the railway used by coal trains, flying coal dust will cause total suspended particle concentrations to rise significantly. And within a distance of 50 m on both sides of the railway, exceedance of transient concentration limits may occur.

[20]Source: Ref. [5].

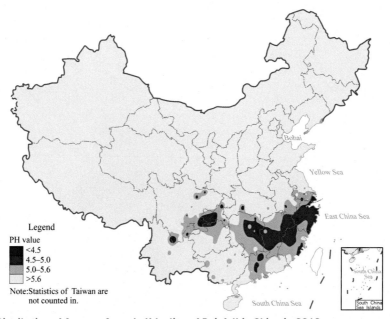

FIGURE 1.34 Distribution of Average Annual pH Isoline of Rainfall in China in 2013

Source: 2013 Environment Bulletin of China, Ministry of Environmental Protection of the People's Republic of China.

In addition, the exploration and utilization of oil and gas will produce pollutants like wastewater, waste gas, and oily sludge, with a negative effect on water, air and soil. The penetration of corroded pipelines can also seriously contaminate soil and underground water sources, causing not only soil salinization and poisoning that lead to soil degradation and destruction, but also harm to human health eventually as a result of the toxic substances entering the food chain through crops or underground waters. The United States has to face possible environmental problems such as damage to the ecosystem, pollution of underground waters, and methane emissions brought about by the recent shale gas revolution.

2.3 CHALLENGES TO ENERGY ALLOCATION

2.3.1 Allocation of Fossil Energy

Global allocation of fossil energy is characterized by large volumes, multiple steps, and long transmission distance. The existing modes of sea, rail, or highway transportation, usually entail a long chain of operations with low efficiency and require the integration of multimodal transport to complete the entire energy–transport process. As a result, this complex transport system is easily subjected to external influences. Particularly in international transport, external factors, geopolitical, and otherwise, will have a significant impact on energy security and prices.

Take oil transportation as an example, global oil trade is currently highly dependent on offshore oil transport corridors such as the Malacca Strait and the Hormuz Strait, which are exposed to high potential risks. In the event of political unrest or war in one of the countries concerned, the lifeline of offshore oil transportation will probably be interrupted, thereby threatening the world's oil supply and transport chain. To ensure the safety of offshore oil transport, the countries concerned have been compelled to launch military escorts, incurring massive military spending. With the steady growth of global natural gas trade, the secure allocation of global natural gas has become increasingly prominent, as the delivery of piped gas is strongly influenced by geopolitical factors.

In recent years, international geopolitical conflicts have been on the rise and increasingly centered on oil and gas resource–rich countries in the Middle East and North Africa. The intensifying local conflicts pose a serious challenge to the security of oil imports for the world' s major energy consumers, including China and Japan. Since the 1970s, the Middle East and North Africa have been facing multiple problems of population, employment, religion, and race. Since 2010, countries in the Middle East and North Africa have seen more political upheavals or even armed conflicts. The complicated geopolitical situation determines that global oil and gas energy supplies are highly vulnerable.

2.3.2 Allocation of Clean Energies

As global energy development is increasingly focused on clean energies, the importance of electricity as an energy resource for long-distance, large-area allocations has come into the spotlight. However, the current level of power allocation capacity is obviously insufficient. To cope with the challenges of global energy supply and the energy environment, clean energy development is the way forward as the world's energy structure is trending toward being clean-energy oriented from being fossil-fuel based. This is being accompanied by a gradual shift in the world's energy demand from fossil energy to clean energy. As hydro, wind, solar, and other clean energies need to be converted into electricity before they can be effectively used, the important role of electric power in the future allocation of global energy is ensured. Currently, the allocation capacity of the world's electric power is limited and can hardly satisfy the future development of clean energy. Compared with fossil energy, the current level of global electricity trade is minimal, representing approximately 80 million tons of standard coal or less than 2% of fossil energy trade.

The current coverage of power allocation is limited and cannot meet the need for worldwide allocation of clean energy. Judging by the global distribution of clean energy resources, high-quality wind and solar energy resources are concentrated in the polar and equatorial regions and the major clean energy bases on each continent and are hundreds to thousands of kilometers away from load centers. The economic transmission distance of existing EHV power grids ranges from 500 km to 1000 km, which cannot meet the need for large-area development and allocation in the future.

To meet the requirement for large-scale development of clean energy, there is a pressing need to establish an efficient global platform for electric power allocation. With the large-scale development of clean energy, an energy allocation structure focused on electric power is in the making. To meet the need for large-scale and long-distance allocation of clean energy, a global platform oriented toward clean energy must be established with the focus on electricity, higher voltage levels, bigger transmission capacity and longer transmission distances. This being the case, ways to improve the capacity of global electric power allocation to meet the power demands of various regions in the world, will become the crux and focus of future energy development.

2.4 CHALLENGES TO ENERGY EFFICIENCY

Currently, the efficiency of exploration, allocation, and application of either fossil fuels or clean energies is still relatively low, with significant scope for improvement.

2.4.1 Exploration

The exploration and utilization rates of energy resources are low. Currently, the average recovery rate of global oil reserves is only 34%, compared with 65–70% for coal, 80–90% for natural gas from constant volume depletion gas fields, and 65–80% for natural gas from gas condensate fields. Problems with the exploration and utilization of coal, oil, and other mineral resources, are rife in some developing countries. Examples are scattered mining of large deposits, disorderly and indiscriminate mining, and selective mining of rich deposits rather than infertile deposits. According to statistics, the average recovery rates of China's coal and oil deposits stand at only 35 and 28%, respectively, far below the world's average.

Energy conversion efficiency is also low. At present, China's average thermal coal consumption is about 330 g of standard coal/kWh. Germany has a relatively higher efficiency at an average of 290 g of standard coal/kWh. In comparison, some countries have a thermal coal consumption of as much as 370 g of standard coal/kWh, indicating much room for improvement. Constrained by exploration and application technologies, the exploration and utilization of wind and solar energy is inefficient. Studies have shown that the integrated efficiency of wind power is around 38%, and the generating efficiency of photovoltaic power ranges from 12% to 18%, again indicating major room for improvement.

2.4.2 Allocation

The allocation of fossil energy involves a multitude of processes resulting in low efficiency. Not just used directly for terminal consumption to a certain extent, significant quantities of fossil fuels, including coal, natural gas, and even fuel oil, are also used to generate electricity. Power generated from fossil fuels usually goes through a series of processes before arriving at a power plant, which results in huge energy losses. In China, energy development is excessively dependent on coal, while electric power development aims to achieve local balance, a situation that leads to strained transport capacity, high costs of thermal power generation, and soaring coal prices. Compared especially with rail transport, it consumes 10 times the energy to carry each tonne of coal per kilometer by highway, which also contributes to traffic jams, and even higher consumption of gasoline. Fig. 1.35 shows the processes involved in coal transport and power transmission.

The key to solving the problems of complicated thermal coal transport processes and heavy loss of energy is to shift the mode of energy development away from being focused on achieving local balance toward a one-stop allocation solution by replacing coal transport with electricity transmission, especially through UHV grids with the advantages of high capacity, long distance, high efficiency, low loss and space economy. In terms of transmission capacity and distance, a 1000 kV UHV AC transmission line is four to five times and three times, respectively, as much compared with a 500 kV AC transmission line, while the loss of energy per unit is only one-quarter to one-third the level sustained by a 500 kV AC line. A ±800 kV UHV DC transmission line has a transmission capacity of 10 GW and a transmission distance of up to 2500 km, 3 and 2.5 times, respectively, as much as a ±500 kV DC transmission line, with the loss of energy per unit at only 73% of a ±500 kV DC transmission line. Table 1.33 shows the correlation between voltage levels, transmission capacity and transmission distance. Compared with coal transport, UHV transmission presents a one-stop power solution by making it possible to deliver, at one go, large quantity of hydro, nuclear, wind, and solar power to load centers. However, the technology edge of UHV grids has yet to be fully leveraged, given the limited scale of UHV application in a world still dominated by extra-high power grids.

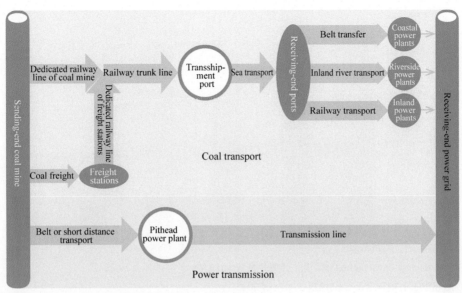

FIGURE 1.35 Diagram of Coal Transport and Power Transmission

Table 1.33 Correlation between Voltage Levels, Transmission Capacity, and Transmission Distance			
Items	**Voltage (kV)**	**Transmission Capacity (10,000 kW)**	**Transmission Distance (km)**
AC	10	0.02–0.2	6–20
	35	0.2–1	20–50
	110	1–5	50–150
	220	10–30	100–300
	330	20–80	200–600
	500	100–150	150–850
	750	200–250	Over 500
	1000	200–600	1000–2000
DC	±500	300	1000
	±800	800	2500
	±1100	1200	5000

2.4.3 Utilization

Energy utilization is inefficient. On the whole, energy utilization in developed countries is generally more efficient than in developing countries. Energy consumption per unit of GDP among OECD countries is only 25% of the non-OECD countries.[21] In 2012, China's energy consumption per unit of GDP was 2.7 times higher than the global average. Fig. 1.36 shows the energy consumption per unit of

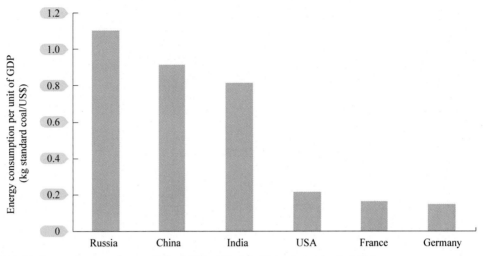

FIGURE 1.36 Energy Consumption per Unit of GDP of World's Major Countries in 2012

Note: Data of Fig. 1.36 is in constant US dollars of 2005.

Source: Refs. [67] and [68].

GDP of the world's major countries in 2012. The overall efficiency of China's energy processing and conversion was 72.4%, which was 10–20% lower than the world average, while the average energy consumption per unit of each product of China's major energy-intensive industries was 10–15% higher than the international advanced level. The energy consumption of heating per unit of floor area in some developing countries is two to three times higher than the level in developed countries with similar climate conditions.

Electric power accounts for a small share of terminal energy consumption. Electricity has a higher efficiency of terminal consumption than other forms of energy. In China, studies have shown that every one percentage point increase in the share of electricity in terminal energy consumption will translate into a 3.7 percentage point decrease in energy intensity.[22] Energy saving at the end-user segment of power consumption has an amplifying effect: every 1% increase in the relative efficiency of terminal consumption will contribute to a relative efficiency 4–5% higher than the energy production sector. A greater share of electricity in end-use energy consumption can help raise economic output and improve overall energy efficiency at the community level. In 2012, electricity accounted for just 18.1% of global terminal energy consumption. The level of electrification urgently needs to be further increased to realize improved energy conservation and efficiency.

On the whole, global energy development is facing serious challenges in terms of resources, environment, allocation, and efficiency, especially in view of the environmental problems like air pollution, climate change, and resource depletion resulting from large-scale exploration and utilization of fossil fuels. Other practical issues remain daunting, including high costs, low efficiency, and difficulties in

---

[21]Source: Ref [67].

[22]Source: State Grid Energy Research Institute, "Study of the Impact of Total Energy Consumption Control on Power Demand," 2011.

energy allocation over long distances. To meet these challenges, intensive efforts are required to promote an energy revolution so as to drive the global development of energy resources that are safe, efficient, clean, and sustainable.

SUMMARY

1. Global energy development has undergone a shift from firewood to other fossil fuels like coal, oil, and natural gas, providing a driving force to the first and second industrial revolutions. The faster growth of clean energy we see today is set to lift the postindustrial revolution and human civilization to a new height.
2. Coal, oil, and natural gas are the world's most important primary energies. However, the overdependence on fossil fuels is unsustainable and has created pressing problems such as resource depletion, higher costs, environmental pollution, and climate change.
3. In a world rich in renewable energy resources, including hydro, solar, and wind power, clean energy is the strategic direction of future energy development. Given the current limited capacity of clean energy development and less favorable economics, technology innovation must be pursued to resolve the existing constraints on clean energy development, in terms of energy conversion, resource allocation, and efficient utilization.
4. The Arctic and equatorial regions are rich in clean energy resources, making them important strategic bases of clean energy development around the world. We need to rely on UHV transmission technology to achieve safe, economical, and efficient exploration and utilization of energy.
5. In the face of global problems such as energy supply, environmental concerns, energy allocation, and energy efficiency, efforts are required to quicken the pace of energy revolution to drive the intensive development of clean energy, and pave way for sustainable energy that is safe, clean, and efficient.

CLEAN ENERGY REPLACEMENT AND ELECTRICITY REPLACEMENT

2

CHAPTER OUTLINE

1 WIND AND SOLAR ENERGY DEVELOPMENT IN THE WORLD

The Earth is abundant in wind and solar energy resources, yet efficient utilization of these energy sources can only be achieved by converting them into electric power. The world's wind and solar energy technologies have undergone a slow process of development since the nineteenth century. Since the twenty-first century, wind, and solar energy development has entered a new phase of large-scale exploration, with more importance being attached to clean energy development and new energy technology advancing rapidly.

1.1 WIND ENERGY DEVELOPMENT

In 1887, Prof James Blyth at Anderson's College, now the University of Strathclyde, in Scotland installed the first power-generating windmill in the world, in his house, marking mankind's entry into an advanced age of wind energy exploration. The United States, Germany, Denmark, the Soviet Union, and France followed suit by commencing wind energy research, development and application, but overall progress was slow. Since the two severe global oil crises in the 1970s and 1980s, the utilization of wind power has gained global recognition and grown rapidly, under the double impact of the tightening supply of conventional fossil energy and the deterioration of the global ecological environment.

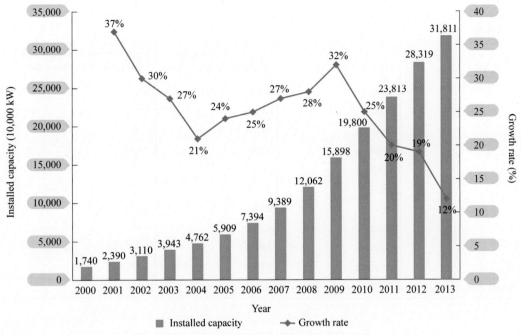

FIGURE 2.1 World Installed Wind Power Capacity and Growth Rates, 2000–2013

Source: Global Wind Energy Council (GWEC), Annual Market Update 2013.

1.1.1 Rapid Growth in Installed Capacity

In the early twenty-first century, Europe and North America were the fastest growing regions in the world in wind energy development. In recent years, Asia has emerged quickly to become the world's major wind power market. In 2013, global installed wind power capacity totaled 320 GW, accounting for approximately 5.6% of the world's total, and wind power generation was estimated at 640 TWh, accounting for approximately 2.9% of the world's total. From 2000 to 2013, both installed wind power capacity and power generation grew 17 times, representing an annual growth rate of 25.0%. See Fig. 2.1 for the world's installed wind power capacity and growth rates from 2000 to 2013.

By the end of 2013, 6 countries in Europe, 2 countries in North America, and 2 countries in Asia were among the world's top 10 nations in terms of installed wind power capacity. Collectively these countries boasted an installed wind power capacity of approximately 85% of the world's total. See Table 2.1 for the basic information on the world's top 10 countries in terms of installed wind power capacity in 2013.

A total of 103 countries and regions around the globe have been developing and harnessing wind energy, with the United States and some European Union countries in particular accounting for a relatively large share of the world's total capacity. Wind power has become the largest source of power supply in Denmark and Spain, representing, 34 and 21%, respectively, of the total electricity consumed nationwide. Wind power represents 20, 16, and 9% of total electricity consumption in Portugal, Ireland, and Germany, respectively.

Table 2.1 Basic Information on Top 10 Countries in Terms of Installed Wind Power Capacity, 2013

Rankings	Countries	Installed Capacities (10,000 kW)	Shares of Total Nationwide Installed Power Capacities (%)
1	China	7548	6.2
2	USA	6109	5.7
3	Germany	3425	19.3
4	Spain	2296	21.8
5	India	2015	8.1
6	UK	1053	11.1
7	Italy	855	6.9
8	France	825	6.4
9	Canada	780	5.8
10	Denmark	477	33.9

Source: Annual Market Update 2013, Global Wind Energy Council (GWEC); Ref. [69].

1.1.2 Rapid Progress in Wind Power Technology

Wind power technology is concerned mainly with wind energy resources assessment and prediction, wind power equipment manufacturing technology, wind turbine testing, and grid interconnection technology. The development of and breakthroughs in equipment manufacturing technology, being the core of wind power technology, hold the key to realizing large-scale commercialization of wind power. Since the mid-1990s, the world's major wind turbine manufacturers have stepped up research and development efforts, leading to increasingly mature equipment manufacturing technology and continued improvements in utilization efficiency, technological standards, and system user-friendliness.

First, the continued growth of single-unit wind power capacity has contributed to higher levels of utilization efficiency, lower unit costs, greater economies of scale, and space economy of wind farms. Wind turbines produced in the 1980s typically had a single-unit capacity of only 20–60 kW. Reflecting the growing trend worldwide of single-unit wind power capacity in recent years, the capacity of the world's major wind turbine models has increased from between 500 kW and 1 MW in 2000 to 2–3 MW in 2013. In 2013, the average single-unit capacity of newly installed wind turbines in the world was 1923 kW. In China, newly installed turbines had an average single-unit capacity of 1720 kW, with 1.5 MW and 2 MW turbines as the mainstream models. See Fig. 2.2 for the International Energy Agency (IEA) report on the changes in the single-unit capacity and hub height of wind turbines in the world.[1]

---

[1]The world's wind turbine technology has experienced very rapid progress in recent years. Onshore turbines commissioned in 2014 have reached a maximum single-unit capacity of 5 MW, and offshore wind turbines a maximum single-unit capacity of 8 MW. The trend towards larger turbines is expected to further accelerate.

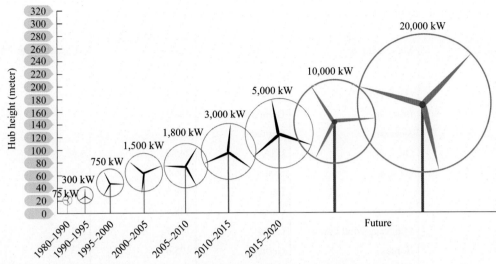

FIGURE 2.2 Changes in Single-Unit Capacity and Hub Height of Wind Turbines in the World

Source: IEA, Technology Roadmap Wind Energy 2013 Edition.

Second, variable blade pitch control technology has seen major progress to contribute further to turbine stability, safety, and efficiency. Typically, random changes in wind speed and direction cause continuous changes in the angle of attack of the blade, leading to fluctuating output power at the expense of power quality and grid stability. With the employment of variable blade pitch control technology, the angle of blade can be changed in line with the random change in wind speed, and the angle of attack of the wind stream can be maintained within an appropriate range. Power output can be kept stable especially when wind speed exceeds rated wind speed. Variable blade pitch control has been applied extensively in wind turbines (especially large turbines) in recent years. Along with variable pitch applications and power electronics developments, most turbine manufacturers have introduced variable-speed constant-frequency technology (VSCF) and invented variable-speed variable-pitch turbines to align rotational speed change with wind speed and further improve turbine efficiency. Currently, 90% of the world's installed turbines are VSCF-supported, and the figure continues to rise.

Third, the rapid development of system-friendly wind farm technology contributes to ever-greater controllability and adjustability and gradually improved coordination with conventional power sources and grids. Given the high stochastic volatility and intermittent nature of wind energy, large-scale wind power integration will pose many grave challenges to load balancing, grid safety, and power quality. Conventional wind farms attach more importance to the turbine's power generation capacity at the expense of the coordination between the turbine, other electrical equipment, grid access for turbines, and wind farms, required to maintain safe and stable grid operations. This produces a serious impact on the maximization of resources and the safety of grid operations. Thanks to the design and control technology employed, a modern wind farm has all the characteristics resembling a conventional power plant, enabling it to fully meet the requirements for generation of performance, as well as stable and safe grid operations. Generally speaking, system-friendly wind farms carry three features. First, they have a wind power prediction system capable of short-term and ultrashort-term projections

Table 2.2 Basic Information on the World's Top 10 Turbine Manufacturers

Rankings	Manufacturers	Countries	Installed Capacities (10,000 kW)	Market Shares (%)
1	Vestas	Denmark	489.3	13.1
2	Goldwind	China	411.2	11.0
3	Enercon	Germany	368.7	9.8
4	Siemens	Germany	277.6	7.4
5	GE	USA	245.8	6.6
6	Gamesa	Spain	206.8	5.5
7	Suzlon	India	199.5	5.3
8	United Power	China	148.8	4.0
9	Mingyang Wind Power	China	129.7	3.5
10	Nordex	Germany	125.4	3.3
	Others		1144.8	30.5
	Total		3747.8	100

Source: Ref. [91].

for dispatching and operation. Second, the turbine is capable of active/reactive power regulation and low voltage ride-through to maintain uninterrupted operation in times of grid fluctuations. Third, there is a focus on optimizing the allocation of active and reactive power control systems to realize remote turbine control.

1.1.3 Rapid Development of Wind Power Generation Equipment Industry

Driven by the continued growth of market demand, the wind power generation equipment industry is developing rapidly. By the end of 2013, global whole-system turbine manufacturing capacity amounted to approximately 55 million kW on an annual basis. Turbine manufacturers were concentrated in China, the United States, Germany, Denmark, and Spain. China alone accounted for 50% of the world's total production capacity. Among the top 10 wind turbine manufacturers in 2013, 3 were from China, 3 from Germany, 1 from the United States, 1 from Denmark, 1 from Spain, and 1 from India. See Table 2.2 for the basic information on the world's top 10 turbine manufacturers in 2013.

1.1.4 Significant Improvement in Wind Power Economics

The economics of wind energy are a function of power generation costs, with investment costs, operation and maintenance costs, resource availability, and grid-based consumption capacity as the major contributing factors. The wind turbine, as the core component of the wind power generation system, accounts for approximately 70% of the total investment in a wind farm. Lower turbine costs are driven by improved economies of scale and technological advancement. The reduction in wind power costs since 2005 has mainly reflected the benefit of growing development capacity, while the impetus to drive costs further down is expected to come more from technological innovation and breakthrough.

Table 2.3 Turbine Prices in Global and Chinese Markets, 2008–2013

Years	Global Market Prices (Yuan/kW)	Turbine Prices in China (Yuan/kW)
2008	11,109	6,300
2009	10,064	5,000
2010	8,500	4,000
2011	7,800	3,800
2012	7,371	3,600
2013	7,130	3,500

Source: Ref. [22]; China Wind Energy Development Report 2014, Renewable Energy Committee, China Association of Circular Economy.

Fierce competition in the wind power generation equipment market over the years has led to lower turbine prices across the board. With rapidly growing demand and fast-expanding manufacturing capacity, turbine prices are falling significantly, especially after the global financial crisis in 2008. In China, turbine prices fell by 37% on an aggregate basis between 2008 and 2010. Following the restructuring and consolidation of the global wind power market, indiscriminate capacity expansion has been kept in check, market competition has witnessed a return to rationality, and the downward trend of turbine prices has slowed. See Table 2.3 for the turbine prices in the world market and mainland China between 2008 and 2013.

Wind power generation costs around the world have shown a steadily downward trend year by year. From 1980 to 2005, wind power generation costs globally went down over 90%. Currently, the investment costs of onshore wind power projects stands at US$ 970–1400 per kW and electricity generation costs at around 10 US cents per kWh. China's wind power generation costs have dropped to RMB 0.45–0.55 kWh. By 2020, the overall production costs of onshore turbines are expected to fall a further 20–25% and offshore turbines a further 40%, bringing with them lower electricity generation costs. Driven by more advanced wind power technology and growing development capacity, wind power may become more price-competitive as costs are likely to fall to a level comparable to or even lower than conventional fossil fuel-fired power generation costs.

1.1.5 Global Efforts to Support Wind Power

To cope with climate change, optimize the energy structure, and foster emerging industries of strategic importance, countries worldwide are gradually expanding the capacity of the wind power market through incentive policies designed to encourage wind energy development. The stage is already set for strong growth in wind power worldwide. Development is expected to accelerate even further with some countries and regions having announced their own wind power planning objectives.

The United States encourages wind power development primarily through policy on production tax credit and quotas for renewable energy development. The production tax credit works as a subsidy calculated on a per kilowatt hour basis, and the renewable energy quotas implemented at the state level have been made into law to set the share of renewable sources, in total electricity consumption at a designated level. In 2008 the United States Department of Energy carried out a feasibility study, 20% Wind Scenario, which found a 20% share of total electricity consumption by 2030 as a practically achievable target for wind power.

European countries encourage wind power development primarily through subsidies on a per kilowatt hour basis. Under this system, either fixed feed-in tariffs set directly by government are provided, at which power purchased by grid operators is calculated or wind farms can participate directly in tenders with the government providing a subsidy based on market tariffs. According to the National Renewable Energy Action Plan submitted by EU nations in 2010, installed wind power capacity and wind power generation in the European Union are expected to reach over 200 GW and 500 TWh by 2020, respectively, accounting for 12.7% of total electricity consumption in that year. The Turkish government, for instance, is planning to achieve a total installed wind power capacity of 20 GW by 2020.

India has set up the National Clean Energy Fund to fund technology research and projects in clean energy. A total of 17 members from 25 Central Electricity Regulatory Commissions have jointly promulgated the Regulations for Renewable Purchase Obligation and 18 "pradeshes" (provinces) have announced a feed-in tariff setting mechanism for wind power. In addition, India has moved to cut import duty on certain turbine components from 10% to 5% and waive a 4% surcharge on the procurement of related raw materials. By the end of 2013, onshore installed wind power capacity had surpassed 20 GW, with the development of offshore wind power expected to accelerate.

In *China*, the Renewable Energy Act of the People's Republic of China sets forth a renewable energy policy system covering priority grid access, benchmark tariffs, and cost apportioning. With effect from 2009, the territory of China has been divided into four different classes of wind resource regions, where four levels of benchmark feed-in tariffs apply, respectively, RMB 0.51, RMB 0.54, RMB 0.58, and RMB 0.61 per kWh, respectively. At the end of 2014, the benchmark feed-in tariffs for Classes I, II, and III wind resource regions were readjusted downward by RMB 0.02 per kWh and the tariffs for Class IV regions have remained unchanged. Also, wind energy development funds were subsidized by means of a renewable energy surcharge imposed on electricity sales to end-users. By 2020, onshore installed wind power capacity is expected to double from the current level to 200 GW.

1.2 SOLAR ENERGY DEVELOPMENT

In 1839, French scientist Henri Becquerel discovered the photovoltaic (PV) effect. The first utilitarian monocrystalline silicon solar cell (mono c-Si) was invented by Bell Laboratories in the United States in 1954, giving birth to a utilitarian PV technology to convert solar energy into electric power. Since the 1970s, solar power generation has gained increasing policy promotion and attention in the world. Under the government-level Sunshine Project launched in the United States in 1973, spending on solar energy research received a significant boost, with the establishment of solar energy development banks to promote the commercialization of solar power products. In 1974, Japan's equivalent of the Sunshine Project was announced, with the program incorporating solar energy R&D technologies such as solar houses, industrial solar systems, solar thermal power generation systems, solar cell manufacturing systems, PV distributed generation (PV–DG), and large PV systems. Germany launched its 2000 Photovoltaic Roofs Project in 1990, followed by the Million Solar Roofs Program in the Netherlands in 1998. From 2009 to 2013, China implemented the Golden Sun Project, providing fiscal subsidies for PV–DG programs in a move that set the domestic PV power generation market in motion.

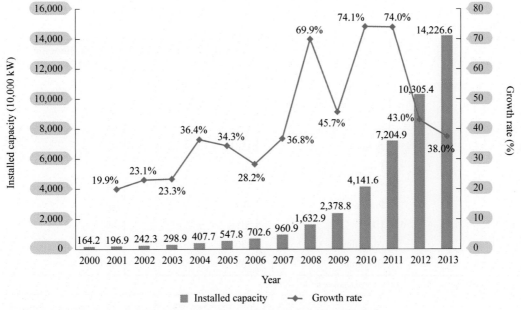

FIGURE 2.3 World's Installed Solar Power Capacity and Growth Rates, 2000–2003

Source: Ref. [90].

1.2.1 Rapid Growth in Installed Solar Power Capacity

In recent years, technological progresses have driven down the costs of PV and solar thermal power generation so quickly, that solar energy has become the fastest growing clean energy. In 2013, the world's total installed solar power capacity came to 142 GW, accounting for 2.5% of the world's total installed power generation capacity, and total solar power generation was 160 billion kWh, accounting for approximately 0.7% of the world's total electricity generation. Between 2000 and 2013, the world's installed capacity and electricity generation from solar power skyrocketed by around 86 times, accounting for an average annual growth of 40.9%. See Fig. 2.3 for the world's installed solar power capacity and growth rates between 2000 and 2013.

Among the world's top 10 countries in terms of installed PV capacity in 2013, 6 were in Europe, 2 in Asia, 1 in North America, and 1 in Oceania. The 10 countries represented 86% of global installed PV capacity. In recent years, China's solar power industry has grown rapidly, as evidenced by the establishment of a megawatt-grade PV power generation base in Qinghai Province. See Table 2.4 for the basic information on the world's top 10 countries in terms of installed PV capacity in 2013.

Europe currently leads the world in terms of PV power generation capacity. In 2013, PV power generation became the second most widely utilized new power source after wind energy for the third consecutive year, meeting 3% of Europe's total electricity needs. In some European countries, PV power generation accounted for an even larger share of maximum load. In 2013, instantaneous PV power output as a share of maximum load amounted to 49% in Germany, 20–25% in Italy and Spain, and even up to 77% in Greece. See Table 2.5 for the PV maximum power output (MPO) in selected European countries in 2013.

Table 2.4 Basic Information on the World's Top 10 Countries in Terms of Installed PV Capacity in 2013

Rankings	Countries	Installed PV Capacities (10,000 kW)	Share of Domestic Installed Capacities (%)
1	Germany	3571	20.1
2	China	1942	1.6
3	Italy	1793	14.4
4	Japan	1381	4.7
5	USA	1373	1.3
6	Spain	534	5.1
7	France	467	3.6
8	UK	338	3.6
9	Australia	330	5.2
10	Belgium	298	14.3

Source: EPIA, Global Market Outlook for photovoltaic 2014–2018.

Table 2.5 PV MPO in Selected European Countries, 2013

Countries	PV MPO (10,000 kW)	System Minimum Loads (10,000 kW)	MPO as a Share of System Minimum Loads (%)
Germany	2349	3480	67.5
France	304	3372	9.0
Italy	1322	2506	52.8
Spain	388	2128	18.2
Belgium	210	688	30.5
Czech	137	493	27.8
Greece	198	222	89.2
Bulgaria	67	273	24.5

Note: The table covers the period from May 2013 to September 2013 when PV power output in Europe was at its highest level.
Source: EPIA, Global Market Outlook for photovoltaic 2014–2018.

More and more countries are actively exploring solar thermal power generation, building and putting into operation a number of typical projects. In 2011, the Sevilla solar thermal farm was completed and commissioned in southern Spain. With an installed capacity of 20,000 kW and molten salt for heat storage, it is the first 24-h operating solar thermal farm in the world. In February 2014, the Ivanpah solar thermal power station was connected to the grid and became operational. With a total installed capacity of 392,000 kW, the power station consists of three concentrated solar power towers with an installed capacity of 133,000, 133,000, and 126,000 kW, respectively. Known as the largest of their kind in the world, the three towers alone accounted for 30% of the total installed solar thermal power capacity then in the United States. In July 2013, the first phase (10,000 kW) of the Supcon (Delingha) 50,000 kW solar thermal farm in Qinghai was successfully connected to the grid, marking a solid step forward in the commercialization of China's proprietary solar thermal power generation technology.

1.2.2 Fast-changing Solar Power Generation Technology

Solar power generation technologies, including solar photovoltaics, solar thermal power, related materials, and processes, are undergoing a continued process of innovation. Crystalline silicon cell (c-Si) technology accounts for over 85% of the market and is expected to remain the mainstream technology in 2025–2035, reflecting a trend toward higher efficiency and thinner wafers. Thin film solar cell (TFSC) technology has also been improving to such an extent that it now accounts for approximately 15% of the market. In terms of energy conversion efficiency, the c-Si cell has been improving at an annual rate of 0.5%. In this respect, the conventional c-Si cell has achieved an efficiency of 16–18%, compared to 20–23% for the heterojunction with intrinsic thin layer (HIT) c-Si cell, and 20–21% for the back-contact c-Si cell. Among the different TFSC cell types, the cadmium telluride (CdTe) cell has achieved an efficiency of 9–11%, compared to 13–15% for the copper indium gallium selenide (CIGS) cell. Both cell types have maintained continuous efficiency improvement at an annual rate of 1.0–1.5%. See Table 2.6 for the efficiency of the major solar cell types around the world.

1.2.3 Rapidly Growing Solar Power Industry

In the polycrystalline silicon (poly-Si) industry, global capacity amounted to 393,000 tons and production reached 227,000 tons in 2013, with an average capacity utilization rate of 57.8%. In terms of

Table 2.6 Efficiency of Major Solar Cell Types in the World

	Solar Cell Types		Energy Conversion Efficiencies (%)
c-Si cell	Conventional cell		16–18
	Specially structured cell	HIT	20–23
		Back-contact	20–21
TFSC cell	Chemical cell	CdTe	9–11
		CIGS	13–15
	Crystallite silicon (c-Si) cell and amorphous silicon (a-Si) cell		9–10
	Amorphous silicon (a-Si) cell	Single-junction	5–6
		Double-junction	6–8
		Multijunction	8–12
Concentrator solar cell (CSC)	Low concentration		30–40
	High concentration		

Source: Wang Sicheng, Latest Breakthroughs in Photovoltaic Power Generation and the New Photovoltaic Package Deal.

poly-Si capacity, China ranked first in the world with approximately 151,000 tons or around 38.4% of the world's total. The United States was second-placed with nearly 76,000 tons or around 19.3% of the world's total. South Korea, Germany, and Japan came third, fourth, and fifth, respectively, with 57,000, 52,000, and 26,000 tons in the same order. In terms of poly-Si production, China ranked first with a total output of 82,000 tons or 36.1% of the world's total. The United States came second with 59,000 tons or 26.0% of the world's total. Germany, South Korea, and Japan produced 46,000, 41,000, and 13,000 tons, respectively, accounting for 20.3, 18.1, and 5.7% of the world's total in the same order.

In the solar cell industry, global capacity was estimated at approximately 78 GW and production at approximately 39.5 GW in 2013, with a capacity utilization rate of around 50.6%. In 2013, global c-Si solar cell capacity reached approximately 69.6 GW. Ranking first globally, China claimed the lion's share of this amount, with 49.30 GW or 70.8% of the world's total. In 2013, global production of c-Si solar cells was estimated at approximately 35.5 GW, divided in the ratio 3:1 between the poly-Si and mono-Si cell types. Production in mainland China totaled 21.5 GW, accounting for around 60.6% of the world's total and ranking first globally. Production in Taiwan totaled 8.5 GW, accounting for around 23.9% of the world's total and ranking second globally. Production in Southeast Asia, Japan, and South Korea was around 2400 MW, 1700 MW, and 1500 MW, respectively, accounting for 6.8, 4.8, and 4.2% of the world's total in the same order. In 2013, global TFSC cell capacity totaled around 8.41 GW, doubling the level in 2010. By technology type, production capacity of Si-based, CIGS TFSC, and CdTe TFSC cells accounted for 50, 22, and 28%, respectively. In 2013, global TFSC cell production totaled about 3.95 GW, up 9.1% from 2010.

In the solar components industry, global production capacity surpassed 76 GW and production reached 43 GW in 2013. Production capacity of c-Si cell components was estimated at around 68.4 GW, compared to around 8000 MW for TFSC components, and 230,000 kW for concentrator solar cell components. China was the largest producer with an output of 27.4 GW in 2013, with c-Si cell components representing 99% of total production. Europe ranked second, producing 3.8 GW, around 20% of which was attributable to TFSC components. Japan produced about 3.5 GW, with c-Si cell components and TFSC components accounting for 71.4 and 28.6%, respectively of the total. Southeast Asia, South Korea, and the United States produced 2.8, 1.7, and 1 GW, respectively.

1.2.4 Steadily Improving Economics of Solar Power Generation

Along with sharply lower prices of PV cells and components, the construction costs of PV power stations around the world have fallen to US$ 1500 per kW. In China, a number of key enterprises have mastered the complete manufacturing processes of poly-Si and c-Si cells up to a production level of 10,000 tons, leading to increasingly lower PV equipment costs. Since 2010, the investment costs of poly-Si cells per 1000 tons have gone down 47%, compared to a more than 55% decline in the investment costs per megawatt of c-Si cells. Silicon consumption has also dropped by 25%. Investment in PV power generation systems had fallen from RMB 25,000 per kW in 2010 to RMB 9,000 per kW in 2013. Investment in PV–DG farms ranged from RMB 9,000 per kW to RMB 11,000 per kW. See Table 2.7 for the investment costs of PV power generation in selected countries in 2013.

PV power generation costs are falling quickly. Generation costs per kilowatt hour are falling significantly due to maturing PV power generation technology, increasing equipment utilization hours, and declining system-manufacturing costs. In solar resource-rich regions like California, Germany, and Italy, generation costs per kilowatt hour have dropped to a level below end-user tariffs and are increasingly

Table 2.7 Investment Costs of PV Power Generation in Selected Countries, 2013

PV Power Type	Australia	China	France	Germany	Italy	Japan	United Kingdom	United States (Unit: $1/W)
Residential	1.8	1.5	4.1	2.4	2.8	4.2	2.8	4.9
Commercial	1.7	1.4	2.7	1.8	1.9	3.6	2.4	4.5
Ground-mounted	2.0	1.4	2.2	1.4	1.5	2.9	1.9	3.3

Source: IEA, Technology Roadmap Solar Photovoltaic Energy 2014 Edition.

drawing closer to conventional power tariffs set based on the most stringent environmental standards. In 2013, Germany achieved a generation cost of US$ 0.11–0.19 per kWh, which could go down to US$ 0.08 per kWh if the annual solar irradiation intensity should exceed 2000 kWh/m^2.[2] Large ground-mounted PV solar power plants in Western China achieved a generation cost of RMB 0.7–0.9 per kWh, and the figure in Eastern China was RMB 0.9–1.2 per kWh.

At US$ 4000–9000 per kW, the investment costs of solar thermal power generation are relatively high around the world, with the unit cost varying greatly with solar resources and the availability and capacity of heat storage facilities. The cost of building the Delingha solar power tower in Qinghai without a heat storage system cost is estimated at RMB 18,500 per kW, compared to RMB 27,800 per kW for a tower with a 2-h heat storage system.

The costs of solar thermal power generation around the world remain higher than those of PV power generation. Take operating projects as an example, 40% of Spain's solar thermal power plants are equipped with a 4-h energy storage system and governed by a feed-in tariff rate of US$ 0.4 per kWh. In Morocco, the Ouarzazate solar thermal power plant, with a 160,000 kW installed capacity and a 3-h energy storage system, is governed by a feed-in tariff rate of US$ 0.19 per kWh. At 110,000 kW, the Nevada Crescent Dunes solar power tower with a 10-h heat storage system is governed by a feed-in tariff rate of US$ 0.135 per kWh, but the actual tariff stands at around US$ 0.19 per kWh after adjustment for the preferential policies implemented. Based on different technology packages and combinations of units, the generation costs of solar power towers in China can be controlled at RMB 1.2–1.5 per kWh.

1.2.5 Worldwide Support for Solar Power Generation

Driven by policy incentives in different countries, PV power generation worldwide has been entered a period of tremendous growth from a low level of development initially confined to a few nations. By contrast, solar thermal power generation is still in the stages of technology development and experimental demonstration. From the perspective of policy trends and planning around the world, solar power is expected to maintain rapid development and even outgrow wind power over time.

The United States provides policy support for the PV industry in terms of technology research and tax refund. An investment tax credit program is available to refund 30% of the total investment in a lump sum or over a specified period of years, which is equivalent to a front-end investment subsidy for PV projects. In 2010, the Senate Committee on Energy and Natural Resources voted through the Million Solar Roofs Initiative, with plans to invest at least US$ 250 million each year as a roof-mounted PV project

[2]Source: IEA, Technology Roadmap Solar Photovoltaic Energy 2014 Edition.

subsidy from 2013 to 2021. Furthermore, many State governments promulgated incentive policies on solar power generation. For example, California officially commenced the California Solar Initiative in 2007, planning to invest approximately US$ 2.2 billion over a 10-year period in front-end investment subsidies or feed-in tariff subsidies for solar power projects with a total capacity of nearly 2 GW. It is anticipated that installed PV capacity will have surpassed 100 GW by 2021.

Europe provides primarily tariff-based stimulus packages for PV power by requiring grid operators to purchase PV power as a priority and pay for the power based on government-mandated fixed feed-in tariffs, or by offering grid operators appropriate subsidies based on market tariffs, a policy similar to that applicable to wind power. Meanwhile, Germany and some other European countries encourage end-use consumption of electricity generated by customer-side PV projects, with any residual electricity to be fed into the grid. The significant fall in PV power generation costs has led the European countries to cut back upon incentives and lower subsidy levels in an orderly manner. It is expected that Germany will begin to scrap its subsidy policy for new PV power generation projects from 2017. According to the National Renewable Energy Action Plan submitted by European Union member nations in 2010, European Union countries will have installed a total solar power capacity of over 90 GW by 2020, including 84 GW of solar PV power. The Turkish government, for instance, has planned to build a total installed PV power capacity of 5 GW by 2020.

India launched the Jawaharlal Nehru National Solar Mission in 2009, with specific principles and directions proposed for building India into the world's major solar energy consumer. A robust policy and management framework was also established to achieve the goal of building 20 GW of installed grid-connected PV capacity and 2 GW of installed off-grid PV and thermal capacity in three stages by 2022. In 2014, Indian Prime Minister Narendra Modi set another goal to bring installed PV capacity to 100 GW, five times the target envisaged in the Jawaharlal Nehru National Solar Mission, by 2020.

China started implementing the Golden Sun Project in 2009 to subsidize approximately 50% of the initial investment for industrial parks and other operators of PV–DG programs, opening up a new era of large-scale development in PV power generation. By the end of 2012, the installed PV–DG capacity of the Golden Sun Project exceeded 6 GW. In 2013, China formally promulgated a policy governing PV power tariffs. Under the policy, China was divided into three solar resource regions according to the availability of solar energy and construction costs. Benchmark feed-in tariffs were set at RMB 0.9, RMB 0.95, and RMB 1 per kWh, respectively for the three regions and PV–DG power generation projects were offered an RMB 0.42 per kWh subsidy. According to plan, China's installed PV power capacity will have reached 100 GW by 2020, approximately 65 GW of which was attributed to ground-mounted PV power stations.

All in all, global wind and solar energy development is growing rapidly, with increasingly sophisticated technology and improved economics. Given government support worldwide, the prospects of this energy sector are bright, and a foundation is laid to resolve the increasingly severe energy and environmental problems.

2 CLEAN ENERGY REPLACEMENT

Clean energy substitution refers to the substitution of clean energy for fossil fuels to move toward a low carbon development path characterized by a gradual transition, from heavy reliance on fossil energy with clean energy as a complement to heavy reliance on clean energy with fossil energy as a complement. Clean energy substitution will basically resolve the problems of resource and environmental constraints that we face in energy supply. It is a strategic move to achieve energy sustainability and an inexorable trend in the world's future energy development.

2.1 THE NECESSITY OF CLEAN ENERGY REPLACEMENT

2.1.1 Energy Supply Security

The Earth is abundant in clean energy resources. Clean energy presents a fundamental solution to the acute shortage of fossil energy sources and ensures mankind's growing demand for energy can be met. With the growing population, urbanization, and rapid industrialization, global energy demand is expected to continue to grow strongly. However, the world's fossil energy sources are limited, and development costs will only continue to increase due to growing technical difficulties. Therefore, we must seek to change a development mode characterized by over dependence on fossil fuels in order to overcome the heavy pressure on global energy security. We have abundant renewable and inexhaustible clean energy resources, such as water, wind, and solar energy, with a theoretical developable capacity of 150,000 trillion kWh, which is far more than enough to meet all mankind's energy demand. Technology innovation can help realize large-scale development of clean energy, solve global energy supply problems, ensure global energy security, and satisfy the needs of economic growth and social progress.

As the fastest-developing source of energy, clean energy will gradually become the world's dominant energy source. From 2000 to 2013, global installed wind and solar power capacity registered an annual growth of 25.0 and 40.9%, respectively, and the share of nonhydroelectric renewable energy rose from 1.8% to 4.8%. In some European nations and the United States, clean energy has now become the key source of electricity. In 2013, Denmark derived 32.1% of its total electricity generation from wind energy, and Germany derived 25% of its total electricity generation from renewable energy. If global wind and solar power generation maintains annual growth of 12.4%, clean energy will be able to meet 80% of the world's total energy demand by 2050, forming a new energy development pattern dominated by clean energy to help fundamentally resolve the various energy concerns we face today. The future energy structure will exhibit a clear trend toward cleaner energy.

2.1.2 Environmental Protection

Clean energy substitution contributes to lower carbon emission, mitigates global climate change resulting from the burning of fossil fuels, and supports sustainable communities. Of all six greenhouse gases (GHGs) identified by the International Panel on Climate Change, carbon dioxide (CO_2) from fossil fuel burning accounts for the largest share of carbon emissions and is the most significant source of manmade GHG emissions. In the combustion process, raw coal, crude oil, and gas produce around 2.77, 2.15, and 1.64 tons of CO_2, respectively. If no concrete action is taken as quickly as possible, the atmospheric concentration of CO_2 will exceed the 450 ppm alert value and the global temperature will have risen more than 4°C by the end of the twenty-first century to pose a major threat to human existence. The development and utilization of coal, oil, and gas, being carbon-intensive energy sources, will inevitably produce substantial carbon emissions to increase the risk of climate change. Therefore, global energy development must not take the conventional path of high energy consumption and high carbon emissions; rather, we need to pursue a new direction of low carbon development to delink economic development from carbon emissions, and realize coordinated and sustainable development on the economic, resource, and environmental fronts. In the United States–China Joint Announcement on Climate Change published on November 12, 2014, the Chinese government proposed to raise the share of nonfossil energy in primary energy production to approximately 20%, with carbon emissions expected to peak in 2030. The United States Government also aimed to achieve an overall reduction of 26–28% in GHG emissions by 2050 from 2005 levels. If global consumption of clean energy (water, wind, and solar energy) climbs to 80% of primary energy consumption by 2050, carbon emissions from the burning of fossil fuels will decline to levels below 12 billion tons, contributing effectively to lower GHG atmospheric concentrations.

Clean energy substitution can help resolve environmental problems, including air, soil, and water pollution, arising from fossil energy development and utilization. Fossil energy is known to be contributing to serious environmental pollution by producing substantial emissions of CO_2, sulfur dioxide (SO_2), NO_x, dust, mercury (Hg), and other toxic metals in the production, transportation and utilization processes. On a whole life cycle basis, clean energy is far less pollution-intensive than fossil fuels in development and utilization terms. Under the current technical and economic conditions, replacing coal power per kilowatt hour with wind power or solar PV power can save 2.2 g of SO_2, 2.0 g of NO_x, and 0.38 g of dust emissions. In 2013, China produced a total of 140 TWh of wind power, contributing to 308,000 tons less SO_2, 280,000 tons less NO_x, and 53,000 tons less dust. The substitution of clean energy for fossil energy can avoid environmental pollution caused by energy development and utilization, improve mankind's living environment significantly, and reduce healthcare spending for the benefit of the community.

2.1.3 Promotion of Economic Development

As a strategic emerging industry, clean energy development has the clear effect of drawing investment with much room for expansion. The clean energy industry is a capital and technology-intensive one, which involves a very long industry chain and numerous other sectors. Its significant technology diffusion effect and multiplier effect have made it an important industry for the world to stimulate investment and create jobs. Based on the statistical information in the Global Trends in Renewable Energy Investment 2014 released by UNEP, global investment in renewable energy was estimated at US$ 1.6 trillion between 2006 and 2013. From development through construction to operation and maintenance, the clean energy industry can create many job opportunities given its heavy manpower requirement.

Clean energy development is a common choice around the world to boost development momentum and generate new economic growth. Recovering from the international financial tsunami that dealt them a heavy blow, all countries have been seeking a new driver of economic growth. To revitalize the economy, the United States and European countries have chosen to develop the new energy industry to boost the national economy. In 2001, the United States invested only US$ 286 million in clean technology in 2010 – the figure rose to US$ 4.6 billion in 2010, representing a more than 15-fold increase. In 2012, 14% of venture capital investments found their way into the clean energy technology sector.[3] As pointed out by the All-of-the-Above Energy Strategy as a Path to Sustainable Economic Growth published by the White House,[4] the energy industry contributed increasingly to United States GDP growth between 2000 and 2013, and the lower energy imports helped cut trade deficits. The world has now resorted to clean energy investment or grid infrastructure upgrading and retrofitting projects, as an important means of boosting economic growth. As forecast by the IEA,[5] global energy infrastructure

---

[3]Source: Ref. [24].
[4]The English title of the report is "The All-of-the-above Energy Strategy as a Path to Sustainable Economic Growth."
[5]Source: Ref. [70].

investment will total US\$ 37 trillion between 2012 and 2035 (based on the United States dollar's exchange rates in 2011), averaging US\$ 1.6 trillion or around 1.5% of the global GDP every year. Developing countries are expected to step up investment notably and non-OECD countries are expected to record a total energy investment equal to 61% of the world's total.

Clean energy development is of strategic significance. In recent years, international energy development has witnessed profound changes. The worsening resource constraints and environmental problems have led to an international consensus on the need to combat global climate change together, with many countries deciding on clean energy as a strategic national goal of energy development. When the world economy has moved into a new round of adjustment aimed at securing a leadership position in the international technology race especially after the financial crisis, the world's major economies, such as the United States and the European Union, have launched policies with an unprecedented focus on clean energy, like the Green Energy Project and the Green New Deal. The European Union's 20–20–20 strategy in 2007 was designed to achieve a 20% reduction in GHG emissions from 1990 levels by 2020, with the share of renewable energy in primary energy consumption to rise from 8.2% to 20% in 2006, and energy utilization efficiency to improve by 20%. In January 2014, the European Commission promulgated the 2030 Framework for Climate and Energy Policies, setting a further goal to cut GHG emissions by 40% and to derive at least 27% of energy output from renewables by 2030. In the United States, the American Clean Energy and Security Act approved in 2009 presented for the first time a national emission control program. Renewables targets at the national level were also formally proposed to meet 20% (15% attributable to wind, solar, and biomass energy) of electricity needs by 2020 through renewable energy development and energy efficiency improvement. In June 2014, the Environmental Protection Agency announced plans to cut carbon emissions from power plants nationwide by 30% by 2030. As for Japan, the government has reassessed the role of nuclear power in electricity supply after the Fukushima nuclear incident, with renewables expected to become the new focus of energy development. Not just the developed countries, the developing countries are also attaching great importance to renewable energy development. Globally, more than 120 countries have so far developed relevant laws, regulations or action plans with the aim of achieving strategic renewable energy goals by statutory or mandatory means. Renewable energy development has become a strategic option for most countries in the world.

2.2 THE FOCUS OF CLEAN ENERGY REPLACEMENT

As clean energy substitution is set to change the longstanding overdependence of the world on fossil fuels, it is particularly important to achieve significant breakthroughs in terms of technology, economics, safety, and policy mechanism.

2.2.1 Key Technologies

Technology for efficient conversion of clean energy. Due to technological constraints, wind and solar energy development has been marred by unfavorable economics and low utilization efficiency. Research has shown that wind power efficiency worldwide has amounted to around 38%. However, typical solar PV power efficiency is within the range of 12–18%, far lower than that of conventional fossil energy, for example, coal and petroleum. Currently, breakthroughs in wind power technology are focused on key turbine component design, while the focus of application technology for solar power is on improving the conversion efficiency of solar PV and solar thermal energy.

Technology for clean energy allocation over large areas. Globally, wind energy resources are distributed primarily in the Arctic, central and northern Asia, northern Europe, central North America, east Africa, and the near-shore regions of each continent. Solar energy resources are distributed primarily in equatorial areas, for example, north Africa, east Africa, the Middle East, Oceania, and central and South America. Water resources are concentrated primarily in the major drainage basins of South America, Asia, North America, and Central Africa. Most of these clean energy-rich areas are remote, sparsely populated, and far away from load centers. Besides, the major clean energy bases on each continent are also distributed inversely with the centers of energy consumption. Therefore, the key to efficient energy utilization lies in how to deploy an intracontinental, intercontinental, and even global energy allocation system to accommodate concentrated development and long-distance delivery of clean energy.

Technology for clean energy grid connection and consumption. Judging by the development trend, wind, and solar power tends to be generated in large bases, consumed on a large scale, developed in a distributed manner and utilized locally. A robust smart grid features a robust grid structure to accommodate the complementary strengths of different power-producing regions, improve system-wide consumption of electricity generated by large renewable energy bases, realize large-capacity, long-distance power transmission, and ensure system safety and stability. A robust smart grid allows concentrated access for random and intermittent power sources and efficient application of distributed power supply systems to support large-scale development of wind and solar energy.

Wind and solar power technology under extreme conditions. Although the Arctic and Equatorial areas are strategic energy bases of global importance and abound in concentrated energy, the fact that these regions are extremely cold or hot gives rise to challenging working conditions, which spells the urgent need to seek breakthroughs in a new series of energy development technology capable of withstanding inclement weather conditions. The Arctic is a high latitude, highly humid area, so major technological innovation is required for the tower and blade materials of wind turbines in order to cushion the impact of the freezing cold on utilization efficiency and improve the tolerance of the equipment to the Arctic's extreme climate. An offshore turbine must be able to withstand strong wind load, corrosion, tidal impact, and other special conditions. Continued improvement in operation and maintenance skills is also necessary. Construction of large solar PV power stations or solar thermal power stations in north Africa, the Middle East, and similar equatorial areas rich in solar energy resources requires important breakthrough technologies and processes to overcome challenging weather conditions, for example, high temperature, temperature difference, wind, and sand.

2.2.2 Economics

Costs of clean energy development: Considering the economics of developing and utilizing different types of power sources, the non-renewable nature and heavy consumption of fossil fuels will continue to drive the development costs of fossil energy higher. As an unconventional fossil fuel, shale gas also faces cost problems. However, given the continued breakthrough in and growing maturity of clean energy generation technology, the development cost of clean energy will trend lower from the current relatively high levels. Wind power technology will continue to see breakthroughs and growing single-unit capacity, while offshore wind power technology will gradually mature. A diversity of generation modes and the significant improvement in energy conversion efficiency will lead to sharply lower costs of solar power generation.

Market competitiveness of clean energy. The cost differentials between clean energy and fossil energy will gradually narrow, with the cost of clean energy generation technology falling, and the generation cost of conventional fossil energy rising year after year. When the cost gap is eventually closed, clean energy will become competitive as grid parity becomes possible. Grid parity includes grid- and user-side parity. The former means that on-grid tariffs for clean energy generation and fossil fuel generation are identical and the latter indicates that the cost of clean energy generation on the user side is on a par with the relevant end-user tariffs. Global wind power and solar energy will achieve on-grid parity in 2020 and 2025, respectively, and user-side parity even earlier as predicted by the IEA and other international organizations.

2.2.3 Safety

Grid safety issues associated with large-scale grid access for clean energy: Large-scale grid integration of wind, solar, and other forms of clean energy will bring about new challenges to grid safety and stability, electric power system planning, and the economical operation and operational management of power systems. The challenges are mainly reflected in a number of areas. First, the volatility of clean energy generation produces a relatively large impact on the power quality and voltage stability of local systems, as well as the stability and transmission efficiency of regional networks, and system-level rotational reserve requirements. Second, the variability of clean power generation at the system level makes economical dispatch of electricity more difficult and increases the demand for auxiliary services. Third, as large-scale grid integration of clean energy (such as wind power) is already having an impact on power system operation on a per-second level, system dispatch capability needs to meet more stringent requirements in terms of responding timely to fluctuations in wind and other clean energy generation, thereby readjusting the way of system operation. All this calls for innovations in technology and management in order to improve the operational technology and management capabilities for large-scale grid access for and consumption of clean energy on a global basis.

Safety issues associated with grid integration of distributed power generation: A high level of grid access for distributed power sources turns distribution networks into active grids, thereby causing a series of technical problems with voltage stabilization, relay protection, short-circuit current, and power quality. When a higher proportion of distributed generation is reached, distribution networks will feed power back to transmission networks in individual time slots, which will change the current flows and distribution of the power grids and significantly increase the complexity of managing grid dispatch operations.

2.2.4 Development Mechanism

Mechanism for innovation in clean energy technology: Technology maturity is a basic requirement for realizing large-scale commercialization of clean energy. Yet technological innovation is time-consuming, capital-intensive, and high risk, with economic returns difficult to achieve in the short-term. At the initial stage of development, clean energy requires special policy support from government. The focus is on establishing the important role of clean energy technology innovation in the country's system for energy technology innovation and equipment manufacturing by formulating a medium to long-term innovation roadmap and conducting major technological research. The dominant role of enterprises in clean energy technology innovation must also be fully leveraged through tax incentives and funding support from the state to encourage the development of R&D centers and demonstration projects for new technologies. In addition, global cooperation, exchange, and sharing of clean energy technologies must be enhanced to provide the world with more, better and quicker access to clean energy research findings.

Full cost accounting for clean energy: The emission of GHGs like CO_2 during the development and use of conventional fossil energies leads to climate change. Air, water, and soil pollution caused by large quantities of pollutants has caused serious ecological damage detrimental to human health. The ecological and environmental benefits of clean energy can be fully manifested and the competitiveness of clean energy in the energy market significantly enhanced if these hidden environmental costs are reflected as part of the real costs of developing and using fossil energy. Therefore, the government should take full account of the environmental costs of fossil energies by implementing pricing and tax measures, such as resource, effluent, and carbon taxes, which can not only accelerate the trend toward cleaner fossil energies, but also create a level playing field for clean energy development.

Mechanism for developing clean energy markets: In clean energy development, the focus is on fostering a competitive market, leveraging the role of market forces in determining the areas and directions of investments, and introducing market plurality to facilitate technological progress and lower costs through competition. In clean energy utilization, the focus is on building an electricity market where electricity tariffs fluctuate with market supply and demand to fundamentally resolve the mismatch between clean energy generation and consumption. Restricted by natural and meteorological conditions, clean energy generation is often out of sync with load characteristics. Generally, more wind power is generated after midnight when demand is at its lowest. As a result, surplus wind power cannot be fully used and can only be abandoned. By building a mechanism to introduce market competition for user-side resources, users are encouraged to consume more clean electricity in periods of high generation and less in periods of low production. Industrial users are also encouraged to come up with a more rational production program so that power loads can be transferred to low ebb periods. At the same time, in response to the market mechanism in force, users can store any clean electricity unused during nighttime ebb periods in an energy storage facility and release it for use in daytime periods of high demand. In this way, economic returns can be obtained and efficient utilization of clean energy achieved.

3 ELECTRICITY REPLACEMENT

Electric energy substitution means the substitution of electric power for fossil energies, including coal, oil, and natural gas, for direct consumption to increase the share of electric energy in energy end-use. As the electrification process accelerates, electric energy will play an increasingly important role in energy end-use and eventually become the most dominant energy option for terminal consumption, which will help realize cleaner, more efficient, and safer energy utilization.

3.1 THE NECESSITY OF ELECTRICITY REPLACEMENT

3.1.1 Improve Energy Efficiency

Electric energy is a clean, efficient and convenient secondary energy source, which provides highly efficient terminal consumption, cleanliness, and zero emission. Compared to other energy options, electricity yields the highest level of utilization efficiency, at over 90%, in the end-use segment. With the proportion of clean energy generation increasing, clean energy generation will gradually replace fossil energy generation, and most of the primary energy sources will be converted into secondary energy, resulting in much lower conversion losses and further manifesting the characteristics of electricity as a clean and efficient energy alternative. In terms of energy efficiency, electrical equipment is far more efficient than appliances powered by direct combustion of coal and oil. For instance, the heat efficiency of coal-fired boilers is only 70% or so, compared to over 90% for electric boilers. As another example, electric trains consume just about 60% of the energy required by diesel locomotives.[6] In Germany, 80% of industrial electricity is used for heating. Electricity used directly in the manufacturing process has very high heat efficiency, but where heating is obtained by direct burning of fuels, only 20% of the heat produced is used in the manufacturing process.[7]

Electric energy substitution can improve energy efficiency in an all-round manner. In terms of utilization, electric energy is convenient and precisely controllable; in terms of conversion, electric energy can be converted to/from various forms of energy, and all primary energy sources can be converted into electric energy; in terms of allocation, electricity can be produced on a large scale, transmitted over long distances and instantly sent to any end user through a distribution system. Driven by industrialization, urbanization, informatization, agricultural and rural electrification, and new technology application, electricity is chosen for its unique characteristics and is extensively used to power economic and social growth in all countries. The advent of electricity has made large-scale and automated production in the agricultural and industrial sectors possible to substantially improve labor productivity and product quality. The electronics and information industries also benefit from the extensive use of electric power. By increasing the share of electricity in energy end-use to promote electric energy substitution in the areas of industry, transport, commerce, as well as urban and rural living, we can not only enhance utilization efficiency, but also increase economic output and improve energy efficiency at the community level. According to data published by China, the economic efficiency of electricity is 3.2 times that of oil and 17.3 times that of coal, meaning that the economic value created by electricity of 1 ton of standard coal-equivalent is comparable to that created by oil of 3.2 tons of standard coal-equivalent, and coal of 17.3 tons of standard coal-equivalent.

3.1.2 Promote Clean Development

Most clean energy needs to be converted into electricity for efficient utilization. An essential requirement for clean energy development and electric energy substitution is the inexorable outcome of implementing clean substitution policy and a necessary condition for developing a new energy system focused on electricity. With the progress of a new energy technology revolution, clean energy will be used on a larger scale and more primary energy sources will be transformed into electricity and delivered to load centers, to provide adequate clean power to meet the massive energy requirements from electrified transport, electric boilers, electric kilns, electric heaters, and electric cookers. This will effectively replace oil, coal, and other fossil fuels, as well as provide more room for developing and using renewable energy like solar, wind, and waterpower.

---

[6]Based on 80% energy conversion efficiency of electric locomotives, 40% coal-fired power generation efficiency, and 20% energy conversion efficiency of diesel locomotives.

[7]Source: Wang Qingyi, Introduction to Electrification of National Economy, Electric Power Technology Economics, 2008(20):12-18.

3.1.3 Improve Electrification Levels

Electrification is an important hallmark of a modern society. It ranked first among the engineering feats with the most significant social implications in the twentieth century, as appraised by the Selection Committee of the American National Academy of Engineering in December 1999. Helping to build a production system driven by industrial mechanization and automation, electrification is maximizing the progress of industrialization, expediting the migration of rural population into cities and promoting urbanization. It has changed the mode of agricultural production, accelerated the mechanization and industrialization of agriculture, and greatly improved human lifestyles and quality of family life. Currently, the level of electrification around the world is steadily increasing.

Implementing electric energy substitution policy is crucial for improving electrification levels. Generally, there are two measures of the level of electrification. One is the share of electricity in primary energy consumption and the other is the share of electricity in terminal energy consumption. The course of development in developed countries fully demonstrates that the level of electrification increases with economic progress and social affluence. As indicated by the progress of power development, electricity as a share of energy end-use has shown a clear upward trend in both developing and developed countries. The share is over 20% in the majority of developed countries. Electric energy is expected to account for more than 50% of energy end-use globally by 2050, with clean energy developing rapidly. Electricity as a share of primary energy consumption in the world increased from 34% in 1990 to 38.1% in 2012. The figure is expected to rise further to nearly 80% by 2050.

3.2 THE FOCUS OF ELECTRICITY REPLACEMENT

As implementation of the policy on electric energy substitution is completely changing the patterns of energy consumption, the key task is to promote the strategy of "replacing coal and oil with clean electricity delivered from afar."

3.2.1 Substitute Electric Energy for Coal

Substituting electric energy for coal refers to the replacement of coal with electricity in energy end-use, in order to significantly alleviate environmental pollution. Burning coal produces large amounts of sulfur dioxide, nitric oxide, smoke, and dust as well as other pollutants that contaminate the air in the form of coal smog. In 2012, approximately 52% of coal production was used for generating electricity in China, with coal for direct combustion and coal used as raw materials each accounting for around 24%. As a result, the power sector produced emissions including 8.83 million tons of sulfur dioxide, 9.48 million tons of nitric oxide, and 1.51 million tons of smoke and dust. Emissions from coal not for power generation included 9.49 million tons of sulfur dioxide, 3.9 million tons of nitric oxide, and 7.15 million tons of smoke and dust. As decentralized burning of coal produces much more pollutants compared to coal-fired generation, most developed countries accord priority to the transformation of coal into electric energy with emissions substantially reduced by managing emission performance at the plant level, and direct consumption of coal in the end-use sector is minimal. In the United States for instance, more than 90% of coal is used to generate electricity.

Electricity technologies like heating, heat pumps, electric kilns, and electric cookers have been well developed and are well placed to replace coal. Take electric boilers for example, the heat supply technology was developed swiftly and used extensively with the improving socialized production and living standards in the early twentieth century. Powered by an adequate supply of electricity, electric boilers provide energy efficiency, cleanliness, safety, and other advantages unrivalled by any other heating equipment. Promoting the development of electric boilers then became an inexorable trend. Electric boilers were first produced and used in Europe in 1926. The United States and Europe began to promote electric heating in 1930. In the 1950s and 1960s, electric heaters became vastly popular. Currently, electricity as a share of heating energy is 90% in Norway, 80% in Japan and the Republic of Korea, 70% in France, and 50% in the United States, Canada, Denmark, and Sweden.

As the world's largest consumer of coal, China still has room to grow to substitute coal with electric energy. Currently, there are about 620,000 electric boilers in operation in China where industrial coal-fired boilers number around 370,000. The sheer number of coal-fired boilers in China means a high level of coal consumption every year, not to mention the problems of high energy consumption, wastage, and environmental pollution still associated with many of these coal-fired boilers. By promoting the use of electricity instead of coal for industrial, commercial and household purposes, such as electric boilers with heat storage capability (electric heating), heat pumps, electric kilns, and distributed electric heating systems, China is expected to be able to cut down approximately 320,000 tons of sulfur dioxide, about 260,000 tons of nitric oxide and 13,000 tons of $PM_{2.5}$ particulates each year by 2020. If clean energy is supplied to meet new demand for electricity, 160 million tons of coal for direct combustion and approximately 320 million tons of CO_2 can be saved. With the technological progress on electric energy substitution, it is anticipated that China's direct combustion of coal will have gone down 60% by 2030 and basically eliminated by 2040.

Substituting electric energy for coal also helps improve people's livelihoods. Serious air pollution and potential safety hazards exist in China and some other less developed countries, where burning coal to stay warm in the winter remains a common practice in rural areas. There are fatalities caused by gas poisoning in northern China's countryside each year. Gas poisoning incidents are actually preventable by promoting distributed electric heating, cooking, and bathing to ensure power safety.

3.2.2 Substitute Electric Energy for Oil

Substituting electric energy for oil refers mainly to the replacement of fuel oil with electric energy in such areas as electric cars, rail transport, and shore power. This can mitigate oil-induced pollution and also reduce dependence on oil. Accounting for one-third of global energy consumption, transport systems are powered predominantly by oil, making them highly reliant on petroleum and also discharging a high level of motor vehicle exhaust as one of the major sources of air pollution. By adopting electricity technology covering electric cars, electrified railways, and shore power, seeking alternatives to oil for energy efficiency's sake has become a common direction in energy utilization for the world's transport industry.

Electric cars are powered by electricity with highly efficient rechargeable batteries or fuel cells. They are clean and pollution-free with the greatest potential for the substitution of electricity for oil. Although conventional internal-combustion engine vehicles are still dominant, clean electric cars are the way forward and set to reshape the automobile industry of the twenty-first century. In terms of energy utilization efficiency, the energy conversion efficiency of fuel oil to power transport systems is 15–20%, with little room for major enhancement. On the contrary, with up to 90% efficiency in the conversion of electricity into kinetic energy, coupled with a 90% recharging efficiency of storage batteries, the efficiency of converting electric energy into motive power can reach 90%. The efficiency of converting natural gas, oil, and coal into electric energy is 55–58, 50–55, and 40–50%, respectively. On this basis, the energy utilization efficiency of electric cars is 1.5–2 times that of oil-powered cars. Electric cars have the advantages of much improved energy utilization efficiency and zero emission. China's car ownership is forecast to exceed 200 million by 2020. Assuming electric car ownership of five million, based on mileage of 20,000 km per car each year, and average oil consumption per 100 km of 10 L, 7.1 million tons of gasoline consumption can be reduced, and 15 million tons of CO_2 emissions saved every year.

Shore power technology means that quayside power supply systems are provided to supply power to berthed ships, which need not rely on oil-fired engines on board for power supply. Shore power that meets the demand of berthed ships for electricity to power up lightning, communication, air conditioning and water pumps, can eliminate exhaust emissions during power generation and noise pollution during the operation of generating units on board. So the use of shore power can dramatically reduce energy costs, as opposed to ship-mounted generators, which are economically unsound, being costly, and with a low generating efficiency.

3.2.3 Electricity from Afar

The contrary distribution of energy resources and load centers determines the basic pattern of "delivering electricity from afar." In China, about 80% of coal and over 70% of clean energy resources are concentrated in western and northern areas but both are scarce in central and eastern areas where load centers are located. As local consumption is limited in large energy bases in northern and western China, electricity generated is mainly exported. By transporting coal together with renewables like water, wind, and solar energy in these areas to central and eastern China, we can ensure electricity supply security for these resource-scare regions and optimize nationwide allocation of energy resources, while avoiding the problems of strained coal transport and environmental pollution caused by moving coal to load centers over long distances.

"Delivering electricity from afar" is crucial for solving environmental issues in areas where load centers are located. China's load centers are situated in the relatively well-developed mid-eastern regions, covering 12 provinces (including municipalities directly under the central government) and accounting for 45% of the country's population, approximately 58% of the national GDP, and 13.5% of the nation's land area. For a long time, numerous coal-fired power plants have been constructed in areas where load centers are located. About 75% of the nation's installed coal power capacity is concentrated in mid-eastern regions, with a power plant built per 30 km along the Yangtze River. The sheer number of power plants has far exceeded the carrying capacity of the local environment, with local population density exacerbating further the risk of health damage and associated losses due to environmental pollution. In eastern China's Yangtze River Delta region, sulfur dioxide emissions per kilometer reaches 45 tons annually (20 times the national average), turning the region into acid rain-afflicted areas with frequent smoggy weather. The transport and storage of coal from western and northern areas to central and eastern regions year after year is the culprit of serious air pollution. In 2013, an Action Plan for the Control of Air Pollution was published by the State Council to curb new coal-fired power capacity in the western and northern regions, replace coal with electricity, and satisfy the growing demand for energy. What this means is that the need for long-distance transport of large quantities of coal will be eliminated, dust problems in the transport process removed, and the mid-eastern regions no longer be exposed to soil, air, and water pollution from the storage and burning of coal.

3.2.4 Clean Electricity

Low carbon electricity supply is a fundamental solution for global climate change. Although electric power supplied from afar can solve the problems of imbalanced power supply and demand and also pollutant emissions, the global CO_2 and pollutant emission problems cannot be fundamentally eradicated if the electricity delivered from afar is not a clean energy but generated from the burning of coal or other fossil fuels. Therefore, from the perspective of global energy sustainability, clean electricity is a basic requirement for combating global climate change.

Clean electricity supply is the inevitable outcome of clean substitution. The shift from fossil energy to clean energy requires a transition to clean energy at the energy development stage and to electricity consumption at the end-use stage. Similarly, the allocation of energy should help realize the transition from delivery of fossil energy to transmission of clean electric power. Large-scale development of clean energy will inevitably result in long-distance transmission of clean electricity and substantial growth in power consumption, making transmission of clean electricity the major form of energy transport in the future.

Clean electricity supply will be a progressive process. Thanks to the UHV power transmission technology, thermal power, wind energy, and solar power in the western and northern parts of China as well as hydropower in southwest China are delivered to eastern and central parts of China on a large scale from a long distance to meet the demand for electrical power and alleviate the strain on the local environment. In the future, along with the large-scale development of clean energy in western and northern parts of China, the proportion of thermal power will gradually go down and transmission of clean electricity focused on wind, solar, and hydroelectric power will play a dominant role in supplying clean electricity continuously to the eastern and central regions. According to State Grid Corporation of China's plan, a UHV AC/DC power grid with a capacity of 0.45 GW for large-area allocation will be fully developed by 2022 to meet the requirement for delivering clean energy of 550 GW and consuming clean energy of 1700 TWh annually, which will save 700 million tons of raw coal together with 1.4 billion tons and 3.9 million tons of CO_2 and SO_2 emissions, respectively.

Looking into the future, along with the development of wind energy in the Arctic and solar energy in the Equatorial region as well as clean energy bases on different continents, a sufficient supply of clean electricity will be delivered to load centers on each continent through UHV and EHV grids, making the transmission of "clean electricity from afar" completely achievable.

4 TWO-REPLACEMENT AND ENERGY REVOLUTION

Two-substitution policy signifies a major change in the mode of energy development, which will revolutionize energy consumption, energy supply, energy technology, and energy systems to become a driving force behind global energy sustainability. The future of global energy development hinges on making the most of the opportunities arising from the new energy revolution and expediting the implementation of the two-substitution policy.

4.1 CLEAN ENERGY REPLACEMENT AND ENERGY REVOLUTION

Clean energy substitution is an immutable law of energy transition. Throughout the history of global energy development, energy transition has exhibited two major trends. One is the trend toward steadily higher energy intensity from firewood and coal to oil, natural gas, and electric power. Despite being low intensity energy sources by nature, clean energy like hydro, wind, and solar power can become high intensity energy after transformation into electricity as an energy source of high quality, accessibility, and efficiency. The intensity of clean energy will be further manifested especially if and when breakthroughs in energy storage technology are achieved. The other trend of energy transition is one toward an increasingly lower carbon content of energy from firewood to coal, then from coal to oil, and further to the rapid development of natural gas, reflecting a direction toward cleaner energy.[8] The development and utilization of zero-carbon clean energy like hydroelectric, wind, and solar energy will naturally become the focus of a new round of energy transition.

Clean energy substitution marks an important direction for a new round of energy revolution. First and foremost, the new round of energy revolution should aim to realize the low carbon development of energy while ensuring energy supply security. Initially, a two-pronged approach should be coordinated and promoted to achieve "clean and efficient utilization of conventional energy" and "development and utilization of clean energy." However, given the increasingly formidable challenges of climate change and resource depletion, plus the lower potential for and higher cost of clean utilization of fossil energy, the energy revolution will be mainly driven by clean energy substitution to eventually develop a new energy supply system oriented toward clean energy.

Clean energy substitution is the key to energy sustainability. In the progression from firewood to coal and further to gas in human history, new energy alternatives have continuously evolved as substitutes for older energy types on the strength of their inherent advantages. The new energy revolution represents a choice made in line with the trend of energy development for the sustainability of human society, amid continued efforts to grapple with the problems of energy supply, climate change, and environmental pollution. As the development efficiency of wind, solar, ocean, and other clean energy sources continues to rise with improving technology economics and market competitiveness, the substitution of clean energy for fossil energy will become an inevitable revolutionary trend. The growing momentum of clean energy is already playing a dominant role in reshaping the world's energy landscape that will fundamentally resolve the energy and environmental problems currently inhibiting human survival and development. The rapid development of clean energy is not only required for energy transition, but also a step that must be taken to meet the progress of human civilization.

4.2 ELECTRICITY REPLACEMENT AND ENERGY REVOLUTION

Electric energy substitution is a necessary condition for achieving terminal consumption of high efficiency, low carbon energy. In terms of energy consumption, the energy revolution is about realizing the efficient utilization and decarbonization of energy. The end-use efficiency of electric energy can be as high as over 90%, compared to 50–90% for natural gas and an even much lower rate for coal. Currently, the structure of terminal energy consumption in some developed countries is relatively reasonable, with a small share of coal in energy end-use compared to a higher share of natural gas, oil, and electricity to provide a correspondingly higher utilization efficiency of energy. Driven by an increasing level of clean energy substitution, the share of electricity in terminal energy consumption is fast growing, contributing significantly to lower consumption of fossil energy.

[8]Source: Ref. [23].

Electric energy substitution is an effective way to solve energy and environmental problems. Electric power is clean and zero-pollution energy. In producing the same amount of heat equivalent to 1 kWh of electricity, raw coal will emit 330 g of CO_2, 5.3 g of SO_2, and 1.6 g of nitrogen oxide, whereas diesel oil will emit around 260 g of CO_2, 0.4 g SO_2, and 0.6 g nitrogen oxide. While thermal power generation also produces SO_2 and nitrogen oxide emissions in the production and utilization processes, centralized management of the emissions at the plant level is possible by means of desulfurization and denitrification. Currently, the desulfurization level of thermal power plants amounts to 90%, compared to a denitrification level of 80%. After desulfurization and denitrification, thermal power generation produces far less sulfide emissions compared to coke, gasoline, diesel, and natural gas. Looking ahead, the environmental advantages of electric energy substitution will be further manifested with the development and large-scale generation of pollution-free clean energy.

Electric energy substitution holds great promise. From 1971 to 2012, the share of electric energy in global energy end-use increased from 8.8% to 18.1%, ranking second only after oil. The figure is expected to further rise to 25% by 2030 and to over 50% by 2050. With the massive increase in clean energy supply, end-use energy demand will largely be met by electricity.

SUMMARY

1. World wind and solar power technology has moved into the fast lane of development, providing an empirical foundation for clean energy substitution, electric energy substitution, and global energy sustainability.
2. Clean energy substitution and electric energy substitution mark an important direction for energy revolution and a path that must be taken to resolve global energy and environmental concerns. Clean energy substitution presents a fundamental solution to mankind's energy supply issues by reconciling energy development and utilization with environmental protection. Electric energy substitution is the inevitable outcome of clean energy substitution and the key to improving energy utilization efficiency and electrification.
3. Clean energy substitution calls for breakthroughs at the technology, economic, security, and policy levels. The focus is on establishing a scientific and workable development mechanism and expediting the search for technology breakthroughs in clean energy development, allocation, and coordinated control, underlined by intensive efforts to resolve problems concerning the economics and safety of clean energy development.
4. Electric energy substitution focuses on replacement of coal and petroleum with electricity as well as the delivery of clean electricity over long distances. In the energy end-use segment, direct consumption of electricity rather than fossil fuels (e.g., coal and petroleum) will contribute to clean energy utilization and an all-round increase in electrification levels to drive socioeconomic development.
5. The impact of clean energy substitution and electric energy substitution on global energy development will be nothing short of revolutionary. It will drive a structural shift from being fossil energy–oriented to being clean energy–focused and help achieve the goal of utilizing high efficiency, low carbon, and clean energy.

A GLOBAL ENERGY OUTLOOK

CHAPTER OUTLINE

1 EVOLUTION OF ENERGY DEVELOPMENT

Energy is an important driving force behind the progress of human civilization. The global energy landscape itself is continually developing and undergoing a profound shift from high carbon to low carbon, from low efficiency to high efficiency, and from local balancing to wider-scale distribution. Recognizing and commanding this law of development is crucial for promoting energy science and implementing the "two-replacement" strategy.

1.1 "HIGH CARBON" TO "LOW CARBON"

Throughout the history of energy development, mankind has been continually seeking more forms of energy to secure energy supply and satisfy the growing energy needs of socioeconomic development. Nevertheless, the dominant energy form has been different in different stages of development. With this variance in demand and the development of technology, dominant energy form continues to move toward lower-carbon territory. The carbon-reduction process is best represented by the replacement of firewood, coal, oil, and natural gas with hydropower, nuclear, wind, solar, and other forms of clean energy sources. Table 3.1 shows the carbon intensity of different energy forms.

Fossil energy features very high carbon intensity, evidenced by the fact that enormous quantities of carbon have been emitted through more than a century of consumption. Meanwhile, the global temperature has increased by 0.74°C in the past century with a soaring carbon intensity. Since 1979, Arctic Sea ice has been disappearing by more than 70,000 km^2 every year.[1] In an attempt to mitigate climate change and ensure sustainable development for mankind, measures must be taken to substitute

---

[1]Adapted from NSIDC.

Global Energy Interconnection. http://dx.doi.org/10.1016/B978-0-12-804405-6.00003-8

Table 3.1 Carbon Intensity of Different Energy Forms (t-CO_2/t-ce)

Coal	Oil	Natural Gas	Hydropower	Nuclear Energy	Wind Energy	Solar Energy
2.77	2.15	1.65	0	0	0	0

Source: BP, Statistical Review of World Energy 2014.

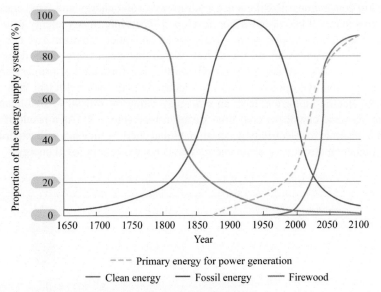

FIGURE 3.1 **Evolution and Development Trends in the Energy Structure**

more fossil energy with clean energy within the energy structure. Mankind has unlimited solar energy and wind energy which promise large-scale development potential and are expected to become the most important clean energy in the future. Under existing and foreseeable future technological conditions, solar energy, and wind energy can be most conveniently exploited through transformation into electrical energy. The development of solar energy and wind energy not only embodies a replacement policy, but also provides clean sources for electricity-based replacement. See Fig. 3.1 for the evolution and development trend in the energy structure.

1.2 LOW EFFICIENCY TO HIGH EFFICIENCY

Technological innovation is the key to improve efficiency in energy exploitation and utilization. In the late eighteenth century, the steam engine as a technological innovation with epochal significance for energy development, promoted the large-scale and highly efficient development of coal, as well as catalyzed the transformation of production from manual labor to large-scale, machinery-aided manufacturing, hence greatly strengthening labor efficiency and social productivity. In the late nineteenth century, the steam engine's potential for further technical improvement increasingly diminished and played a lesser role over time in promoting the coal-dominated energy structure. Appearance and broad application of the internal combustion engine and the electric motor, led to the emergence of new energy forms represented by oil and electricity as epoch-making inventions, which worked positively on energy efficiency and labor productivity. However, fossil energy efficiency now demonstrates increasingly lower potentials for further improvement. Currently, gasoline engines have a direct fuel efficiency of around 30% and the coal-fired generator has a maximum energy efficiency of around 50%.

As opposed to conventional fossil energy, hydropower, solar energy and wind energy are nonstorable energy forms that will be wasted if not developed. Therefore, many energy institutions, such as IEA and BP, take wind power, solar power, hydropower, and other transformative energy forms as the primary energy, while making statistical analyses. Given that clean energy may dominate the energy structure, energy efficiency will depend primarily on the transformation of primary energy, i.e., wind power, solar energy, and hydropower. Currently, electricity is far more efficient than directly applied fossil energy. An electric motor can have an efficiency rating of over 90%, significantly higher than a steam engine, gasoline engine, or coal-fired steam turbine (Fig. 3.2). As a result, if clean energy is developed in a large scale for electricity generation, global energy efficiency will experience a dramatic leap. This is an inherent attribute of clean energy that is not existent in fossil energy.

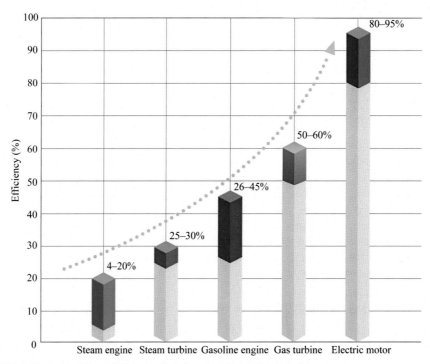

FIGURE 3.2 The Change in Energy Efficiency from Low Levels to High Levels

1.3 LOCAL BALANCING TO WIDER OPTIMIZATION

As a product of modern societal development, the network has been the most significant technological innovation in human history and has contributed dramatically to social development. Networks link up scattered points, areas, and nodes to realize transmission, reception, and sharing of resources. If the locations of development and production with regards to raw materials and the location of demand for the finished product are far apart, a network can be established to optimize resources distribution, maximize efficiency, and minimize costs. This has given rise to the development of transport networks, power grids, railway networks, etc. From the view of energy development, the production end and the consumption end of fossil energy are notably characterized by reverse distribution at a global level. For example, oil and natural gas are delivered from South America and the Middle East to Asia, from the Far East and Siberia to Europe, over distances of several thousand kilometers. At present, energy distribution has assumed a clear networking trend from point-to-point transmission to widespread use of bulk transmission and pipeline networks. It is predictable that power grids will become the main channel of energy distribution in the future in the context of clean energy-dominated and electricity-centered energy development, as shown by Fig. 3.3. On the other hand, distribution of clean energy is highly uneven in the world. Except for a small number of clean energy schemes that can be distributed and utilized locally, most large hydropower plants, wind farms, and solar ranches in regions near the Arctic, the equator and the major continents are located hundreds or thousands of kilometers from the load centers. Therefore, power grids need to be constructed to transmit electricity to the load centers in the most economical and convenient way. With the global expansion of large-scale clean energy development efforts, the coverage of power grids is expected to further widen worldwide, forming a globally linked energy network.

2 A GLOBAL ENERGY OUTLOOK

A global energy outlook requires investigating and addressing the world's energy development concerns from a global, historical, differentiated, and open perspective and stance, with more importance attached to coordination between energy and politics, economy, society, environment, as well as overall development of the various centralized (or base type) and distributed energy facilities. A global energy outlook requires to follow the new trend of the "two-replacement" strategy within the global energy interconnection so as to integrate the development, distribution, and utilization of global energy resources, and ensure a safe, clean, efficient, and sustainable supply of energy. The outlook also requires compliance with the inherent law and adaptation to the new "two-replacement" strategy. Fundamental theories on global energy sustainable development are explored, summarized as basic principles according to which, the overall goal, strategic direction, fundamental principle, development trend, and strategic direction are formulated for future world energy development. This theoretical system is illustrated in Fig. 3.4.

2.1 BASIC FRAMEWORK

The overall goal is sustainable development. The primary task is to transform the development model from overdependence on fossil energy, remove the long-term threat to human existence from massive

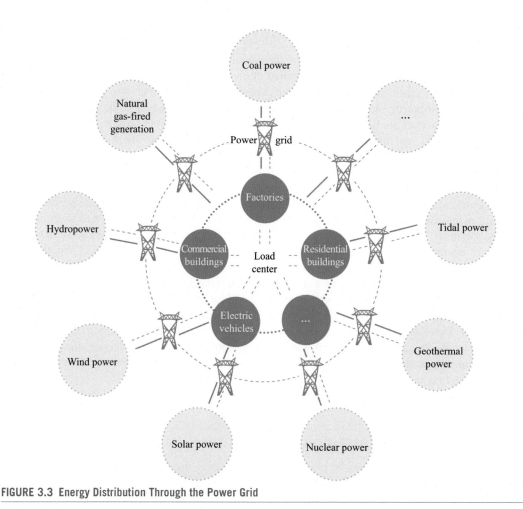

FIGURE 3.3 Energy Distribution Through the Power Grid

carbon emissions, and ensure sustainable development of human society. Since clean energy is an inexhaustible, carbon-free, and pollution-free form of energy, developing clean energy and increasing the proportion of electricity in the energy end-use structure on a global scale may be leveraged to relieve pressure on energy security due to increasing depletion of fossil energy. In addition, direct end-use of fossil energy may be reduced for lower carbon and pollutant emissions, and for a sustainable environment.

The strategic direction is the "two-replacements." The transformation from high carbon to low carbon dictates a clean energy-guided trend in energy production. This is the theoretical rationale for the development of clean energy replacement at a global level. In this connection, the supply of clean, low-carbon energy shall start with the development at source of clean energies. The more widely clean energy replacement are applied and developed, the more the potential for further development. As clean energy replaces fossil energy as the main energy source, global development efforts on a wider scale

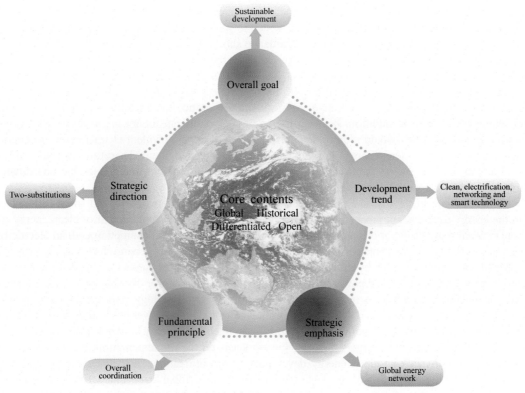

FIGURE 3.4 The Theoretical System Enshrined in Global Energy View

are expected. *The transformation of energy use from low efficiency to high efficiency determines the electricity-centered trend in energy end-use and is the theoretical rationale for electricity replacement and energy efficiency improvement.* As is known to all, electricity has been so far the most efficient energy form, and can contribute to green and efficient energy use. If clean energy is transformed in a large scale into electrical energy in the most economical and convenient way, more electricity will be available for end-uses, it will promote the creation of an electricity-centered energy structure. *The transformation of energy distribution from local balancing to wider-scale network development determines the grid-based energy transmission trend, which serves as a solid foundation for implementing the two replacements.* As for electricity generation with conventional fossil fuels, electricity may be generated at the place of origin before transmission to the load center or primary energy may be transmitted to the load center for electricity generation. Since it is only possible for clean energy to be locally transformed into electricity before transmission to the load center, extensive and large-scale transmission facilities have to be constructed. Therefore, global clean energy development involves the construction of a global power transmission network linking up clean energy bases and load centers as a platform whereupon clean energy and electricity is globally developed, distributed, and utilized in a new structure.

A global energy view involves the fundamental principle of overall coordination. As energy issues are global involving many aspects of socioeconomic development, energy development should be integrated into and coordinated globally with the politics, economy, society, and the environment. Technological innovation and policy guidance should be in place to lower the dependence of socioeconomic development on fossil energy, increase the use of clean energy, alleviate, and ultimately eliminate the political confrontations and conflicts which result from the quest for limited fossil energy. In brief, a sufficient supply of clean energy is expected to contribute to a politically harmonious, economically collaborative, environmentally sound, and socially beneficial new world development pattern. In addition, because the energy development model is closely related to resource endowment, this would require due consideration of the endowment characteristics of energy resources (especially clean energy) in the course of energy development. Therefore, various centralized and distributed energy facilities must be developed in an efficient, integrated, and coordinated manner to secure a well-balanced energy supply system.

The development trend is one toward cleanness, electrification, networking, and smart technology. In the energy development process, a clean energy policy emphasizes replacement of clean energy for conventional fossil fuels for clean energy development. At the energy consumption end, an electricity-centered policy is implemented to substitute electricity for other end-use energy forms as the principle mode of energy consumption by distributing electricity to all power consumers. In the distribution process, a globally interconnected power grid will bring about flexible power distribution by grid transmission; innovations and advances in such fields as electric power technology, widespread application of information communication, and the interconnected grid would make power generation, grid operation, power consumption for clean energy safer, smarter, as an intelligent energy system is put in place.

The global energy view targets the strategic emphasis on the implementation of a global energy interconnection by linking clean energy development with utilization, such that clean energy from every corner of the world is distributed and made available at a global level. Characterized by an innovative system structures inspired by interconnected grids, the energy interconnection integrates many elements to a very high degree, e.g., energy, markets, information and services, and shares the typical Internet features in equality, interactivity, openness, and resources sharing.

The global energy interconnection is open to all forms of centralized or distributed power facilities, users, and grids, regardless of size or national origin. All users may interconnect to the grid, access the market, choose transaction counter parties on an equal footing, and realize fair exchange of energy, information, and service and open sharing of the network. The interconnected grid enables interaction between the energy producer and the user. The user may select the appropriate producer according to demand, and the producer may provide pertinent, differentiated, and individualized products or services for the user based on an analysis of mass consumption data. In addition, there will no longer be any more rigorous distinction between the energy producer and the user; one can assume the dual or, where necessary, interchangeable roles of producer and user by not just producing and consuming, but also sharing products and services with others. The final result is shared development on all sides on the global energy platform.

2.2 CORE CONTENT

The core content of the global energy view is to study and resolve energy problems from a global, historical, differentiated, and open perspective.

2.2.1 Global

The global energy view is focused on understanding and addressing energy issues from a global perspective. *First, the nature of energy development is global.* Due to the low energy intensity of the future energy structure that is focused on renewable energies such as solar and wind power, the availability of local renewable energy can hardly meet the energy requirements of developed regions with substantial energy needs. As a result, energy resources must be developed on a large or even global scale in order to ensure energy supply and sustainable-energy development, while meeting the practical development needs of modern society and population growth. With the benefits of scale, efficiency, and economy, centralized energy development is an important foundation of energy supply whereas distributed energy development is a key complementary source of energy, given its advantages of local resource- operation flexibility, and shorter distance to loading centers. These two energy development modes play an equal role in global energy development, with due regard for both centralized large-scale clean energy bases and distributed clean energy developments. *Second, energy distribution is global.* There is a global disequilibria in energy distribution with great diversities in energy type, availability, quality, and development difficulty in various regions. At the beginning of an industrial society, with a limited energy requirement, energy imbalance could be resolved locally, so energy distribution disequilibria was not a problem. However, continued economic and social progress demands even higher levels of energy development and utilization. As it is difficult for energy-consuming centers to meet their energy shortage with local supply, energy imports are increasingly required. The increasingly obvious reverse distribution of energy resource–rich regions and energy-consuming centers points to the objective need for large-scale energy distribution at a global level, to optimize energy distribution by building a mutually supportive and complementary structure in terms of geographical spread, energy type, and features. *Third, energy security is global by nature.* With the growing globalization of economies, the energy development among different countries is characterized by an interdependent and close-knit relationship. Energy security is a global rather than national or regional issue. Any significant change in the local energy situation will lead to volatile energy prices globally, resulting in supply tension. There is no absolute energy independence, even for an energy self-sufficient country. *Fourth, environmental impact is global.* The ecological environment is a dynamic system subject to different influences. Any change locally will have an impact on the overall situation. Modern energy development has a significant influence on the global ecological environment, causing geological damage, environmental pollution, and climate change. These are threatening the survival and development of human society. Improvement of the ecological environment depends on taking a global perspective and coordinating and integrating energy development and distribution at an international level, with efforts on all fronts to protect the global environment.

2.2.2 Historical

The global energy view was developed through long years of energy development, with an element of historical linkage. *First, energy development is closely linked to the history of social development.* The history of social development is also the history of energy progress. The low-level social development in primitive and agricultural society meant limited energy demand, with low energy efficiency. Animal power and firewood were primary energy sources. In an industrial society with rapid acceleration of productivity and social development, energy development has moved up toward electricity, nuclear power, and renewable energy from coal and oil. Our society is moving from industrial civilization toward ecological civilization. *Second, energy development is closely linked to the progress of technological innovation.* Along with the technological progress from hand-crafted technology to mechanization, automation, electrification, information, and network technology, the scale, efficiency, and economy of energy utilization are constantly increasing, facilitating the energy development mode to be transformed from one of low efficiency, extensive scale, high pollution, and high emission to one of high efficiency, energy conservation, cleanness, and low carbon emission. *Third, all aspects of energy development are moving from lower to higher levels.* It appears in the form of a shift from firewood and animal power to high quality coal, oil, and other forms of fossil energy and clean electricity. The *energy development* mode is transformed from exploitation of nonrenewable fossil fuels to clean renewable energy. The *energy distribution* mode is also changing from one of long-distance, low-efficiency railway, road, and pipeline transport, to instant power grid transmission. And *energy utilization* is changing from inefficient direct burning to efficient terminal use of electricity.

2.2.3 Differentiated

The global energy view emphasizes win–win partnerships and coordinated development by taking into account the differences in energy endowments, social development, and political and economic conditions between different countries and regions. *First, energy endowments are different by nature.* The energy resources distribution among different nations is uneven and the distribution of fossil energy and clean energy is different. Subject to the political, economic, environmental, and other factors of development, clean energy is gradually replacing fossil fuels as the dominant energy source. The focus of energy development, utilization, as well as the distribution of energy will change from being dominated by some countries and regions to a globally interconnected network with changes effected on an as-needed basis and the impact of energy endowment variability on energy development gradually removed. *Second, the levels of energy development are different,* depending basically on overall national strength. The stronger the overall natural strength, higher is the level of energy development achieved, and more advanced is the energy technology, the stronger is the resulting control over energy resources, which leads to a more rational energy structure with higher efficiency in energy production, utilization, and enhanced distribution capability. The global energy view emphasizes closer cooperation and shared development at an international level to promote the balanced development of energy worldwide. *Third, the geopolitical conditions are different* from country to country, as a political manifestation of a lack of fossil energy. The geopolitical condition of an international fossil energy-dominated energy landscape is highly complicated. A country exercises greater control over energy resources and it will have a greater say in the international political community. This situation results in a fierce competition for energy resources and a major transport route all over the world, with lasting tension seen in a number of energy-rich countries and regions. With a large-scale development and dominant position of clean energy, an adequate supply of energy will eventually get realized and the international energy situation will become lesser tensed. Geopolitically, one will see the energy situation shifting from one of confrontation to the achievement of a win–win outcome based on mutual benefit and cooperation.

2.2.4 Open

Global energy development is a dynamic process, with the continued evolution of energy types, structures, characteristics, and markets. *First, energy resources are open by nature.* Fossil fuels are characterized by scarcity and regional differences, with a close relation with territorial sovereignty, national security, and diplomacy. In contrast, clean energy is inexhaustible with an endless supply and unrestricted access. As a result, coordinated development efforts and sharing of resources on a global scale will become an inevitable trend in clean energy. *Second, the clean energy system is characterized by openness.* An open system is also a safer and more dynamic system. With a growing number and coverage of energy types, the energy system will move from simple energy development and supply toward developing multiple functions in terms of information, service, and interconnection. As a result, the integration of energy technology with information, material, and Internet technologies will also become an inevitable trend. The future energy system will be a global energy interconnection being fully open and with a global reach and interconnectivity capability. *Third, the energy market is an open one.* Ensuring an adequate supply of energy shall be an important basis and precondition for reviving the characteristics of energy as a common commodity, thereby building a fair and open energy market. The future energy market will be based on a global energy interconnection with open access for energy suppliers and consumers worldwide to achieve global power trading in accordance with established market rules.

SUMMARY

1. Global energy development has experienced a course from high carbon to low carbon, from low efficiency to high efficiency, and from local balancing to wider-scale optimization. The two replacements are an inevitable trend in this sphere.
2. The global energy view is the principle theory for global energy development. The key point is that global energy resources analysis and solution should be based on a global, historical, differentiated as well as open perspective, and stance.
3. The global energy view defines a theoretical system for future energy development with an overall objective for sustainable development and a strategic direction toward "two replacements." The fundamental principle is an emphasis on coordination, underlined by a trend toward cleanness, electrification, networking, and smart technology with a strategic focus on establishing a globally interconnected energy network.

SUPPLY AND DEMAND OF GLOBAL ENERGY AND ELECTRICITY

CHAPTER OUTLINE

1 MAJOR FACTORS

Nowadays, energy development is closely associated and aligned with socioeconomic development. It determines the close link that energy supply and demand has with economy, society, the environment and resources. Generally speaking, socioeconomic development, energy resource endowment, energy environment constraints, technological progress, and energy policy regulation are the five major factors that influence energy supply and demand (See Fig. 4.1). In particular, socioeconomic

Global Energy Interconnection. http://dx.doi.org/10.1016/B978-0-12-804405-6.00004-X
Copyright © 2015 China Electric Power Press Ltd. Published by Elsevier Inc. All rights reserved.

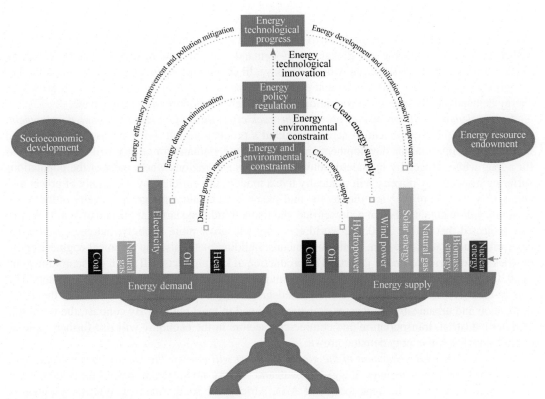

FIGURE 4.1 Schematic Diagram of Factors Influencing Energy Supply and Demand

development, including economic and population growth, industrialization, and urbanization, determines the growth trend of gross energy demand and regional distribution. The reserves and distribution of energy resources determine the supply potential, structure, and distribution of energy. Energy and environmental constraints restrict the overall level, structure and way of mankind's utilization of energy resources to meet energy demand. As the Earth's ecological environment deteriorates, its impact on energy consumption, supply structure, and structural readjustment has become increasingly prominent. Energy technological advancement and energy policy regulation mirror the combined effect of science and technology and government regulation on economy, energy, and the environment. Energy technological progress is a major decisive factor in directly determining the efficiency of energy production and consumption as well as environmental emissions. Energy-related policy and regulation play a role in guiding and regulating socioeconomic development, energy resource supply, and energy consumer behavior.

In-depth analysis and comprehensive assessment of the previous factors will have important implications for learning the development trend of global energy, studying and evaluating the mode of global energy development in a scientific manner, and formulating in the right way a roadmap for energy development.

1.1 SOCIOECONOMIC DEVELOPMENT

Energy demand reflects the ups and downs of economic development. Looking ahead, development will remain the major theme and global energy demand will continue to rise, despite the crippling impact of the financial crisis on the world economy. In developing countries (regions) in particular, fast-growing populations and accelerated economic expansion will bring about catch-up growth in energy demand and the energy consumption gaps between different regions will narrow.

Energy demand will continue to grow on the back of relatively fast socioeconomic development. It is driven by the requirement to satisfy mankind's production and living needs. With continued improvements in economic development and quality of life, demand for energy will continue to soar for a long while, especially in underdeveloped countries and regions where most of the populations suffering from energy poverty will gradually live a modern life enriched by the supply of power as a commodity. We are of the view that by the mid-twenty-first century, the gross capacity of the world economy will maintain, albeit at a slower rate, the rising trend that started at the end of World War II. It is projected that before 2030, the world economy will grow more strongly, but energy-intensive enterprises in the steel and iron, nonferrous metals, building materials, and chemical engineering industries will experience slower growth. The production of key products will reach the saturation point in 2030 or thereabouts, with energy demand maintaining an average growth rate of 1.6%. After 2030, energy-intensive industries will see themselves degenerating into "sunset industries," with the industrialization and urbanization processes drawing to an end in most regions. By contrast, the weighting of other industrial, transportation and commercial sectors in the economy will rise further, amid an overall slowdown in energy demand growth.

A more balanced development of the global economy will narrow the gap in energy demand per capita between different regions. Sluggish growth and poverty are the major reasons for the long years of war and social unrest in some locales of Asia, Africa, and South America. With the progress of the globalization process, the gaps between the underdeveloped regions of Asia, Africa, and South America and the developed regions of Europe, the Americas, and Oceania will be significantly narrowed in the future, with an improvement in the imbalanced development between north and south. By 2030, the role of Asia as the engine of world economic growth will be further strengthened. China will overtake the United States as the world's largest economy; Africa and South America will occupy a more important position in the world economy. The developed countries in Europe, North America, and Oceania will continue to lead the world in technology, finance, and education, but economic growth will be relatively lower due to the impact of population ageing, heavy government debts, and an excessively exuberant virtual economy. By 2050, the currently developing countries (regions) will rise to developed status following the completion of the industrialization process. The gross capacity of emerging economies will account for over 50% of the world economy by then. The gap in energy demand per capita between different regions will close gradually along with the narrowing differences in economic development. Energy consumption per capita in the underdeveloped regions of Asia, Africa, and South America will increase substantially.

The marked difference in population growth will produce a greater impact on energy demand. According to United Nations projections, under the scenario of medium fertility rates, the world population will continue to increase, albeit at a steadily lower rate. The world population is forecast to reach 9.55 billion by 2050. The difference in population growth will become increasingly obvious among different regions, with the African population growing fastest, followed by the Oceanian population;

the Asian and South American populations will fall sharply from a high level; the North American population will maintain a low and steady rate of growth while the European region will see negative population growth. In the mid-twenty-first century, along with a slowdown in economic and population growth, energy demand in developed countries will only post marginal growth in the long term. By contrast, the rapid economic expansion, rising populations and faster urbanization among developing and emerging markets will continue to drive global energy consumption higher.

Global economic policy is shifting in the direction of globalization, balance, and low-carbon development. Over the past 20 years, the global economy has exhibited a growing trend of "moving east," with emerging economies like China, India, and Russia becoming the new drivers of global economic growth. This trend will remain intact and spread to underdeveloped countries and regions. Driven by a fundamental view of concerted and sustainable development on a global basis, all countries are working hard to create a new international economic order characterized by globalization, multipolarization, and mutual coordination, guided by the shared objective of combating global climate change and eliminating war and poverty. Moreover, the emerging low carbon economy is providing an important impetus to global economic growth in the future. In view of the global impact of carbon emissions, countries around the world are finding themselves competing and working with each other to promote global low-carbon development together.

1.2 ENERGY RESOURCE ENDOWMENTS

Judging by the supply of energy resources around the world, the long-term development and utilization of fossil energy has imposed rigid growth restrictions on fast-growing energy demand. In contrast, renewable energy sources are inexhaustible with great development potential. The development and utilization of energy resources is limited by regional resource endowments and affected by the economics of technology.

The supply of fossil energy is subject to rigid resource restrictions, with limited room for development and utilization in the future. Fossil energy sources like coal, oil, and natural gas supported the progress of human civilization and socioeconomic development for 200 years in the nineteenth and twentieth centuries. In spite of a steady year-on-year rise in proven reserves of global fossil energy thanks to fast-developing exploration technology, the reserves of global fossil energy remain limited. Unless mankind can stop relying on fossil fuels, this source of energy will eventually and inevitably be exhausted as a matter of reality and natural restriction.

The abundant renewable energy resources around the world will become the dominant energy source in the future. Given their abundance and thanks to the growing maturity of development and application technologies, hydropower, wind, and solar energy resources worldwide can meet energy development requirements. The global demand for energy in 2050 can be satisfied simply by developing just a fraction, at 0.05 percent (5/10,000), of the developable capacity of global wind and solar energy. Moreover, the Earth also possesses other abundant energy resources, like ocean, biomass, and geothermal energy. If these renewable energy resources can be developed on a large scale, the energy problems facing mankind will be fundamentally eradicated.

The imbalanced distribution of energy resources calls for efforts to coordinate energy allocation on a global basis. Regionally, remaining recoverable coal reserves are distributed mainly in Europe and Eurasia, Asia Pacific, and North America. Remaining recoverable, conventional oil reserves are concentrated in the Middle East, Central and South America, and North America. Remaining recoverable natural gas reserves are mainly located in the Middle East, Europe, and Eurasia. Given the abundance of hydropower resources available and a low degree of development and utilization, Asia, Africa, and South America will become the focus of hydropower development in the future. The world's wind and solar energy resources are mostly concentrated near the Arctic and equatorial regions, making them ideal locations for development and construction of large energy bases. Historically, the uneven distribution of conventional energy sources has propelled the development of global fossil energy trade and traditional energy markets. The global development of electricity-oriented clean energy in the future will lead to the formation of global electricity trade and electricity markets and invoke new requirements on the global allocation of electric power.

1.3 ENERGY ENVIRONMENT CONSTRAINTS

With the substantial growth of energy consumption, environmental issues arising from energy development and utilization are becoming increasingly prominent and commanding widespread attention. Global energy environment issues are reflected mainly in greenhouse gas emissions, environmental pollution and ecological damage caused by burning fossil fuels.

Global efforts to combat climate change have accelerated low-carbon development of energy. Putting a spotlight on the growing urgency of work on energy efficiency and emission control, global climate change has been driving the development and improvement of relevant energy technology, energy policy and global energy management systems. The massive emission and build-up of CO_2 caused by fossil fuel combustion has increased the concentration of this pollutant gas in the atmosphere and further aggravated the greenhouse effect, resulting in abnormal climate events and global ecological imbalances. To avoid severe disaster and achieve the goal of limiting global temperature rise to no more than 2°C in this century from the temperature recorded before industrialization, the major developed countries and emerging market economies are expected to introduce mandatory measures to control emissions in the future. The global action to combat climate change will bring improvements, making global energy technology, energy policy and global energy management systems more comprehensive, in-depth, and synergistic.

Pollutant discharges caused by energy utilization are attracting growing attention. Long before the Industrial Revolution, firewood was the main source of energy for consumption. Basically this did not create environmental concerns at a time when consumption was limited and timber resources harvested could be replenished through revegetation, not to mention the fact carbon emissions from burning could be offset by the carbon fixation of the growing plants, and the smoke pollution caused by combustion was within an environmentally acceptable limit. The Industrial Revolution triggered rapid growth of energy consumption, making fossil fuels the dominant energy source, with a tremendous impact on the environment. For example, China now faces the severe challenge of compound air pollution caused by soot and vehicle exhaust emissions. With the growing prominence of environmental problems and the ever-higher demand for environmental quality, more importance is being attached to the pollutant discharges from fossil fuel burning as a restrictive factor for energy development.

The current development and utilization of energy resources is ecologically and environmentally unsustainable. Energy development and utilization focusing on fossil energy has brought about increasingly conspicuous damage to the ecological environment. For instance, land subsidence and water contamination caused by coal mining as well as the serious heavy acid rain caused by particulate, sulfur dioxide, and nitrogen oxide emissions in the coal burning process have led to soil acidification, forest deterioration, and other ecological damage. The current over-reliance on fossil fuels for energy development and utilization is unsustainable, which demands the completion an energy transition as soon as practicable in order to reduce destructive exploitation and promote clean energy development.

1.4 ADVANCEMENT OF ENERGY TECHNOLOGY

By taking advantage of better energy technology to improve demand-side energy efficiency, we can reduce energy supply. By enhancing supply-side capacity while lowering the cost of energy supply, we can significantly alleviate the environmental impact of energy development and utilization and also ease energy environment constraints.

Technological progress has led to improved energy efficiency and lower energy demand. Around the world, energy end-use and intermediate conversion efficiency have improved to various extents, thanks to technology process improvement, energy efficiency technology development, and energy management enhancement. For example, the world's advanced-level aluminum electrolytic AC power consumption decreased from 14,400 kWh/ton in 1990 to 12,900 kWh/ton in 2012, whereas comprehensive energy consumption of ethylene also went down from 897 kg of standard coal/ton to 629 kg of standard coal/ton over the same period. Energy consumption in the construction and transport sectors has shown significant improvement along with growing technology development in motors, electronic information, materials, and energy gradient utilization. Thanks to the advanced gas turbine and coal-fired power generation technologies, such as extra supercritical coal-fired power generation, integrated gasification combined cycle power generation, and circulating fluidized bed, the efficiency of fossil-based power generation has improved exponentially. Improvement in energy development, energy conversion, and utilization efficiency has not only reduced the production of primary energy required to meet the same level of demand, but also provided conditions for restructuring the energy mix and mitigating energy and environmental problems.

Technological progress has enhanced energy supply capacity and lowered energy supply costs. The oil crisis of the 1970s triggered a far-reaching structural change in the world energy market, prompting global efforts to actively develop energy efficiency technology and seek alternative energy sources to ensure supply security. For example, along with the rapid development of exploration technology, the proven reserves of fossil energy around the world have increased from year to year. The successful development of horizontal well technology, multilayer fracturing technology, hydraulic fracturing technology, refracturing technology, and simultaneous fracturing technology has made the commercialized mass production of shale gas in North America possible. Technological advancements in nuclear energy utilization and renewable energy generation have led to ever-higher levels of clean energy utilization, steadily lowering costs, and growing capacity expansion. As a result of technological development, grid parity for photovoltaic energy generation will be reached in 2016–2017 in Europe and America. However, parity between on-grid photovoltaic power tariffs and residential sales tariffs in China is not expected to be reached until 2020, given the country's relatively low power price benchmarks. In line with technological progress, generating costs based on the same type of power generation show a steadily downward trend, allowing the electricity industry to meet increasing power demand at increasingly competitive system costs.

Pollutant emissions are reduced and impact on energy and the environment mitigated through technological advancement. As a unique high-quality energy source, electric power is not only highly efficient (over 90% in general), but also pollution-free in consumption and conducive to high-precision control. It is also a good substitute for fossil energy at the end-use level. Amid the dwindling supply of fossil energy resources and the growing concerns about the contribution of fossil energy development and utilization to environmental pollution and climate change, application technologies for renewable energy, such as wind and solar energy, have become the focus of competition in the technologies for global primary energy development and the way forward in energy technology development around the world. With technological progress, the share of electric power in energy end-use and primary energy consumption can be improved and the structure of energy demand optimized to reduce the environmental impact brought about by energy development and utilization and alleviate energy and environmental constraint.

1.5 ENERGY POLICY REGULATION

Energy policy can be described as the regulator and controller of energy development, providing a guiding tool at the macro level and a management tool at the micro level to enable the private and public sectors to adjust the relationship between the energy system and the socioeconomic/environmental systems.

Energy policy drives innovation in energy technology. Progress and innovation in energy technology is an important pillar of energy development. Driven by a consensus on global sustainability, governments around the world hold in high regard the development of energy technology, as evidenced by the strong financial, policy and taxation support provided for the development of energy efficiency technology and clean energy technology. For instance, a US$150.7 billion investment plan for the years 2009–2014 was put forward in the United States Recovery and Reinvestment Act by the Obama Administration to provide direct investment, tax incentives, and loans or loan guarantees for clean energy technology. Of this funding amount, 74% is dedicated to promotion and application of clean technologies, 18% to the research and development and demonstration of clean technologies, and the remaining 8% to the provision of financial subsidies for clean technology manufacturers. In recent years, the Chinese government has been working vigorously on the research and development and promotion of energy technologies and creating conditions for energy technology development by improving the market mechanism, technical standards and policy environment.

Policy-guided energy production and utilization. Energy is an important physical foundation for socioeconomic development. With the fast expanding energy demand and the growing scarcity of resources over the past decades, meeting the energy demand arising from socioeconomic development has become the top priority of policy-based regulation. To achieve this goal, measures such as technological progress, market regulation, and system guidance have been taken to support the policy objectives of promoting energy development and ensuring a more adequate supply of energy, while encouraging energy conservation and efficiency and controlling rapid energy demand growth. For example, to improve energy conservation and rein in unreasonable energy demand, the Chinese government has proposed a target of approximately 4.8 billion tons of standard coal for total primary energy consumption and of approximately 4.2 billion tons of standard coal for total coal consumption by 2020 so as to strengthen control of the coal-dominated energy mix. In addition, plans have been proposed to vigorously develop renewables as well as clean energy like nuclear energy and natural gas, with the share of nonfossil energy in primary energy consumption set at 15% and 20% by 2020 and 2030, respectively, to improve the level of substituting clean energy for fossil energy.

Energy policy supports energy and environmental improvements. For a long time, the continued growth of fossil energy consumption has been responsible for environmental issues like ecological damage, environmental pollution, and global climate change. Against this background, the resolution of these energy and environmental problems has figured more prominently in policy terms in some countries. For instance, environmental standards governing energy utilization have been developed, covering coal consumption and pollutant discharges at thermal power plants, environmental well-being and energy efficiency, vehicle exhaust emissions, and the economics of fuels. Intensive efforts have also been exerted to promote the application of environmental management technologies, covering clean energy, high-efficiency power generation, and high-efficiency desulfurization and denitrification.

2 ENERGY DEMAND

Energy supply and demand is affected by various factors, such as socioeconomic development, energy resource supply, environmental constraints, technological progress, and regulatory policy. This reality gives rise to great uncertainties surrounding the future development of energy. Among the many probable scenarios of global energy and power demand growth, we focus particularly on the impact of shared global prosperity and an active response to climate change on the future demand for energy and electricity. In terms of socioeconomic development, the global economy is expected to maintain relatively steady growth between 2010 and 2050, with annual growth of approximately 3%,[1] and the world's population will continue to grow from 6.92 billion to 9.55 billion[2] in the same period. In response to the restrictions caused by climate change, a global consensus has been reached with aggressive efforts to attain the achievable target of limiting global temperature rise to 2°C. This is expected to lower global carbon emissions from energy consumption by 40–70% to within 12 billion tons[3] by 2050, representing a dramatic 50% reduction from 1990 levels and nothing short of a revolutionary challenge to the way of global energy development in the future.

Taking an integrated view of energy demand and environmental constraints, among other factors, an energy system analysis model of "end-use energy demand – energy processing and conversion – demand for primary energy" is adopted for analyzing scenarios of global demand for energy and electricity. The general idea is to project the end-use demand for different types of energy (e.g., coal, oil, natural gas, electricity, and heat) based on the levels of economic activity and historical energy consumption among different energy end-use sectors, while also taking into account trends of fossil energy consumption, electric energy substitution, and carbon constraints. Based on the end-use demand so projected, an integrated view is then taken of the conversion efficiency of different energy segments, including raw coal-fired generation, heating, coking, crude oil refining, power generation, natural gas-fired generation, liquefaction, and nonfossil fuel–based power generation and heating, as well as the availability and technology economics of different resources for power generation. This will form the base for projecting the demand for primary energy resources such as coal, oil, natural gas, and nonfossil energy. See Fig. 4.2 for the reasoning behind the global energy and electricity analysis model.

---

[1]Refer to the research findings of the United Nations, World Bank, Standard Chartered Bank and WEC.
[2]United Nations forecasts based on a scenario of medium fertility rates.
[3]Findings based on a 450 ppm scenario of limiting global temperature rise to no more than 2°C, according to IPCC AR5.

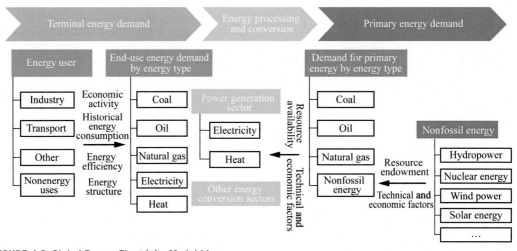

FIGURE 4.2 Global Energy Electricity Model Idea

Based on the previously mentioned reasoning, a scenario analysis of global demand for energy and electricity is performed for the target years of 2020, 2030, 2040, and 2050 in order to forecast the total volume, structure and distribution of global primary energy demand as well as the total volume and distribution of electricity demand in the future. The analysis provides fundamental support for global energy development and the building of a globally interconnected energy network.

2.1 TOTAL ENERGY DEMAND

Primary energy demand worldwide is expected to continue growing, albeit at a steadily lower rate. In 2013, global GDP totaled US$74 trillion; with a global population of approximately 7.2 billion, and total primary energy demand of approximately 19.5 billion tons of standard coal.[4] With the steady economic and population growth worldwide and continued improvements in energy efficiency, global GDP is expected to reach US$220 trillion and the global population to increase to around 9.55 billion by 2050. We have adopted here a model of "end-use energy demand – energy processing and conversion – primary energy demand" by taking into full account global socioeconomic development, energy supply, energy environment constraints, energy technological progress, and regulatory energy policy. Based on a scenario of quickening clean energy development, global primary energy demand is expected to increase to 30 billion tons of standard coal (see Fig. 4.3) by 2050, which is in line with the scenario proposed in the World Energy Council's World Energy Scenarios: composing energy futures to 2050. Under this scenario, global energy consumption is expected to grow 1.2% on a yearly basis in 2010–2050, representing an increase of 11.2 billion tons, or the aggregate energy consumption of China, the United States and the European Union combined in 2010. Global energy consumption per capita will grow by around 15% from 2.7 tons of standard coal to 3.1 tons of standard coal, or 46% of the coal consumption per capita of 6.7 tons of standard coal among OECD nations in 2000. This level of per capita consumption is necessary to support the economic growth

---

[4]Energy demand referred to in this chapter includes energy of a noncommodity nature.

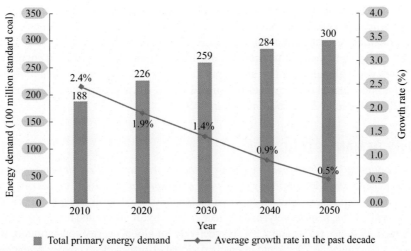

FIGURE 4.3 Total Volume and Growth Rate of Global Primary Energy Demand

of developing countries. At the same time, the progress of energy technologies will contribute to a 50% reduction in energy consumption per unit of GDP from 2.7 tons/US$10,000 to 1.4 tons/US$10,000, indicating a significant improvement in the efficiency of energy utilization.

In the future, the elasticity coefficient of energy consumption is expected to fall steadily, with faster economic growth being supported by relatively low energy growth. Between 1990 and 2000, average annual global economic growth stood at 2.8%, compared with an average annual growth of 1.4% in global energy consumption, indicating an elasticity coefficient of energy consumption at about 0.5. Between 2000 and 2010, average annual global economic growth was estimated at 2.7%, compared with an average annual growth of 2.4% in energy consumption. The elasticity coefficient of energy consumption climbed to 0.9 as a result of the growing energy consumption in non-OECD countries. As regards future energy development, average annual global economic growth is expected to be about 3.0% between 2010 and 2050, compared with energy demand growth of about 1.2%, indicating an elasticity coefficient of energy consumption at about 0.4. On a phased basis, between 2010 and 2020, average annual global economic growth is estimated at about 3.0% and annual energy demand growth at about 2.0%, indicating an elasticity coefficient of energy consumption at about 0.6. Between 2020 and 2030, average annual global economic growth is expected to rebound to 3.2%, fuelled by the growing emerging economies, and the quickening pace of promoting and applying green energy technology (like the industrial internet, smart buildings, and transport electrification) in the industrial, construction, transport, and other major energy-consuming sectors, is expected to drive global energy demand growth down to 1.4% and the elasticity coefficient of energy consumption down to 0.4. With a slowdown in global economic growth, average annual global economic growth is expected drop to 3.0% in 2030–2040, while average annual energy demand growth will decline to 0.9% and the elasticity coefficient of energy consumption will be about 0.3. With more intensified global efforts in energy efficiency and GHG emission control, average annual world economic growth is expected to fall to 2.8% in 2040–2050, with average annual energy demand growth easing to 0.5%, indicating an elasticity coefficient of energy consumption of approximately 0.2.

The afore-mentioned are projections of world energy consumption growth based on a historical view of global energy development and scenarios of faster clean energy development. But objectively there are uncertainties, as total primary energy demand growth is inextricably linked with global economic growth, changing industry structures, urbanization, population growth, and energy policy. In the event of a slowdown in global economic and population growth, or significant breakthroughs in energy efficiency technology, global primary energy demand may reach approximately 23 billion tons of standard coal by 2030, to rise to 25–27 billion tons in 2050.

2.2 ENERGY DEMAND STRUCTURE

The continued optimization of the primary energy demand structure will result in a fundamental shift of focus from an energy system dominated by fossil fuels and supplemented by clean energy, to one oriented towards clean energy and supplemented by fossil fuels. In 2013, coal, oil and natural gas accounted for 30.1%, 32.9%, and 23.7%, respectively, and nuclear energy, hydropower and nonhydropower renewable energy accounted for 4.4%, 6.7%, and 2.2%, respectively, of global energy consumption, with the share of fossil energy standing at 86.7%. Amid intensified global efforts in clean energy development, the development of favorable hydropower resources will be basically completed by 2030, with continued rapid growth in various nonhydropower energy resources. Two-thirds of new energy demand will be met by renewables, with over 50% of this new demand to be satisfied by nonhydropower renewable energy, such as wind and solar power. That said, coal, oil, natural gas, and other fossil energy sources will still account for two-thirds of total primary energy demand by 2030. After 2030, we will see more mature technology for development and utilization of wind, solar, and other renewable energy, with continued improvements in conversion efficiency and economics to expedite the substitution for fossil energy (especially coal and oil). Consumption of coal, oil, and natural gas will experience negative growth and all new demand for energy will be met by renewables. Based on a scenario of faster clean energy development, clean energy is expected to represent 80% of total energy supply by 2050, replacing fossil energy as the dominant energy source, while the share of fossil energy will drop to around 20%. See Fig. 4.4 for the global demand for different forms of primary energy between 2010 and 2050.

The substitution of electricity for fossil energy, together with improved electrification, has become the dominant trend in the changes in the world's end-use energy structure. With population expansion and improving quality of life, mankind's demand for available energy[5] will continue to grow. The efficiency of energy utilization will rise substantially, driven by the growing conversion of clean energy into electricity to gradually replace fossil energy in the end-use sector. Based on meeting equivalent level of demand for available energy, a higher share of electricity in the end-use energy structure will lead to lower energy demand at the end-use level. Higher demand for available energy and improved efficiency of end-use energy will cause end-use energy demand globally to peak around 2030. From 1980 to 2010, global end-use energy consumption increased by 1.6% on an annual basis. Average annual growth in 1980–1990, 1990–2000, and 2000–2010 was 1.6, 1.1, and 2.1%, respectively. Global end-use energy demand is expected to post average annual growth of 0.4% in 2010 to 2050. See Fig. 4.5 for the global end-use energy demand and global demand for available energy from 2010 to 2050.

Judging by the trend of end-user, energy utilization has been marked by a shift from a direct, low efficiency consumption model to an indirect, high-efficiency model, with demand for electricity

---

[5]Available energy means the energy actually usable net of energy loss in the end-use segment.

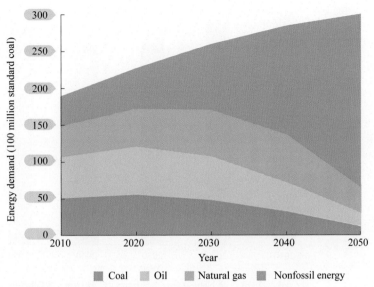

FIGURE 4.4 Global Demand for Different Forms of Primary Energy, 2010–2050

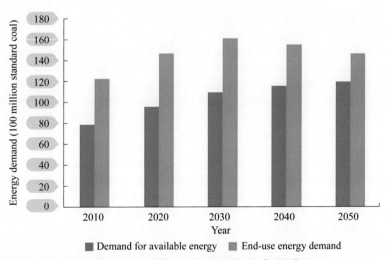

FIGURE 4.5 Global End-Use Energy Demand and Global Demand for Available Energy

continuing to grow. In 2010, electricity was responsible for 17.7% of global terminal energy consumption, 2.2 percentage points higher than that in 2000. Electricity consumption of OECD countries accounted for 21.9% of end-use energy consumption, and non-OECD countries accounted for 15.7%. It is expected that from 2010 to 2030, the substitution of electricity for coal will be carried out rapidly in the world's industrial and construction sectors, electric vehicles will gradually be used

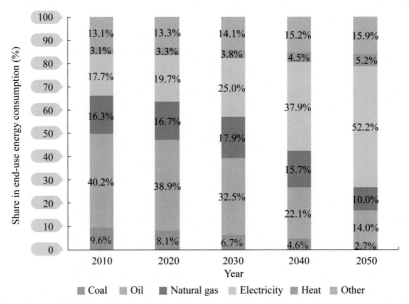

FIGURE 4.6 Structure of Global End-Use Energy Consumption, 2010–2050

for commercial application, railway electrification will grow rapidly, coal and oil will have a decreasing share in end-use energy consumption, and electricity will have an increasing share in the energy structure. By 2030, electric energy will have accounted for 25.0% of end-use energy consumption, 7 percentage points higher than that in 2010. See Fig. 4.6 for the world's end-use energy consumption structure from 2010 to 2050.

Because the major emerging economies and developing countries in Asia, South America, and Africa will complete the industrialization process successfully after 2030, the electric furnace will gradually take the place of traditional converters and blast furnaces as the major iron smelting equipment. In other industries (including construction), renewables-generated electricity, and heat will be utilized on a larger scale. In transportation, electric vehicles will replace traditional petrol vehicles at a faster rate to such an extent that oil will be edged out of its current position as the dominant fuel in transportation. Under a scenario of the "two-replacement policy" gathering speed, electricity is expected to account for over half (52.2%) of end-use energy demand by 2050, or a doubling of its share in 2030.

Given the stringent carbon emission controls, the share of fossil energy in total energy consumption is expected to be limited to around 20% by 2050, with more than half of the coal consumed and around 45% of the natural gas consumed going into power generation. The remainder will be used mainly by certain segments of the industrial sector or for nonenergy purposes. By that time, oil-fired generation will basically be nonexistent, with oil used primarily for water transport, air transport, and nonenergy uses. Of all nonfossil energy sources, around 88% will be utilized in the form of electricity and the remaining in the form of heat.

The uncertainty of the energy demand structure is derived mainly from clean energy substitution on the supply side, and electricity substitution at the end-use level. If clean energy substitution and electricity substitution should proceed at a slower-than-expected speed due to technological, cost, policy and other reasons, fossil energy will still have accounted for one-quarter to one-third of the primary energy demand structure by 2050, and electricity will still accounted for less than 50% of end-use energy consumption.

2.3 ENERGY DEMAND DISTRIBUTION

As Asia, South America, and Africa have moved into or completed the industrialization and urbanization processes, accompanied by relatively fast population growth, the share of these regions in global energy consumption is expected to increase quite strongly. By contrast, Europe and America will see a declining share, but nonetheless will remain the world's most energy-intensive regions in terms of per capita consumption and total energy consumption. Traditionally, the developed nations of Europe and America have been the world's largest energy consumers. But in recent years, Asia has also become one of the largest energy consumers, reflecting strong demand growth in China and India. In 2013, Asia, North America and Europe were responsible for 39.4, 21.9, and 23.0% of global primary energy consumption, respectively, and South America, Africa, and Oceania accounted collectively for 15.7% of the world's primary energy consumption. In the future, with the progress of economic globalization and balanced development, Asia, Africa, and South America are expected to experience faster economic growth and account for a higher share of the world economy from 34% in 2010 to around 45% in 2030, rising further to over 50% by 2050. As the north–south divide narrows, underdeveloped regions will enjoy higher living standards and energy consumption. From 2010 to 2050, annual energy consumption per capita in Asia will increase from 1.9 tons of standard coal to 3.1 tons of standard coal, reaching the world average level; annual energy consumption per capita of Africa and South America will grow from 0.6 and 1.8 tons of standard coal to 1.7 and 3.0 tons of standard coal, respectively. Although the energy consumption per capita of North America, Europe and Oceania is notably higher than the world average, it is expected to show a downward trend after reaching its peak under the pressure to reduce greenhouse gas emissions in absolute terms. See Fig. 4.7 for the energy consumption per capita of the world and different continents between 1990 and 2050.

The rapid growth in energy consumption per capita and population size is expected to make Africa the fastest-growing region in terms of total energy consumption before 2050. Africa's population will grow from 1.03 billion in 2010 to 2.39 billion in 2050, accounting for a higher share of the world population from 14.9% to 25.1% during this period. The continent's share of global energy consumption will also rise from 3.0% in 2010 to 13.7% in 2050. Between 2010 and 2050, global energy demand is expected to grow by 11.2 billion tons of standard coal, with Asia, Africa, and South America contributing to the growth. Given their massive population base, these three regions will be able to further consolidate their position as the world's largest energy consumers. These three regions are also expected to take up a 74.1% share of total global energy demand by 2050. Africa and South America will rise to a more prominent position as the world's major energy consumers. See Fig. 4.8 for the share of primary energy in global energy consumption by continent between 2010 and 2050.

North America, Europe, and Oceania, regions of traditionally high energy consumption, are expected to experience slower growth in energy consumption, with a correspondingly lower share of world energy consumption to 11.7, 13.4, and 0.8%, respectively, by 2050, as a result of slower economic and population expansion and high energy efficiency. However, annual energy demand per capita in these

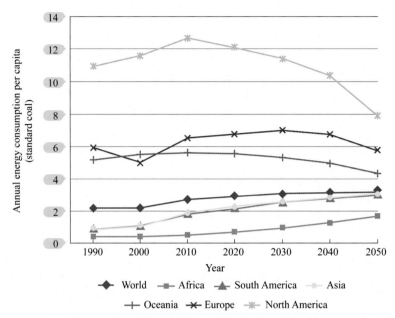

FIGURE 4.7 Energy Consumption Per Capita in the World and Continents, 1990–2050

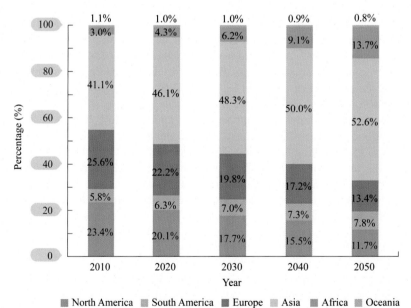

FIGURE 4.8 The Share of Primary Energy in Global Consumption by Continents, 2010–2050

Table 4.1 Energy Demand Scenario Analysis by Continent

Region	Energy Demand (Billion Tons Standard Coal)					Share (%)		Growth Rate (%)
	2010	2020	2030	2040	2050	2010	2050	2010–2050
Asia	7.7	10.4	12.5	14.2	15.8	41.1	52.6	1.8
Europe	4.8	5.0	5.1	4.9	4.0	25.6	13.4	−0.4
North America	4.4	4.6	4.6	4.4	3.5	23.4	11.7	−0.6
South America	1.1	1.4	1.8	2.1	2.3	5.8	7.8	1.9
Africa	0.6	1.0	1.6	2.6	4.1	3.0	13.7	5.0
Oceania	0.2	0.2	0.3	0.3	0.3	1.1	0.8	0.0
World	18.8	22.6	25.9	28.5	30.0	100	100	1.2

regions will expectedly remain higher than the world average by 1.5, 0.8, and 0.4 times. See Table 4.1 for scenarios for energy demand by continent.

Economic globalization has produced the greatest impact on the distribution of global energy demand. If economic growth is lower than expected, the share of Asia, Africa, and South America in energy consumption may be lower than the levels described in the previous scenario. In Africa in particular, local industrialization and urbanization will remain low, dragging down regional energy consumption growth and weakening its position in the world's energy demand structure, if economic development has not significantly improved while war and turbulence persist. Against this background and as opposed to the previous scenario, Africa's total primary energy consumption in 2050 may fall to 1.5–2.0 billion tons of standard coal, thereby bringing down global energy consumption to around 27 billion tons of standard coal. By 2050, the share of Africa in global energy demand will maintain at around 5%, and annual energy consumption per capita at 0.6 tons of standard coal, little changed from 2010 levels.

Currently, energy consumption per capita in Africa, where 600 million people are still without access to electricity, is less than one-quarter of the world's average. Judging by the fast-growing energy and electricity consumption in the developed countries of Europe and America and also China since it launched a policy of reform and opening-up, it can be envisaged that Africa will also see rapid growth in energy and electricity demand after embarking on a modernization drive. Expanding the supply of modern energy and improving energy consumption per capita in Africa is one of the important objectives towards eliminating energy poverty and achieving joint development globally, and also the major driving force behind global electricity demand growth.

3 ELECTRICITY DEMAND

Electricity demand is an important component of the energy demand structure. Given the continued economic and population growth worldwide, along with improved electrification, electricity demand will experience continued growth at a relatively fast speed. Under the framework of electricity substitution and clean energy substitution, the share of electricity will continue to rise in the end-use energy consumption structure and the primary energy supply structure. The share of electricity in energy demand will also rise substantially. Electricity will play a more important role in supporting socioeconomic development and helping to achieve greater economic expansion with lower energy demand growth.

3.1 TOTAL ELECTRICITY DEMAND

Global electricity demand is expected to maintain relatively rapid growth. Over the past few decades, global electricity demand has been growing at a steadily faster rate, with average annual growth of electricity demand worldwide being 2.6% between 1990 and 2000, rising to 3.3% between 2000 and 2010. Judging by the growth trend in global energy demand and considering the accelerating efforts in implementing the two-replacement policy, global electricity demand is expected to soar from 21,400 TWh to 73,000 TWh in 2010–2050, representing annual growth of 3.1%. Annual electricity consumption per capita will shoot up from 3096 kWh to 7654 kWh, representing a 1.5-fold increase and annual growth of 2.3%. On a phased basis, electricity demand growth will slow down, as the major developed countries will have reached the saturation point by 2020 and renewable energy development and electricity substitution are still in their nascent stage. With the energy demand among emerging economies and underdeveloped countries growing rapidly while electricity substitution in developed countries has moved into a stage of fast advancement, electricity demand growth will pick up between 2020 and 2040. With the growing base of electricity demand and less room for substitution, electricity demand growth will slow by 2050. See Fig. 4.9 for the world's total electricity demand and growth rates between 2010 and 2050.

Electricity demand growth is comparable to economic growth. Over the past two decades, the elasticity coefficient of global electricity consumption has hovered around 1. In 1990–2000, the elasticity

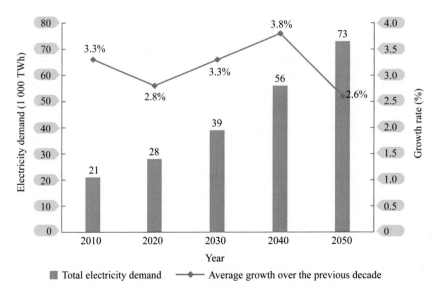

FIGURE 4.9 World's Total Electricity Demand and Growth Rates, 2010–2050

coefficient stood at 0.9, increasing to 1.3 in 2000–2010. It is expected to fall to 1.0 in 2010–2050. By 2020, electrification in the industrial, construction, and transportation sectors will gain steady momentum, contributing to constantly rising electricity demand. However, with the subdued consumption growth among the major energy-intensive industries impacted by the global financial crisis, average growth of global electricity demand is expected to be 2.8% between 2010 and 2020, down 0.5 percentage points from the first decade of the twenty-first century, with an elasticity coefficient of electricity consumption of 0.9. After 2020, the substitution of electricity in various end-use segments for conventional fossil energy, especially coal and oil, will progress at an increasingly noticeable pace. The generating capacity of wind, solar, and other forms of renewable energy will experience significant expansion, with global electricity demand accelerating steadily, on an average annual basis, at expected rates of 3.3 and 3.8% in 2020–2030 and 2030–2040, respectively, with the elasticity coefficient of electricity consumption of 1.0 and 1.3, respectively. After 2040, due to the significantly expanding basis for comparison, electricity demand growth will slow to 2.6%, with the elasticity coefficient of electricity consumption easing to 0.9, in 2040–2050.

The fact that electricity demand growth has surpassed energy demand growth indicates the increasingly dominant role of electricity in the energy structure. It is statistically shown that global electricity demand will post annual growth of 3.1% between 2010 and 2050, 2.6 times the rate of energy demand growth and slightly higher than the average rate of economic growth. Average energy demand growth is expected to be 1.2%, indicative of the saturation of demand and growth "delinked" from the total economy. The fact that electricity demand growth has markedly surpassed energy demand growth reflects a gradual strengthening of the central position of electricity in the energy system and the need to prioritize electric energy development. See Fig. 4.10 for the total world economy and growth in energy and electricity demand between 2010 and 2050.

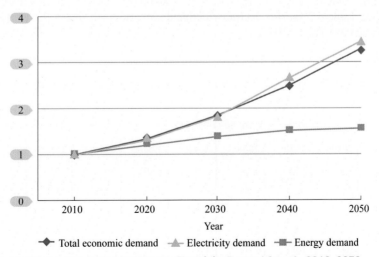

FIGURE 4.10 The World's Total Economy and Energy, Electricity Demand Growth, 2010–2050

Note: The initial values of the total economy, electricity demand, and energy demand are set at 1 for the purpose of nondimensionalization.

In developed countries and regions, electricity demand has moved into a period of steady natural growth, with future growth potential coming primarily from electricity substitution. Asia, Africa, and South America are now the world's major centers of electricity demand growth where demand is expected to maintain relatively fast growth for a while. If the saturation point of electricity demand sets in ahead of time in Asia, Africa, and South America, or electricity substitution in Europe, America and Oceania proceeds more slowly than expected, total global electricity demand may drop to 50,000–60,000 TWh as opposed to the above-mentioned scenario.

3.2 ELECTRICITY DEMAND DISTRIBUTION

Fundamental changes in the world's electricity demand situation are expected to occur in line with economic adjustments. The share of developed economies in Europe and America in total global electricity demand will decline substantially, in contrast with a significantly higher share of Asia, Africa, and South America in a rise in total global electricity demand. For a long period, the developed economies in Europe and America have been the world's major electricity consumers. In 1990, the OECD countries accounted for about 65% of the world's total electricity consumption and non-OECD countries, only about 35%. More recently, and driven by the fast-growing electricity consumption among emerging economies, non-OECD countries accounted for 51% of total global consumption in 2010, rising to 53% in 2013. In 2010–2050, Asia, Africa, and South America will contribute to over 80% of new global demand growth on account of their population and economic growth, representing shares of 56.7, 17.1, and 7.8%, respectively. Asia will continue to see improvements in electrification in the industrial, construction and transportation sectors, with annual electricity demand per capita rising from 2088 kWh to 7361 kWh (equivalent to the consumption per capita of Europe in 2010), and its share of electricity demand increasing from 41% to 52% of the global total. From very low levels of electrification, Africa and South America will see a sharp increase in electrified operations, as evidenced by the expected growth of annual electricity demand to 3971 and 6547 kWh, or 13 and 7% of the global total, respectively. See Table 4.2 for the electricity demand scenario analysis for the world and individual continents between 2010 and 2050.

Table 4.2 Electricity Demand Scenario Analysis for the World and Individual Continents, 2010–2050

Region	Electricity Demand (1000 TWh)					Share (%)		Growth Rate (%)
	2010	2020	2030	2040	2050	2010	2050	2010–2050
Africa	8.7	12.8	18.8	28.9	38.0	40.7	52.0	3.8
Europe	5.4	6.2	7.8	9.4	9.5	25.0	13.0	1.4
North America	5.3	6.2	7.6	9.3	10.2	24.9	14.0	1.6
South America	1.1	1.6	2.3	3.7	5.1	5.0	7.0	4.0
Africa	0.6	1.0	2.0	4.5	9.5	3.0	13.0	6.9
Oceania	0.3	0.4	0.5	0.6	0.7	1.4	1.0	2.2
World	21.4	28.2	39.0	56.4	73.0	100	100	3.1

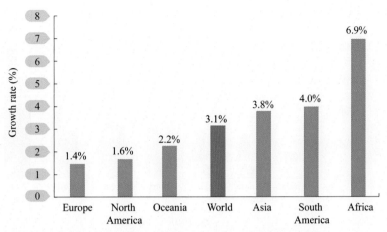

FIGURE 4.11 Annual Growth Rate of Electricity Demand by Continents, 2010–2050

Despite a very high base of per capita electricity demand, North America, Europe, and Oceania will continue to experience growing power demand as the substitution of electricity for traditional fossil energy continues. By 2050, electricity demand per capita in North America, Europe and Oceania will have increased to 22,927, 13,398, and 12,835 kWh, respectively. In 2010–2050, the total electricity demand of the three regions is expected to grow at a yearly rate lower than the world average, accounting for a sharply lower share of the world's total power demand.

In terms of growth rates, Europe, North America, and Oceania will see relatively low demand growth after the completion of the industrialization process, with annual growth of 1–2% between 2010 and 2050. In contrast, Asia, South America, and Africa will see faster growth in electricity demand at a yearly rate of more than 3% between 2010 and 2050, due to their population numbers and ongoing industrialization. Average annual growth in electricity demand in Africa, the least industrialized continent, is expected to almost double the level recorded in Asia, reflecting the combined effect of fast-growing industrialization and population. See Fig. 4.11 for the annual growth of electricity demand by continent between 2010 and 2050.

On a per capita basis, the world's electricity demand is expected to be around 7650 kWh in 2050, slightly higher than Europe's consumption per capita in 2010 and equivalent to that of the United States in 2010. By that time, following decades of relatively strong growth, Asia and South America will reach or come close to the world average in terms of per capita electricity consumption. Despite its fast growing demand, Africa's electricity consumption per capita will only reach 52% of the world average by 2050, given the very low base it started from.

In Europe, Oceania, and North America, consumption per capita will continue to grow driven by a large consumer base and the potential of electricity substitution, but the margin with the world average will become significantly narrower. In 2010, per capita electricity consumption in Europe, Oceania, and North America was 2.3, 2.7, and 5.0 times the world average, respectively. In 2050, the figures are expected to fall to 1.8, 1.7, and 3.0 times, respectively, indicating a narrowing gap in per capita consumption around the world in the future. See Fig. 4.12 for the electricity consumption per capita in the world and individual continents in 2010 and 2050.

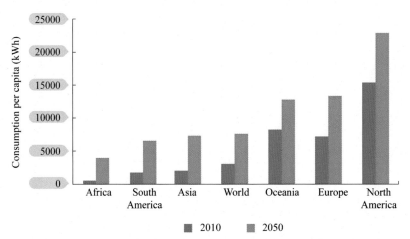

FIGURE 4.12 Electricity Consumption Per Capita of the World and Individual Continents in 2010 and 2050

As an overall trend, the narrowing gap in electricity demand among different continents is apparently in line with the globalization process. Although the speed of narrowing varies with the natural growth of regional demand and the substitution rate of electricity for other energy forms, the overall share of developed countries and regions in global electricity demand will maintain a downward trend.

The factors affecting regional electricity demand are wide and varied, and there are divergent expectations as to the future electricity demand in selected regions. For example, the IEA is relatively conservative about the level of electrification improvement in Africa, expecting that by 2040, over 500 million people on the continent will still be denied access to electricity.[6] In 2012, over 620 million people in Sub-Saharan Africa went without electricity, accounting for half of the world's total population being denied access to electric power. Furthermore, with the benefit of increased power supply offset by rapid population growth, Sub-Saharan Africa will become the world's only region with a population living without electricity. See Fig. 4.13 for information on Africa's "powerless" population.

A lack of power infrastructure has become a major hindrance to the development of Africa, amid its rapid economic expansion. According to IEA estimates, an investment of more than US$300 billion in electricity infrastructure is required for Sub-Saharan Africa to completely resolve the problem of power shortage by 2030. In recent years, China as well as the United States and European countries and regions have strengthened their investments in Africa's electricity infrastructure. For example, China offered a US$20 billion loan to Africa in March 2013, with the greater part of the amount planned to go into building electricity infrastructure in support of the continent's economic development. During his visit to Africa in July 2013, United States President Barack Obama proposed the Power Africa program to resolve the power shortage problem in Africa through an US$7 billion investment. With improvements in the global economy and the governance structure over the next few decades, Africa will see greatly improved popularity of power supplies and a significant reduction in its "powerless" population, with the growing development of electricity infrastructure.

---

[6]Source: Ref [70].

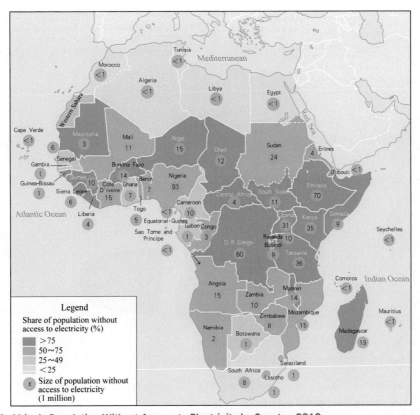

FIGURE 4.13 Africa's Population Without Access to Electricity by Country, 2012

Source: Ref. [70].

4 FUTURE GLOBAL ENERGY DEVELOPMENT STRUCTURE

In 2050, global primary energy demand is expected to surge to 30 billion tons of standard coal, including electricity demand of 73,000 TWh. The need to meet such massive power and electricity requirements spells major changes to the development of energy worldwide. Particularly in a low-carbon development environment in response to climate change, the pace of clean substitution will quicken with a rising share of fast developing and utilized renewables in the energy mix to gradually replace fossil fuels, which will see increasingly lower levels of development and utilization, as the dominant energy source of the future. By 2050, global supplies of fossil energy are forecast to decrease to 6.3 billion tons of standard coal, down 57% from 2010, while global supplies of nonfossil energy are expected to skyrocket to 23.7 billion tons of standard coal, up 480%. As the dominant energy source, renewables will see a new pattern of global development focusing on the construction of bases supplemented by distributed generation, with accelerated efforts in building large bases of hydropower, wind, solar, and other renewable energy in the Arctic and equatorial regions and on each continent.

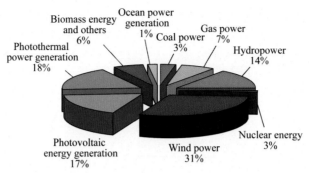

FIGURE 4.14 Global Power Supply Structure in 2050

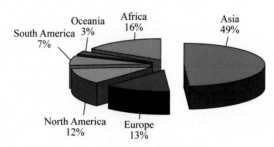

FIGURE 4.15 Share of Clean Energy Generation by Continents in 2050

4.1 OVERVIEW OF ENERGY SUPPLY

As the dominant energy source of the future, clean energy will see accelerated development and it is expected that the generating capacity of global clean energy will reach 66,000 TWh (accounting for 90% of the world's total generating capacity) in 2050. Here is a breakdown of clean energy as a share of global generation: solar and wind power (66%), hydropower (14%), biomass energy and others (approximately 10%), and ocean and nuclear energy (approximately 10%). Judging by the conditions of clean energy development on each continent,[7] Asia's share of global clean energy generation is expected to reach 49% by 2050, followed by a 16% share attributable to Africa as the most important region of solar energy development. See Fig. 4.14 for the global power supply structure in 2050 and Fig. 4.15 for each continent's share of global clean energy generation in the same year.

Fossil energy generation will experience a sharp decrease. Fossil energy, mainly natural gas and coal-fired generation, is expected to account for approximately 10% of global electricity generation in 2050. In anticipation of the need to accommodate the system operational requirements of grid-connected, large-capacity wind, solar, and other renewable energy generation, a certain level of natural gas generation capacity will be retained while efforts continue to develop pumped storage capability. Dictated by its own level of development and the capacity of renewable energy generation, Asia will retain a relatively high level of natural gas and coal-fired generation in 2050. North America, owing to

---

[7]Includes nuclear energy, but not gas-fired electricity generation.

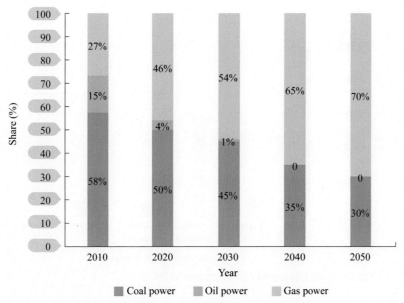

FIGURE 4.16 Changes in the Share of Fossil Energy Generation in 2010–2050

the mature technology and relatively low costs of developing shale gas and other unconventional gas resources, will retain a relatively high level of natural gas generation. See Fig. 4.16 for the change in the share of power generation attributable to fossil energy in 2010–2050. A breakdown of fossil energy generation by continent in 2050 is shown in Fig. 4.17.

Distributed generation is an integral part of energy supply. Distributed generation worldwide is expected to reach 11,000 TWh in 2050, accounting for 15% of total power generation. Based on each continent's development of distributed generation with reference to renewable energy resources, population and other factors, Asia's share of global distributed generation is expected to reach 41% in 2050. Africa, blessed with favorable conditions for developing distributed generation of solar, hydro-power, biomass, and other renewable energy, is expected to have a 27% share, ranking second among all continents. See Fig. 4.18 for a breakdown of distributed power generation by continent in 2050.

4.2 LARGE-SCALE CLEAN ENERGY BASES ON EACH CONTINENT

World continents are basically abundant in clean energy resources, such as hydropower, wind, solar, and ocean energy. The development of large-scale power generation bases in regions with favorable resource conditions can render strong support to the requirements of energy sustainability at the intracontinental or transcontinental level.

4.2.1 Asia

Asia, covering an area of over 44 million km^2 (30% of the global land mass), is the largest continent in the world where water, wind, solar and other clean energy resources abound. Favorable water resource conditions are found mainly in the upper reaches of the Yangtze River and the Yalong Zangbo River

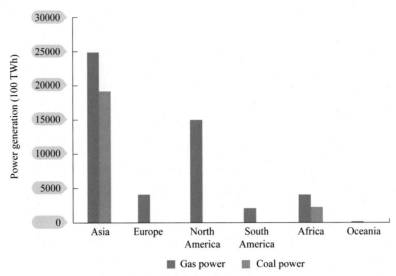

FIGURE 4.17 Fossil Energy Power Generation by Continents in 2050

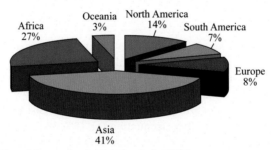

FIGURE 4.18 Share of Distributed Generation by Continents in 2050

in China as well as the Yenisei, Rivers Ob and Lena in the Russian far-east and Siberia. Wind power resources are mainly distributed in Mongolia, Central Asia, the "Three North" region (northwestern China, northern China, and northeastern China), as well as the Kara Sea, the Bering Strait, and the Kamchatka Peninsula in the Arctic region. Solar energy resources abound in Mongolia, Central Asia, the Middle East, and north-west China.

4.2.1.1 Renewable Energy Generation Bases in China

1. *Hydropower bases in southwestern China*. The technologically developable capacity of the country's hydropower resources is estimated at about 570 GW, the great majority (82%) of which is concentrated in southwest China. As at the end of 2013, conventional hydropower stations in China boasted a total installed capacity of 280 GW, with the remaining technologically developable capacity of 290 GW concentrated in Sichuan, Yunnan, Tibet, and other areas. Large hydropower bases are to be built along the Jinsha River, the Yalong River, the

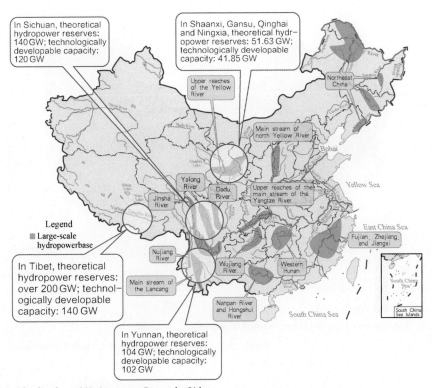

In Sichuan, theoretical hydropower reserves: 140 GW; technologically developable capacity: 120 GW

In Shaanxi, Gansu, Qinghai and Ningxia, theoretical hydropower reserves: 51.63 GW; technologically developable capacity: 41.85 GW

Upper reaches of the Yellow River

Main stream of north Yellow River

Northeast China

Yalong River

Dadu River

Jinsha River

Upper reaches of the main stream of the Yangtze River

Yellow Sea

Legend
■ Large-scale hydropowerbase

In Tibet, theoretical hydropower reserves: over 200 GW; technologically developable capacity: 140 GW

Nujiang River

Main stream of the Lancang

Wujiang River

Western Hunan

East China Sea

Fujian, Zhejiang, and Jiangxi

Nanpan River and Hongshui River

South China Sea

In Yunnan, theoretical hydropower reserves: 104 GW; technologically developable capacity: 102 GW

FIGURE 4.19 Distribution of Hydropower Bases in China

Dado River, the Lanchang River, the Yalong Zangbo River, and the Nujiang River, carrying a total installed capacity of over 260 GW. See Fig. 4.19 for the distribution of hydropower bases in China.

2. *Wind power bases in the "Three North" region.* In China, onshore wind resource potential at 80 m height with a wind power intensity of above 150 W/m^2 is estimated at 10,200 GW; wind resource potential over Grade 3 in near-shore areas with a water depth of 5–25 m and at 50 m height (wind power intensity ≥ 300 W/m^2), at 200 GW. The "Three North" region accounts for approximately 80% of China's national total of onshore wind resources. The distribution of China's wind power resources is shown in Fig. 4.20.

3. *Solar energy generation bases in northwestern China.* The annual solar radiation falling on China's land surface is estimated at the equivalent of 4.9 trillion tons of standard coal. In particular, the Qinghai–Tibet Plateau, northern Gansu, northern Ningxia, southern Xinjiang, the Gobi Desert, and other desert areas are blessed with the most abundant solar resources, with strong development potential estimated at over 85,000 TWh per year or approximately 75% of the nation's total solar energy reserves. With favorable conditions for development, large-scale solar energy generation bases can be established to transmit power to load centers in eastern and central China. The distribution of China's solar energy resources is shown in Fig. 4.21.

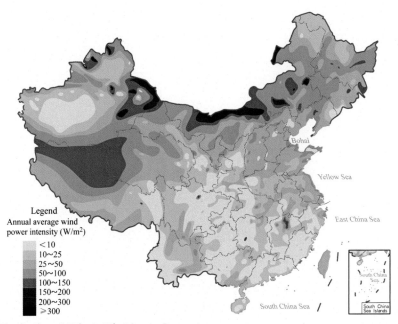

FIGURE 4.20 Distribution of China's Wind Power Resources

Source: SGCC, Wind Power Development Promotion White Paper by SGCC, 2011.

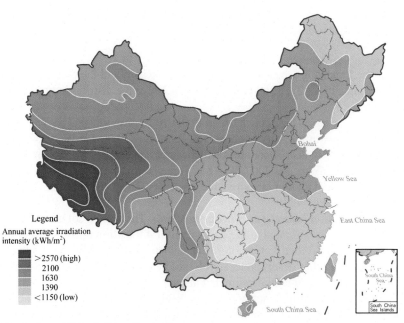

FIGURE 4.21 Distribution of China's Solar Energy Resources

Source: CMA Wind and Solar Energy Resources Center.

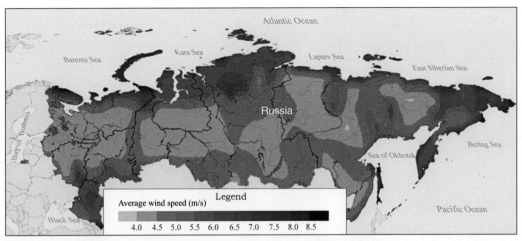

FIGURE 4.22 Wind Speed Map of Russia (Based on Satellite-Based Observations)

Source: Карты ветровых ресурсов России с комментариями.

4.2.1.2 Wind Power Bases in North Russia and Hydropower Bases in Siberia, the Russian Far-East

1. *Wind power bases in north Russia.* Russia is endowed with abundant wind power resources, largely along the coast of the Arctic Ocean in the Arctic region. Technologically developable wind power capacity in the region from the Kamchatka Peninsula to the Bering Strait is estimated at more than 7000 TWh per year,[8] compared with an estimated 3400 TWh per year[8] for the Kara Sea and its coastal regions, based on approximately 4000 h of utilization. With its huge potential, the Arctic region in Russia holds great promise as Asia's major base of wind power to be. See Fig. 4.22 for a wind speed map of Russia (based on satellite-based observations).

2. *Hydropower bases in the Russian far-east and Siberia.* The water resources in Russia available for large-scale development are located chiefly along the Lena River, the Yenisei, the River Ob, and the Amur River (the section in China is known as the Heilongjiang River) in the Russian Far East and Siberia. Out of an economically developable capacity of over 700 TWh/year, 500 TWh/year has yet to be developed. See Fig. 4.23 for the distribution of Russia's major river basins.

4.2.1.3 Wind, Solar, and Hydropower Bases in Central Asia

1. *Wind power bases in Central Asia.* Wind power resources in Central Asia are mainly distributed in Kazakhstan, with a technologically developable capacity of approximately 1800 TWh. Wind power resources are most abundant in Atyrau and Mangistau near the Caspian Sea, centrally located Astana and Karaganda, and regions in the south, where large wind power bases can be built in the future. See Fig. 4.24 for the distribution of wind power resources in Kazakhstan.

---

[8]Source: SGCC, fact-finding report on a visit to Russia concerning cooperative opportunities in wind power development in the Arctic region.

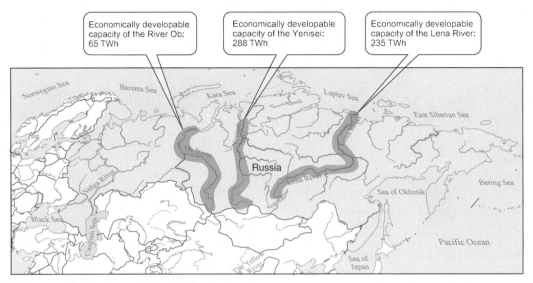

FIGURE 4.23 Distribution of Russia's Major River Basins

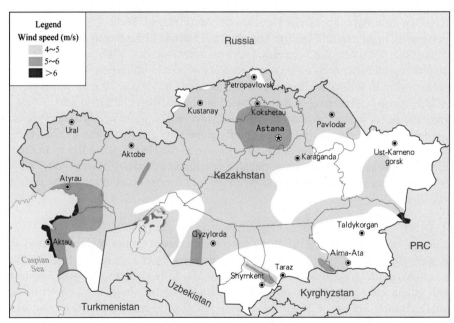

FIGURE 4.24 Distribution of Wind Power Resources in Kazakhstan

2. *Solar power bases in Central Asia.* Central Asia has an annual irradiation intensity of 1300–1800 kWh/m². In the eastern and southeastern parts of Turkmenistan where the terrain is flat, an annual irradiation intensity that often exceeds 1800 kWh/m² is recorded, reaching 70–80% of the comparative figure for the Sahara Desert in Africa. Solar energy resources abound in South Kazakhstan, Kyzylordinskaya, the coastal regions of the Caspian Sea, and southeastern Uzbekistan, where conditions are favorable for the development of solar power stations. See Figs. 4.25–4.27 for the distribution of solar energy resources in Turkmenistan, Kazakhstan, and Uzbekistan.

3. *Hydropower bases in Central Asia.* Hydropower resources in Central Asia are chiefly distributed in Kyrghyzstan and Tajikistan, the former with a technologically developable capacity of about 150 TWh/year, compared with the latter's approximately 260 TWh/year. Large hydropower bases can be built there in the future to supply power to surrounding countries.

4.2.1.4 Wind and Solar Power Bases in Mongolia

In Mongolia, the technologically developable capacity of wind power is estimated at 2500 TWh/year, while solar power potential is estimated at 3400 TWh/year. These resources are concentrated in Mongolia's southeastern regions. Wind power resources on over 10% of Mongolian land receive an "excellent" rating, while wind power resources on over 40% of Mongolian land are rated between "average" and "good." The country's southern and eastern regions see the highest concentration of wind power potential, whereas solar energy resources are mostly located in the central and southern regions

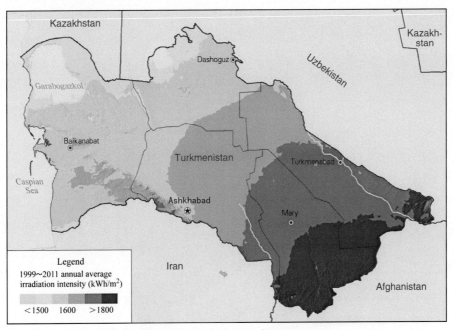

FIGURE 4.25 Distribution of Solar Energy Resources in Turkmenistan

Source: http://solargis.info/doc/free-solar-radiation-maps-GHI.

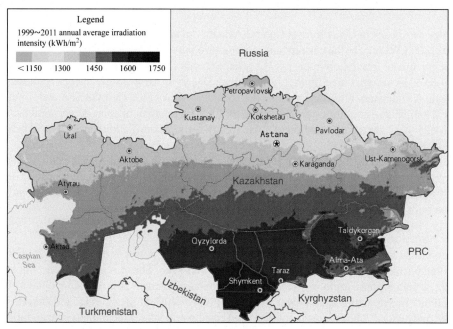

FIGURE 4.26 Distribution of Solar Energy Resources in Kazakhstan

Source: http://solargis.info/doc/free-solar-radiation-maps-GHI.

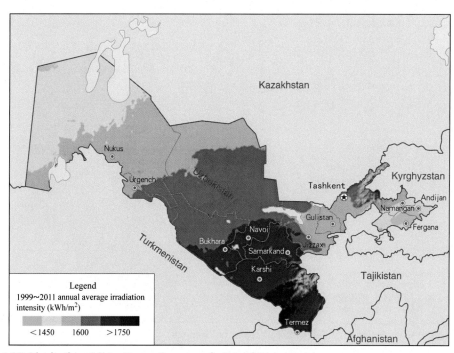

FIGURE 4.27 Distribution of Solar Energy Resources in Uzbekistan

Source: http://solargis.info/doc/free-solar-radiation-maps-GHI.

of Gobi, where conditions are suitable for building large solar power bases with an annual irradiation intensity of 1200–1600 kWh/m^2. Given the country's relatively low power load, virtually all wind and solar energy to be developed locally can be exported. Renewable energy produced in Mongolia is mostly delivered to Northeast Asia, with a transmission distance of less than 2500 km. See Figs. 4.28 and 4.29 for the distribution of wind and solar energy resources in Mongolia.

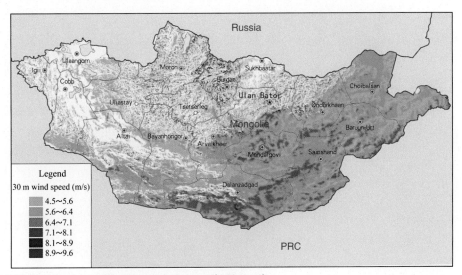

FIGURE 4.28 Distribution of Wind Energy Resources in Mongolia

Source: Wind Energy Resources Map by the US National Renewable Energy Laboratory; data from AWS TruePower.

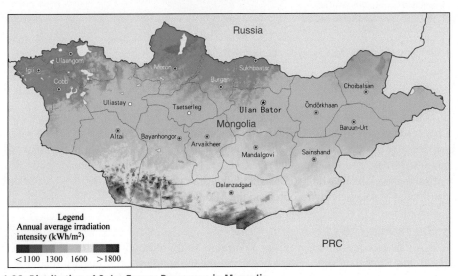

FIGURE 4.29 Distribution of Solar Energy Resources in Mongolia

Source: 3TIER Wind and Solar Energy Resources Evaluation Company.

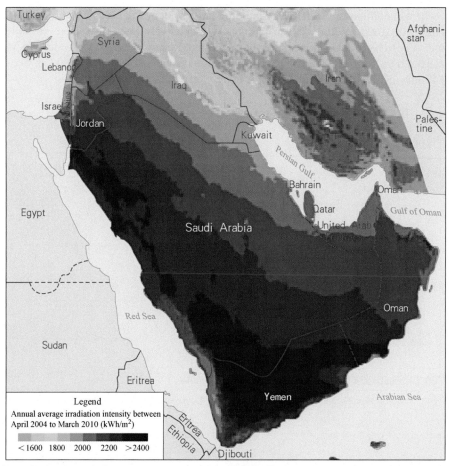

FIGURE 4.30 Distribution of Solar Energy Resources in the Middle East

Source: http://solargis.info/doc/free-solar-radiation-maps-GHI.

4.2.1.5 Solar Power Bases in the Middle East

Solar energy resources abound in the Middle East, especially in Saudi Arabia and Yemen with an annual irradiation intensity of over 2500 kWh/m². The annual irradiation intensity in Iran, Oman, the United Arab Emirates, Jordan, and other countries exceeds 2100 kWh/m². The technologically developable capacity of solar energy in the Middle East is initially estimated at over 100,000 TW/year. See Fig. 4.30 for the distribution of solar energy resources in the Middle East.

4.2.1.6 Renewable Energy Bases in India

A southern Asian country, India provides favorable sunlight conditions for solar power generation. Gujarat and Rajasthan in the west boast the most abundant solar energy resources, with an annual solar irradiation intensity of over 2100 kWh/m², followed by central and southern regions with a sunlight intensity of 1850–2100 kWh/m². These regions are expected to become home to

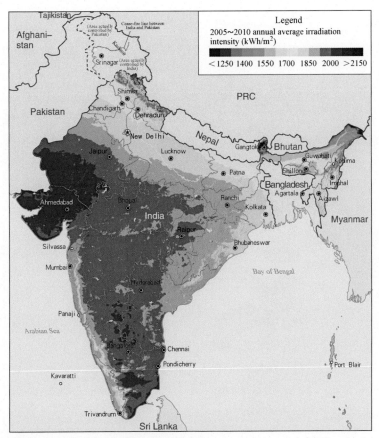

FIGURE 4.31 Distribution of Solar Energy Resources in India

Source: http://solargis.info/doc/free-solar-radiation-maps-GHI.

large solar energy bases in the future. The distribution of solar energy resources in India is shown in Fig. 4.31.

India's total wind power resources are estimated at about 100 GW. Quality resources are located predominantly in the western states of Gujarat and Rajasthan, as well as the southern region and eastern coastal areas, usually with a wind power intensity of 250 W/m². The best wind energy resources are found near the southern state of Kerala, with a wind power intensity of over 350 W/m². The distribution of wind power resources in India is shown in Fig. 4.32.

4.2.2 Europe

With an area of over 10 million km², Europe boasts relatively rich wind and solar energy resources. Wind resources are predominantly located near the North Sea, Greenland and its surrounding waters further north, the Norwegian Sea, and the Barents Sea. Solar resources are mainly distributed along the coasts of the Mediterranean Sea in the south.

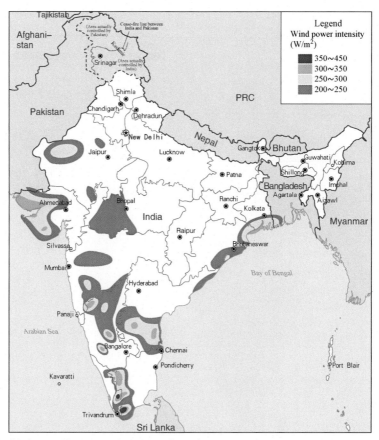

FIGURE 4.32 Distribution of Wind Energy Resources in India

Source: http://www.mapsofindia.com/maps/nonconventional/windresources.htm.

4.2.2.1 Wind Power Bases in Greenland

Greenland and its surrounding waters occupy an area of 2.63 million km^2, with a wind power intensity of over 300 W/m^2 at 70 m height,[9] and a technologically developable capacity of wind energy of about 32.5 TW.[10] The annual average wind across the Greenland Sea south of Greenland carries a maximum speed of 12–14 m/s; the annual average wind speed in the north is relatively low, at 7–10 m/s. The annual average wind speed in the northwest is about 5–7 m/s; the wind speed near the sea is quite low. The wind power intensity in the west, east and south of Greenland is high, while that along the coast is quite low. In winter, the wind power intensity of over 50% of Greenland is higher than 400 W/m^2. Wind power is stronger in the west in spring and also stronger in the east in autumn.

[9]The standard height for wind measurement is 50–80 m.
[10]Source: SGCC, a research report on the environmental characteristics of and evaluation of wind power resources in the Arctic region, July 2014.

Seasonally, average wind speed and wind power intensity move in the descending order of winter > autumn > spring > summer.

4.2.2.2 Wind Power Bases in the Norwegian Sea and the Barents Sea

Lying to the east of Greenland in the Arctic Ocean, the Norwegian Sea, and the Barents Sea are regarded as the second windiest place in the Arctic region, with an annual average wind speed of 9–10 m/s, second only to that of the Greenland Sea.

4.2.2.3 Wind Power Bases in Europe's North Sea Region

In the North Sea region, annual average wind speed at 60 m height is estimated at 8 m/s, with the region's wind power resources amounting to 30,000 TWh/year. Coastal wind power resources are especially abundant in the United Kingdom, Denmark, Germany, and the Netherlands. Based on an installed capacity of 6000 kW/km^2, the United Kingdom's highest offshore wind power potential reaches 986 TWh/year, compared with 24 TWh/year for Belgium, 136 TWh/year for the Netherlands, 237 TWh/year for Germany, and 550 TWh/year for Denmark.[11] Large wind power bases can be built offshore in the future to supply power to the domestic markets and countries in central-southern Europe. See Fig. 4.33 for the wind power development in the North Sea.

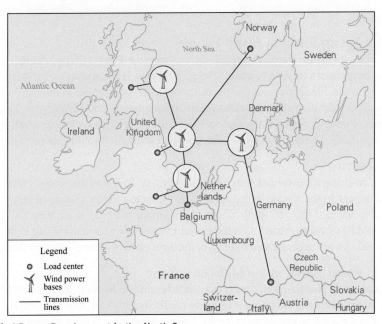

FIGURE 4.33 Wind Power Development in the North Sea

Source: Friends of the Supergrid, the first phase of the European supergrid.

---

[11]Source: *Offshore Wind Energy in the North Sea, Technical Possibilities and Ecological Considerations – A Study for Greenpeace*, October 2000.

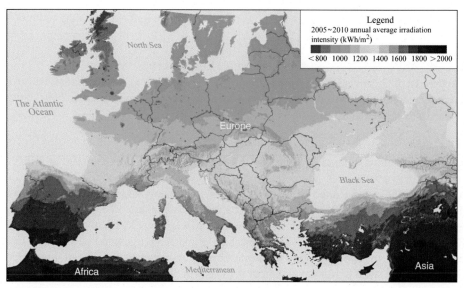

FIGURE 4.34 Distribution of Solar Energy Resources in South Europe

Source: http://solargis.info/doc/free-solar-radiation-maps-GHI.

4.2.2.4 Solar Energy Bases in South Europe

South Europe consists of the Iberian Peninsula, the Apennine Peninsula and south Balkan Peninsula, covering 17 countries on a total area of about 1.66 million km^2. It is also known as Mediterranean Europe for its proximity to the Mediterranean. It has an annual average irradiation intensity of 2000 kWh/m^2, with a technologically developable solar capacity of approximately 2600 TWh/year. Teeming with solar energy resources, Portugal, Spain, Italy, Greece, and Turkey provide favorable conditions for building large solar energy bases. The distribution of solar energy resources in South Europe is shown in Fig. 4.34.

4.2.3 North America

North America, covering a territory of over 24 million km^2 or 16.2% of the world's total land area, is the third largest continent in the world. Its topography is characterized by mountain ranges lying far apart in the west and in the east that run north–south in alignment with the coastlines. The Great Plains are situated in the middle of North America. Wind power resources are most abundant in the Midwest while solar energy resources are concentrated mainly in the American Southwest and northern Mexico.

4.2.3.1 Wind Power Bases in the Midwest

Wind power resources in North America are concentrated mainly in North Dakota, South Dakota, Montana, and Wyoming in the Midwest. Based on a capacity coefficient of over 30% at 80 m height, the technologically developable wind power capacity in the US is estimated at approximately 33,000 TWh/year.[12] Offshore wind resources in the United States are mainly distributed along the coastal areas in the east and west. Based on 830,000 km^2 of offshore areas within 50 nautical miles from the coast and with a wind speed of over 7.0 m/s at 90 m height, the technologically developable wind power capacity is estimated at

---

[12]Wind energy resources development potential data, National Renewable Energy Laboratory, 2010.

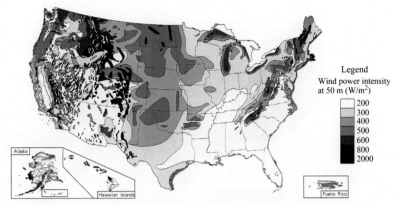

FIGURE 4.35 Distribution of Annual Average Onshore Wind Power Intensities in the United States

Source: Wind power data and map formulated by Pacific Northwest National Laboratory, courtesy of the National Renewable Energy Laboratory.

FIGURE 4.36 Distribution of Annual Average Onshore Wind Speeds

Source: Wind Energy Resources Map by National Renewable Energy Laboratory; data from AWS TruePower.

approximately 17,000 TWh/year. The flat and open terrain of the Central United States, with an annual average wind speed of over 7 m/s and abundant wind resources, creates ideal conditions for the development of large wind power bases. See Figs. 4.35 and 4.36 for the annual average onshore wind power intensities and the distribution of average wind speeds in the United States.

4.2.3.2 Solar Power Bases in the American South-West

Abundant solar energy resources in the United States are predominantly located in the south-west, covering Arizona, New Mexico, California, South Nevada, and other states. Considering the ratio of land suitable for development and also generation efficiency, technologically developable capacity is estimated at approximately 254,000 TWh/year. See Fig. 4.37 for the distribution of solar energy resources in the US.

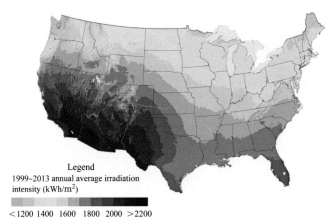

FIGURE 4.37 Distribution of Solar Energy Resources in the United States

Source: http://solargis.info/doc/free-solar-radiation-maps-GHI.

4.2.3.3 Solar Power Bases in Mexico

The annual solar irradiation falling on over 80% of Mexican territory has an intensity of over 2,000 kWh/m^2, providing suitable conditions for building large solar energy generation bases with a technologically developable capacity of about 78,000 TWh. The Baja California Peninsula west of the Gulf of California enjoys the highest annual irradiation intensity at above 2300 kWh/m^2. The distribution of solar energy resources in Mexico is shown in Fig. 4.38.

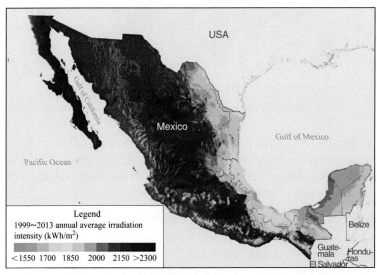

FIGURE 4.38 Distribution of Solar Energy Resources in Mexico

Source: http://solargis.info/doc/free-solar-radiation-maps-GHI.

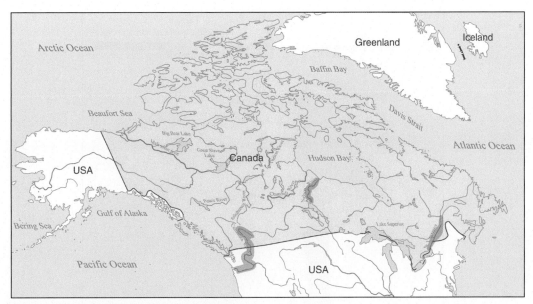

FIGURE 4.39 Distribution of Canada's Major Rivers

4.2.3.4 Hydropower Bases in Canada

In Canada, the technologically developable capacity of hydropower resources is estimated at about 262 GW,[13] with 160 GW under development planning in Quebec in the east (44 GW), Ontario (10 GW), British Columbia (33 GW), Alberta (12 GW), the Yukon (GW), and the Northwest Territories (16 GW). The country's main rivers include the St. Lawrence River, the Nelson River, and the Columbia River. Hydropower development and bases are mainly located in the provinces and regions near the United States border. The distribution of Canada's major rivers is shown in Fig. 4.39.

4.2.4 South America

Occupying over 17 million km², South America is situated in the southern part of the Western Hemisphere, separated from North America by the Panama Canal. In this region, the Andes Mountains, several kilometers high, tower over the west, with plains lying to the east, including the Amazon rainforest. Abundant solar energy resources are distributed west of the Andes Mountains, covering countries such as Peru, Chile, and Bolivia. Water resources are concentrated in the Amazon and other river basins in Brazil.

4.2.4.1 Solar Power Bases Along the East and West Coasts

Solar energy resources in South America are found mainly along the east and west coasts of the continent, especially in the Atacama Desert.[14] Reputed as one of the world's most solar-resource abundant regions, South America has an annual solar irradiation intensity of over 2,300 kWh/m² with a technologically

---

[13]Canada Hydropower Association, report of activities 2013–2014. Canada hydropower data from Canada Hydropower Association, report of activities 2013–2014.

[14]Stretching about 1000 km from north to south between the Andes and the Pacific Ocean, covering a total area of about 180,000 km². Much of this desert lies in Chile, with some areas located in Peru, Bolivia and Argentina.

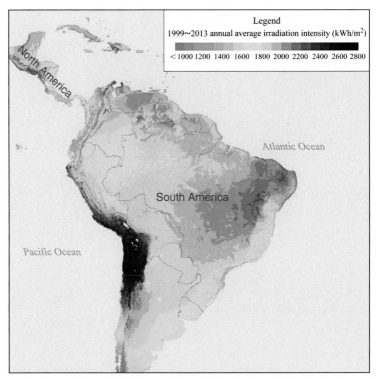

FIGURE 4.40 Distribution of Solar Energy Resources in South America

Source: http://solargis.info/doc/free-solar-radiation-maps-GHI.

developable capacity of some 15,000 TWh/year. Chile and Peru own the most abundant solar energy resources, where large solar power bases can be built. The distribution of solar energy resources in South America is shown in Fig. 4.40.

See Figs. 4.41 and 4.42 for the distribution of solar energy resources in Peru and Chile.

4.2.4.2 Hydropower Bases in the Amazon and Parana River Basins

Hydropower resources in South America are concentrated mainly in major rivers in Brazil, including the Amazon and the Parana River. Brazil's technologically developable hydropower capacity is estimated at 245 GW.[15] In 2012, it boasted an installed hydropower capacity of 83 GW, accounting for 1/3 of the technologically developable capacity, with about 162 GW yet to be developed. With a technologically developable potential of about 120 GW, the Amazon and Parana River Basins offer favorable conditions for building large hydropower bases to supply electricity to the load centers in the east and meet intracontinental power demand. Opportunities are available for joint development of solar, ocean, hydropower and other renewable resources to improve capacity utilization of renewables. See Fig. 4.43 for the distribution of hydropower resources in Brazil.

---

[15]Source: IEA, World Energy Outlook, 2013.

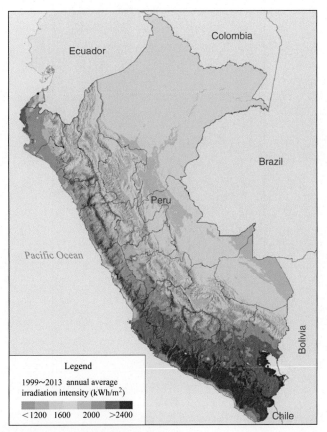

FIGURE 4.41 Distribution of Solar Energy Resources in Peru

Source: http://solargis.info/doc/free-solar-radiation-maps-GHI.

4.2.4.3 Wind Power Bases in Northern and Southern Regions of South America

With an average wind speed of about 8–9.5 m/s at 80 m height, wind power resources are more abundant in Venezuela and the island states of Cuba and Dominica in the Caribbean north of South America, as well as Argentina and southern Chile near the Antarctic region. See Fig. 4.44 for the distribution of wind power resources in South America.

4.2.5 Africa

Africa, measuring over 30 million km^2 or 20% of the world's total land area, is the second largest continent, stretching 8000 km from south to north and 7403 km from east to west. The continent's abundant solar energy resources lie mainly in the north, east and south. Its water resources are also among the world's richest, distributed mainly along the Congo River, the Nile River, and the Zambezi River. Wind power resources are concentrated mainly on the eastern and northwestern coasts of the continent.

FIGURE 4.42 Distribution of Solar Energy Resources in Chile

Source: http://solargis.info/doc/free-solar-radiation-maps-GHI.

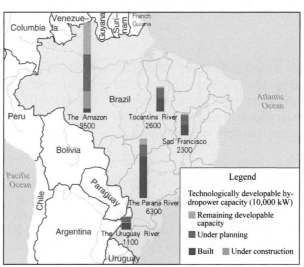

FIGURE 4.43 Distribution of Hydropower Resources in Brazil

Source: IEA, World Energy Outlook 2013.

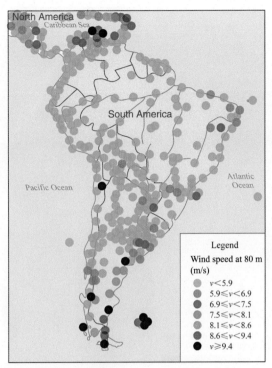

FIGURE 4.44 Distribution of Wind Energy Resources in South America

Source: http://www.geni.org/globalenergy/library/renewable-energy-resources/index.shtml.

4.2.5.1 Solar Power Bases in Africa

Africa[16] has one of the world's highest solar irradiation intensities. The potential for solar energy generation is huge in Morocco, Algeria, Tunisia, Libya, and Egypt. In Algeria, Morocco, and Egypt, the solar irradiation intensity is 2700, 2600, and 2800 kWh/m^2, respectively, compared with over 2500 kWh/m^2 for Tunisia and Libya. North Africa's technologically developable solar energy capacity is expected to reach as much as 141,000 TWh/year, compared with about 187,000 TWh/year for East Africa. Among others, Sudan, Ethiopia, Kenya, and Somalia have abundant solar energy resources, with an annual solar irradiation intensity of over 2200 kWh/m^2. Namibia, South Africa, Botswana, Angola, Zimbabwe, and other countries also teem with solar energy resources, with an annual solar irradiation of above 2400 kWh/m^2.

Favorable conditions exist in North Africa, East Africa, and the southern region of Africa for developing large solar power bases. The distribution of solar energy resources in Africa is shown in Fig. 4.45.

See Figs. 4.46 and 4.47 for the distribution of solar energy resources in Ethiopia, Kenya, Sudan, and Tanzania in East Africa.

---

[16]According to an IRENA Report, North Africa covers Morocco, Algeria, Tunisia, Libya, Egypt, and Mauritania. East Africa covers Sudan, Djibouti, Ethiopia, Tanzania, Kenya, Uganda, Rwanda, Burundi, Eritrea, and Somalia. The southern region of Africa covers Angola, Botswana, Lesotho, Madagascar, Malawi, Mauritius, Mozambique, Namibia, Reunion Islands, Seychelles, South Africa, Swaziland, Zambia, and Zimbabwe.

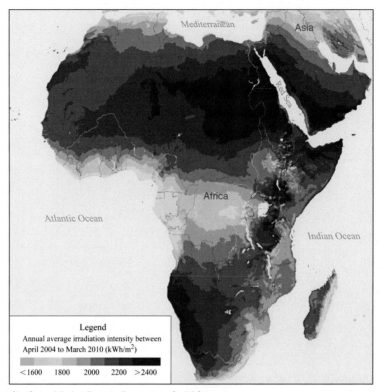

FIGURE 4.45 Distribution of Solar Energy Resources in Africa

Source: http://solargis.info/doc/free-solar-radiation-maps-GHI.

The distribution of solar energy resources in Namibia and South Africa in the southern region of Africa is shown in Fig. 4.48.

4.2.5.2 Hydropower Bases Along the Congo River

Africa has abundant hydropower resources, with a technologically developable capacity of about 1840 TWh/year or about 12% of the global total, surpassed only by Asia, South America and North America. Among different regions of the continent, North Africa and Southern Africa have seen a relatively higher level of hydropower development, while the central African region where water resources are abundant is least developed, with a less than 2% level of development and utilization along the Congo River. In 2011, the total installed capacity of established hydropower stations in Africa amounted to about 27 GW,[17] representing 6% of its technologically developable capacity. There is huge potential for hydropower development, especially along the Congo River, the Nile, and the Zambezi River. The theoretical hydropower reserves along the whole Congo River amount to 390 GW, ranking first among the world's major rivers.[18] In the short to medium term, development efforts are focused on the Congo

---

[17]Source: IEA, World Energy Outlook, 2013.
[18]Source: Jiang Zhongjin, African water resources rank second globally: http://www.geo-show.com/ChannelHY/SN/Content/20146/13430.shtml.

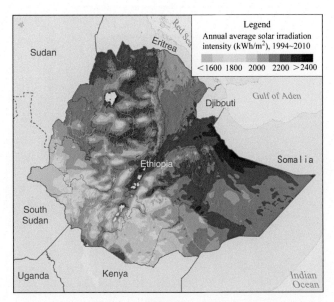

Ethiopia

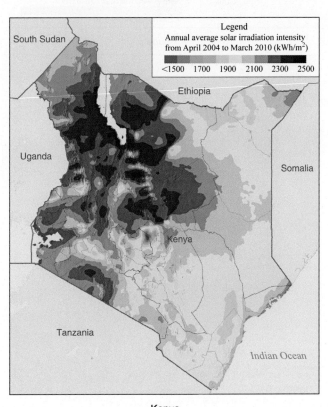

Kenya

FIGURE 4.46 Distribution of Solar Energy Resources in Ethiopia and Kenya

Source: http://solargis.info/doc/free-solar-radiation-maps-GHI.

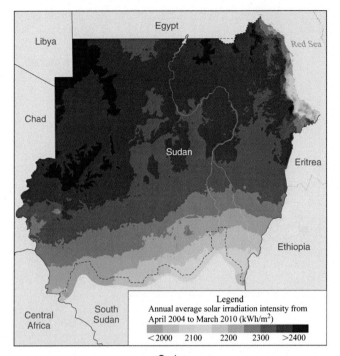

Sudan

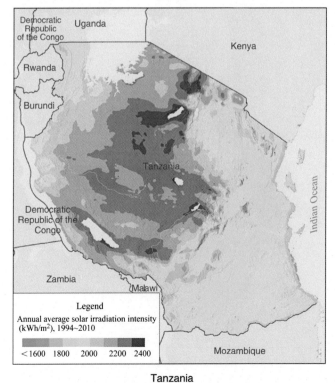

Tanzania

FIGURE 4.47 Distribution of Solar Energy Resources in Sudan and Tanzania

Source: http://solargis.info/doc/free-solar-radiation-maps-GHI.

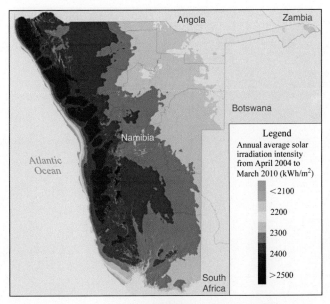

Namibia

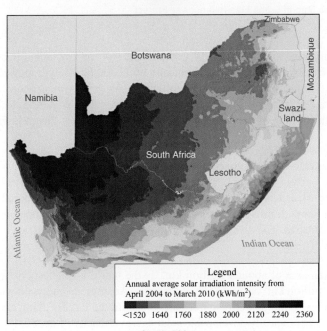

South Africa

FIGURE 4.48 Distribution of Solar Energy Resources in Namibia and South Africa

Source: http://solargis.info/doc/free-solar-radiation-maps-GHI.

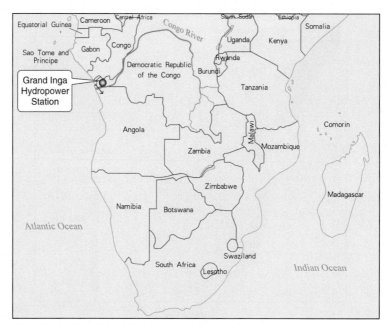

FIGURE 4.49 Location Map of Grand Inga Hydropower Project

River where 52 generating units of 750 MW under Phase 4 of the key Grand Inga hydropower project are being planned to provide a total installed capacity of 39,000 MW. Hydropower generation in Africa is intended mainly for local consumption, with room for joint development and consumption with solar and wind power projects in North Africa and East Africa as well as solar and ocean energy projects in southern Africa. The location of the Grand Inga hydropower project is shown in Fig. 4.49.

4.2.5.3 Wind Power Bases in Eastern and Northwestern Africa

African wind power resources are concentrated mainly in the east and northwest. The annual average wind speed is over 7 m/s in Somalia (and its coastal areas), Ethiopia and Kenya in the east, as well as Western Sahara and Mauritania in the northwest. In near-shore areas, an annual average wind speed of over 10 m/s is recorded, providing favorable conditions for the development of large onshore or offshore wind power bases. The distribution of wind power resources in Africa is shown in Fig. 4.50.

4.2.6 Oceania

Oceania, covering an area of about 9 million km², is the world's smallest continent. Accounting for approximately 7.7 million km² or over 85% of Oceania, Australia is endowed with evenly distributed solar energy resources. Wind power resources are concentrated largely along coastal areas.

Australia has a technologically developable solar energy capacity of over 250 trillion kWh/year. The low-lying areas at the center and the vast expanse of sparsely populated highlands in the west are all arid deserts that account for about 20% of the continent's total area, with ideal conditions for large-scale development of solar energy resources. Category-1 solar-resource areas are mainly distributed in the north (accounting for about 54% of Australia's total land area); Category-2 areas are concentrated in the center

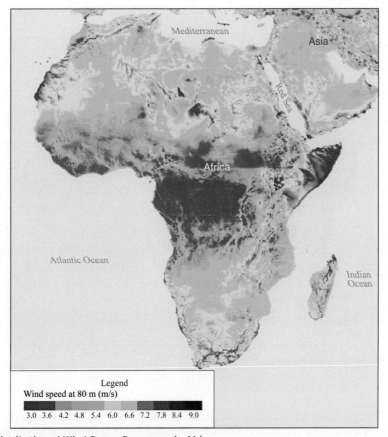

FIGURE 4.50 Distribution of Wind Power Resources in Africa

Sources: 3TIER Wind and Solar Energy Resources Evaluation Company; http://www.geni.org/globalenergy/library/renewable-energy-resources/index.shtml.

(accounting for about 35% of the country's total land area). Category-1 and Category-2 areas combined represent 90% of Australia's territorial land. See Fig. 4.51 for the distribution of solar energy resources in Australia. Australia's solar-resource areas and annual solar irradiation intensities are described in Table 4.3.

Australia's wind power resources are mainly distributed in its northeastern and southeastern regions as well as coastal areas in the southwest. The annual average wind speed at 80 m is over 8–9 m/s in many coastal areas, where favorable conditions exist for the construction of large-scale offshore or near-shore, land-based wind power bases. The distribution of offshore wind power resources in Australia is shown in Fig. 4.52.

4.3 DISTRIBUTED ENERGY DEVELOPMENT

Distributed energy refers to a system capable of power production/storage and also heat production/utilization while at the same time providing integrated utilization and control of energy. Distributed

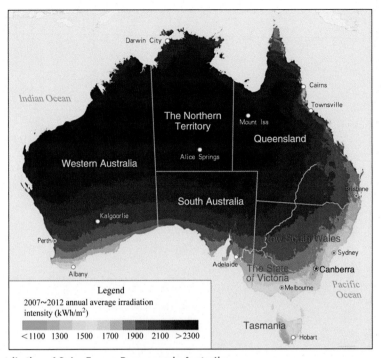

FIGURE 4.51 Distribution of Solar Energy Resources in Australia

Source: http://solargis.info/doc/free-solar-radiation-maps-GHI.

Table 4.3 Australia's Solar-Resource Areas and Annual Solar Irradiation Intensities

Resource Area Classification	Coverage	Annual Irradiation Intensity (kWh/m²)
Cat-1	Northern region, 54%	2100–2400
Cat-2	Central region, 35%	1800–2100
Cat-3	Southern region, 8%	1500–1800
Cat-4	Other regions, 3%	<1500

energy is generally located on the customer side to meet user demand. Normally integrated into or connected to a distribution grid or operated as a standalone unit, distributed energy represents an integrated energy system covering energy production, storage and control. Currently, the major developed nations of America and Europe are taking the lead in distributed energy development. Energy resources in the United States are distributed evenly across different regions, a positive factor for distributed energy development that has contributed to the rapid growth of distributed generation in the United States in recent years. In 2012, small hydropower, wind and solar photovoltaic projects

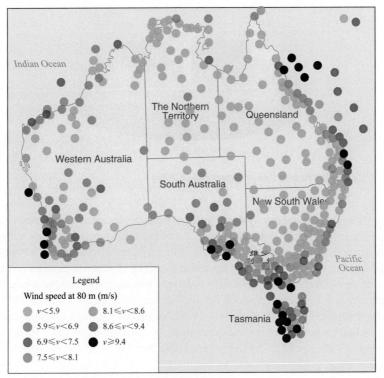

FIGURE 4.52 Distribution of Wind Energy Resources in Australia

Source: http://www.geni.org/globalenergy/library/renewable-energy-resources/index.shtml.

(each under 1 MW) in the United States provided installed capacities of distributed generation of 120, 260, and 550 MW, respectively, demonstrating noticeable year-on-year growth. On average, distributed generation in European Union countries currently accounts for a 10% share of the electricity market. With limited room for growth due to land acquisition, environmental protection and other constraints, onshore wind power development in Europe is gradually giving way to focused development of offshore wind power. In the developed nations of Europe, most residential accommodations are detached, low to medium rise buildings, which create favorable conditions for the development of rooftop photovoltaic projects.

The distributed energy system is valued internationally for its gradient utilization capability and high efficiency. Operating on a small scale, it mainly targets residential customers. In the future, distributed energy can provide a useful backup for large grids by taking advantage of the availability and economics of resources in load centers. Future distributed generation is expected to take the form of microhydropower projects, distributed wind and solar power systems, biomass energy generation, and energy storage systems. In particular, Europe, Asia, North America, and South America are well-placed to develop distributed energy systems based on biomass generation on account of their dense populations as well as the massive availability of urban garbage and forestry and agricultural wastes. In solar energy development, as solar energy resources are extensively distributed across

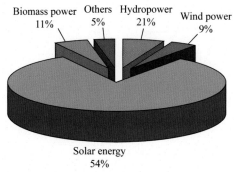

FIGURE 4.53 Structure of Distributed Power Generation in 2050

continents, distributed solar energy systems will become the focus of development against a background of sophisticated smart grid development, major breakthroughs in energy storage technology, and rapid urbanization. Taking into account land utilization, resources, and other factors, distributed wind power generation will form part of a distributed energy system incorporated with solar, energy storage and other power generation capability to supply electricity to remote areas. In 2050, annual distributed generation is expected to reach 3.5 billion tons of standard coal, or 15% of the world's total electricity generation and approximately 11.5% of primary energy consumption. Solar energy is the most important means of distributed generation, accounting expectedly for about 54% of total distributed generation, followed by hydropower (21%). See Fig. 4.53 for the structure of distributed power generation in 2050.

4.4 DEVELOPMENT AND UTILIZATION OF FOSSIL ENERGY

The development of fossil energy will be strictly controlled in a bid to address climate change and reduce carbon emissions. Oil and coal production is expected to peak in 2020 or thereabouts. Natural gas production will reach a peak around 2030. Around 2050, the production levels of oil, gas and coal will be at one-third, one-half, and one-fifth of their peak values, respectively. In the meanwhile, fossil energy trade at the transcontinental level will ease back after an initial rally.

4.4.1 Oil

The focus of oil production is shifting to countries in the Western Hemisphere, accompanied by a sharp decline in North America's demand for imported crude oil from the Middle East and Africa. The Western Hemisphere's increasing unconventional oil reserves have provided solid support to oil production in the hemisphere. Traditionally, resource-rich countries, including Russia in the Eastern Hemisphere[19] and nations in Central Asia, the Middle East, and North Africa, have been part of a major north–south hub of oil and gas supply focusing on conventional resources. In contrast, the reserves and production levels of conventional oil and gas in the Western Hemisphere are far lower compared

---

[19]The Eastern Hemisphere includes most of the Asian continent, eastern parts of the South Pole, most of Europe and Africa, a small part of northwestern North America, most of Oceania and many other islands. The Western Hemisphere mainly covers most of North America and central and south America, and a small part of Asia lying furthest east.

with the Eastern Hemisphere. As at the end of 2011, the Western Hemisphere's oil accounted for a 32.7% share of the world's total remaining proven reserves, compared with 9.3% for natural gas. The hemisphere's oil production accounted for 25.5% and natural gas of 31.65% of the world total. In recent years, advancing technologies have brought both economic and technological breakthroughs to unconventional oil and gas resources, turning oil sands in Canada, shale oil in the United States, extra-heavy oil in Venezuela, deep-water oil in the Gulf of Mexico and Brazil, and shale gas and coalbed methane in North America into an important source of backup for oil and gas resources. Benefiting from the production of unconventional oil resources, oil production in North America is expected to show average annual growth of 0.9% between 2012 and 2030, slightly above the global average of 0.8%. In the meantime, due to sharply higher oil sands and shale oil production levels, Canada's oil production is expected to grow 2.1%, far above the global average of 0.3%. Benefiting from Brazil's deep-water oil production and Venezuela's extra-heavy oil production, South America's oil production is expected to show average annual growth of 2.5%, driven by 4.5% growth in Brazil's deep-water oil production and 2.5% growth in Venezuela's extra-heavy oil production. The Middle East will remain the world's largest oil supplier, expected to account for one third of the global total in 2030, rising to 60% in 2012–2030.

The center of oil consumption is shifting to Asia and the Middle East. Asian countries will continue to be the major driver of oil demand by 2030, expected to account for two thirds of the world's new oil demand, with China as the most important factor. However, growth of China's oil consumption will slow over time, from average annual growth of 3.7% before 2020 to 1.3% between 2020 and 2030. The Middle East has one of the world's lowest oil and diesel prices, where subsidies of petroleum product prices have contributed to fast growing oil demand, expected to reach 1.6%. As Central and Southern America and Africa remain essentially in the developing world, oil demand is expected to grow fast before 2030, then slowing down and finally reaching peak consumption around 2050 before starting to fall again. Contrary to this trend, oil demand in Europe and North America in 2030 is expected to drop dramatically from 2012, reflecting subdued energy demand in transport because of improved car efficiency, high taxes on oil, and slow population growth.

In 2030, North America and Europe will see a falling share of the Middle East's oil exports while the Asia Pacific region's share will go up. In the same year, the world's major oil importers include Europe and the emerging economies of China and India, while the major oil exporters are the Middle East, Russia, and central and southern America, reflecting transcontinental oil trade equivalent to 300 million tons of standard coal. By 2050, global oil demand will have fallen to one-third of 2030 levels, reflecting the impact of large-scale development and utilization of clean energy. In addition, due to the growing development and utilization of unconventional oil resources, local energy demand can basically be met by local production, which will drive global oil trade further down. See Fig. 4.54 for the global flows of oil.

4.4.2 Coal

The Asia–Pacific region has been fuelling global coal consumption, which will peak around 2020 before gradually slowing down. The region's vibrant economy has triggered enormous demand for energy and compared with oil and gas resources, the region is more abundant with coal resources. In particular, China, India, and Australia rank among the world's top nations in terms of coal reserves. With the progress of clean coal technology, countries in the Asia–Pacific region face the realistic option of raising coal-mining capacity to meet higher energy demand. China and India will remain two major

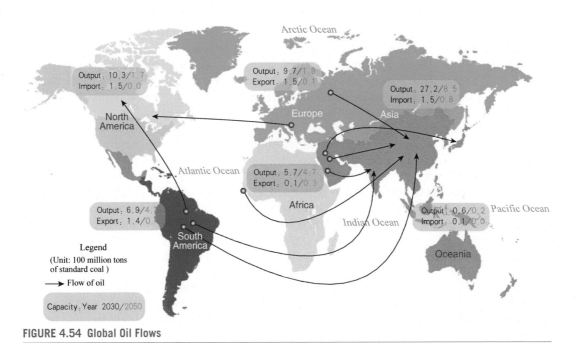

FIGURE 4.54 Global Oil Flows

coal consumers, collectively accounting for 64% of the world's total coal demand in 2030, up from 58% in 2011. China will see coal production increment before 2020, due to the impact of the country's economic restructuring process. In comparison, coal consumption will peak sometime later in India where local coal demand cannot be satisfied by domestic supply and has to be met by imports. After 2020, India is expected to surpass China and become the world's largest coal importer, importing three times as much coal in 2030 as now. Benefiting from growing coal consumption and import in China and India, other Asia–Pacific countries like Australia, Vietnam, Mongolia, and Indonesia will expand coal production substantially. By 2030, Australia's coal production is expected to jump 50%. The Asia–Pacific region's overall share of global coal production will remain steady at 70%.

Based on regional coal demand forecasts and the future scale and speed of clean energy development, global coal trade is expected to peak by around 2020. Among the major coal importers are Asia's emerging market countries and Europe, expected to import the equivalents of 700 million tons and 100 million tons of standard coal, respectively. Among the major coal exporters are Indonesia, Australia, the United States, Russia, and South Africa. Coal trade will shrink gradually after 2020. See Fig. 4.55 for the global flows of coal.

4.4.3 Natural Gas

Overall, North America will switch from the position of a net natural gas importer to one of a net natural gas exporter. The LNG originally intended for export to the United States will then be shipped to the Asia–Pacific region and Europe. As North America has already secured a leadership position in terms of technology and management capability in shale gas development, the United States and Canada will account for a substantial 80% share of the world's total unconventional natural gas production.

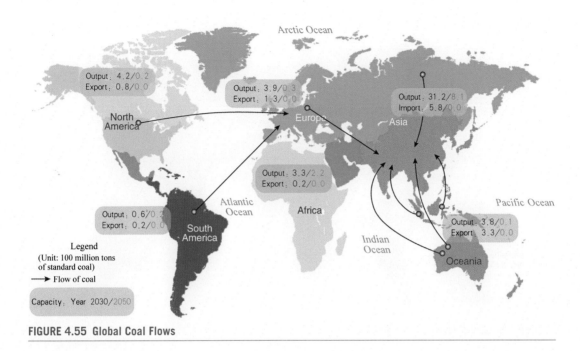

FIGURE 4.55 Global Coal Flows

However, in response to public concerns about the possible environmental pollution caused by hydro-fracturing, a technology used to extract shale gas, the United States will see slower growth in the pro-duction of unconventional natural gas to stabilize at around 600 billion cubic meters per year after 2020. Encouraged by the maturity and application of mining technologies, more countries will jump on the natural gas bandwagon after 2020, ushering in a period of strong growth in this area of production in China, Argentina, and Australia. Currently, the United States natural gas market is suffering from oversupply and high inventory. As the United States shifts gradually to the position of LNG exporter, traditional LNG exporters will refocus its export efforts on Europe and the Asia–Pacific region instead of the United States.

The Asia–Pacific region will become the world's most important natural gas market, with China, Japan and India as major consumers and Australia, Indonesia, the Middle East, and Russia as the major exporters. Over the coming decades, Asia–Pacific will see the most profound changes in the natural gas market. To address the need for energy supply diversification and environmental protection, Asia–Pacific countries, other than Japan and South Korea, will see great upside potential in natural gas consumption, with average annual growth of about 4% expected before 2050. China and India will each witness growth of around 5%. Limited by insufficient natural gas reserves, the Asia–Pacific region has an urgent and significant demand for gas imports from Australia and Indonesia in the region, and the Middle East, Russia, South Africa, South America, and North America elsewhere.

Global natural gas trade is expected to peak around 2030, with trade at the transcontinental level amounting to approximately 500 million tons of standard coal before easing back with declining demand. In 2050, a balance of supply and demand at the intracontinental level will materialize on limited global trade volume. See Fig. 4.56 for the flows of natural gas around the world.

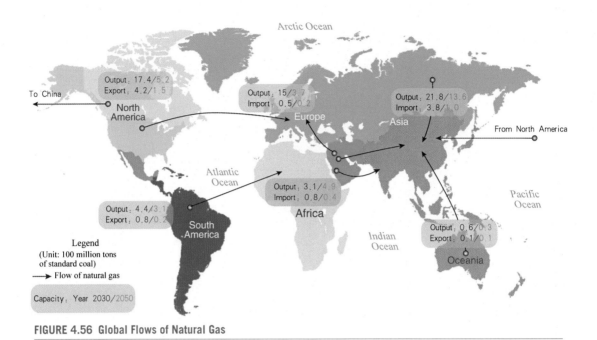

FIGURE 4.56 Global Flows of Natural Gas

5 GLOBAL ELECTRICITY FLOW

Through execution of the "two-replacement" policy, particularly through the development of large-scale energy bases in the Arctic and equatorial regions as well as five continents, a globally-interconnected clean energy system focusing on electricity is gradually taking shape, leading to significantly expanded electricity flows and longer transmission distances around the world.

5.1 REASONING AND PRINCIPLES BEHIND GLOBAL ELECTRICITY FLOW

Global electricity flows carry technological, economic, resource and environmental implications, involving energy production, consumption, distribution, and other factors. Any study of global electricity flows should be guided by "four principles" and a "coordinated view at three levels."

The four principles, the first being the principle of low-carbon development. A consensus has basically been reached among governments around the world on the potential threat of climate change to human sustainability. As low-carbon and clean development has evolved into an inexorable trend in global energy development, major countries have been setting low-carbon targets oriented towards clean energy. Any study of electricity development across continents and global electricity flows should take low-carbon development as a nonnegotiable constraint. *The second is the "locals first" principle.* Every continent is abundant with renewable energy resources, including water, wind, solar, biomass, and ocean power, developed on a centralized or distributed basis to meet power demand locally or from load centers near the source of generation. In terms of reliability and economic performance, this mode of development and utilization compares favorably with the delivery of electricity across nations or continents. These resources should therefore be accorded priority for development and utilization. *The third is the principle of economic benefit and high efficiency.* Any plans to develop major renewable energy bases in the Arctic/equatorial regions and elsewhere as well as long-distance transmission capability, must take account of both development and transmission costs, by comparing the cost required to bring electricity to a receiving region with the cost of local generation and supply. The result of this comparison should form the basis for optimizing a decision on the capacity and direction of transcontinental transmission. *The last is the principle of technological feasibility.* In the design of global interconnections, care should be exercised to avoid building transnational/transcontinental interconnections and transmission channels over high mountains or long undersea distances. Generally, transcontinental interconnections are supported by UHV DC grids and intracontinental/domestic interconnections, by UHV AC or UHV DC grids.

Coordinated view at three levels: the first is a coordinated view of centralized and distributed clean energy development. Depending on the level of resource endowment, global clean energy can generally be categorized into two major types. One is quality resources, abundantly available per unit area, with long utilization hours, low costs, and positive economics. The regions with quality resources are usually regions with strong solar radiation or high annual average wind speeds, relatively sparse population, and far from load centers. For these regions, it is common to adopt centralized development and provide large-scale grid transmission and allocation over extensive areas. The other type of clean energy is resources of a general nature, mostly distributed around load centers, with mild sunlight conditions and wind speeds, dense population, and agreeable climate. In this case, distributed development is usually adopted to provide electricity for local supply and consumption. In the future, electricity demand may partly be satisfied by energy generated from rooftop photovoltaic, biomass, small hydropower, and other distributed sources. But more energy will come mainly from large clean energy bases located far away from load centers but with good resources. Such bases are an integral part of global electricity flows. *At the second level, a coordinated view is required of clean energy development locally and in remote areas.* Any study of the future capacity and transmission direction of global electricity flows must take into account the electricity demand from each continent and major countries, renewable resource endowment and development goals, the feasibility of energy channels, and the cost of transmission technology. Currently, renewables-based generation like solar and ocean power is not as competitive as fossil energy in terms of economic benefit and efficiency. Only the economics of wind power are comparable to those of conventional fossil-fueled generation. In order to reduce the total social costs of electricity usage, priority should be given to renewable resources with good development conditions and short transmission distances. With technological advancement and continued environmental and carbon constraints, the development and environmental costs of fossil energy utilization will increase while, in contrast, the advantages of clean energy generation will manifest themselves more strongly. As a result, the scope of clean energy development will expand to cover renewable resources in remote locations. Electricity flows will also extend gradually from a transnational to transcontinental basis and further to a global level. *At the third level, a coordinated view should be taken of energy balances at the intracontinental level and mutual support in energy flows across continents.* Firstly, the space and demand for renewable energy development on each continent should be analyzed based on the present situation of electricity development and future demand as well as clean energy development goals of the major countries on each continent in order to clearly define the major power-receiving regions and markets around the world. This should be followed by a further study of the development potential of large energy bases in the Arctic and equatorial regions

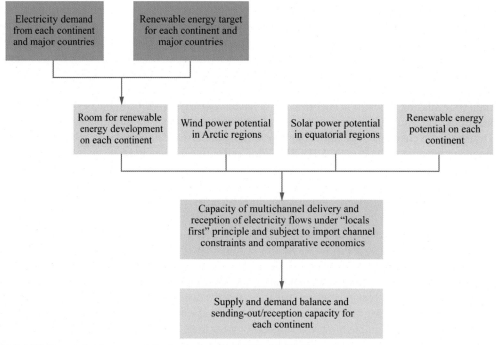

FIGURE 4.57 Reasoning Framework for Analyzing Global Electricity Flows

and elsewhere, with reference to the supply economics based on transmission costs, so as to define the supply capacity and costs of clean energy bases in the future. On this basis, subject to the objectives of low-carbon development and optimizing supply economics, a number of clean energy options are available but priority should be accorded to developing and utilizing renewable energy resources with favorable conditions for local development. Consideration should then be given to clean energy access for large-scale renewable energy bases, with reference to transmission channels and supply economics, so as to determine the capacity of global electricity flows through multichannel delivery and reception and work out the supply and demand balance and the levels of sending-out/reception capacity for each continent.

See Fig. 4.57 for the reasoning framework for an analysis of global electricity flows.

5.2 OVERVIEW OF GLOBAL ELECTRICITY SUPPLY

In the future, the internal and external costs of developing and utilizing conventional fossil energy will show an upward trend, with the growing development costs and ever-more stringent low-carbon clean energy and safety requirements. The generating costs of wind, solar, and other renewables will decline rapidly with higher levels of utilization. After the intersection of the generating cost curve of fossil energy with that of renewable resources, the cost of renewable energy generation will be lower than that of fossil energy generation. The falling costs resulting from improved economies of scale will in turn lead to more capacity being developed in a virtuous cycle of growing capacity and lowering costs, thereby creating enormous room for renewable energy growth. In the foreseeable future, renewable energy will become the driving force behind installed capacity building to meet new electricity demand around the world and fill the capacity shortfall arising from the decommissioning of fossil-fuelled generating units.

Judging by the changing trend of generating costs among different energy options, the generating cost of onshore wind power in regions with better resources is competitive with fossil fuel generation. As generating costs fall further with improved economies of scale, wind power may become even more competitive. Despite its generally better resource conditions than onshore wind power, offshore wind power involves higher construction and maintenance costs. With the progress of generation technologies, ever-falling materials and manufacturing costs, and improving conversion efficiency, the cost of solar power generation will maintain a downward trend, making solar power the most important source of renewable energy generation in the future. Currently, hydropower generation is more cost-competitive than other energy resources. However, as quality resources are being depleted in regions with favorable conditions for development, accompanied by a continued shift of hydropower development into remote areas, the cost of developing and transmitting hydropower resources will continue to rise, depending on infrastructure and development conditions. In addition, ecological problems in regions with hydropower facilities will continue to haunt hydropower development in the long run. Driven by rapid growth, hydroelectric power is expected to maintain its position as the dominant source of renewable power generation until 2030. After 2030, hydropower will move into a period of sluggish growth. Currently, the cost of coal-fired generation is higher than that of most renewables. In the future, however, the internal as well as external environmental costs of coal-fired generation will increase and lead to steadily higher supply costs, due to more stringent emission control requirements governing coal-fired plants, the internalization of external costs, and coal resources of average quality beginning to be developed and utilized. Currently, ocean energy is more expensive, at a generating cost of approximately RMB 6 per kWh. Subject to restrictive development conditions and a lack of technology breakthroughs, the cost of ocean power generation will consistently be higher than that of nonhydropower renewables and conventional fossil fuels. For this reason, ocean energy holds limited promise for the future. As an important clean energy alternative, nuclear power is well-placed to play a vital role in low-carbon energy development. But given the nature of nuclear power as an energy source with a very low risk of accidents but potentially huge damage if something goes wrong, the cost of safety investments will rise continuously, translating into spiraling generating costs. Dictated by the way organic materials are collected and supplied and also the use of these materials for power generation in competition with other purposes, biomass generation offers limited room for technological improvement. As costs are unlikely to fall, the cost of biomass energy generation is expected to stay relatively high. As regards wind energy, the supply (including generation and transmission) costs of wind power in the Arctic region are expected to become more economically competitive in power-receiving regions after 2035, given the anticipated breakthroughs and upgrades of wind generation, materials and transmission technologies in extremely cold regions.

In the long run, solar energy and wind power will assume a dominant role in the electricity supply structure, made possible by a sharp decline in supply costs. Judging by the change in cost trends and with the progress of the "two-replacements" policy, solar energy (including photovoltaic and photothermal) is expected to share a dominant role in the world's generation mix with wind power by 2050, accounting for 35% and 31%, respectively. Restricted by the availability of resources, hydropower is expected to account for 14% or so of total electricity generation by 2050, compared with a 6% share

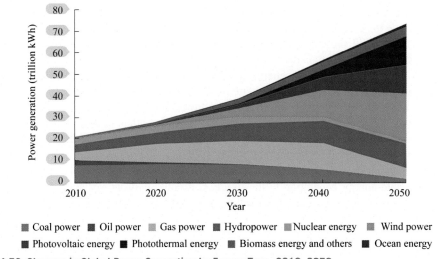

FIGURE 4.58 Changes in Global Power Generation by Energy Type, 2010–2050

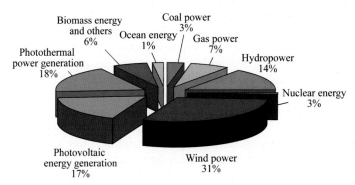

FIGURE 4.59 The Global Power Generation Mix in 2050

for biomass energy and others (mainly including geothermal energy). It is expected that natural gas and coal-fired generation amounting to approximately 10% of total power generation will also be required to meet the operational requirements of power systems and cater to the affordability of some of the underdeveloped regions of Asia, Africa, and South America.

See Figs. 4.58 and 4.59 for the power generation by energy type in future target years and the generation mix in 2050.

It is expected that by 2050, renewable energy generated on a centralized basis, like hydropower, wind, and solar energy, will account for 55% or so of total generation, compared with 15% for distributed generation. Wind power in the Arctic region and solar energy in the equatorial regions will account for 16% of total generation. Changes in the structure of global power supply between 2010 and 2050 are shown in Fig. 4.60

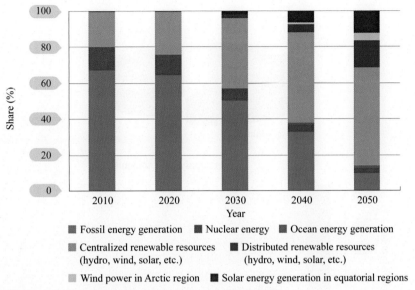

FIGURE 4.60 Changes in Global Power Supply and Demand, 2010–2050

5.3 ELECTRICITY SUPPLY AND DEMAND BALANCE ON EACH CONTINENT

Judging by the supply/demand situation of each continent, Asia, Europe and North America are electricity importers; Africa and Oceania are electricity exporters. South America is geared towards a self-balancing of power supply and demand. See Fig. 4.61 for the supply of power globally and the supply/demand situation by continent in 2050.

5.3.1 Asia

Asia's demand for electricity has maintained relatively strong growth. As an electricity importer, the continent will see a growing supply shortfall. The most densely populated among all continents, Asia has the highest demand for electricity in the world. Fueled by the socioeconomic growth of densely inhabited regions like China, India, and Southeast Asia, Asia's power demand will continue to increase, chalking up the highest growth rate among all continents. By 2050, power demand will reach 38,000 TWh, or 52% of total global demand. Electricity consumption per capita will amount to 7360 kWh/annum, up 2.5 times from 2010.

Asia is rich in renewable energy resources, such as solar, wind, and hydroelectric energy. In the future, the capacity of renewable energy development will be subject to the levels of resources and reserves, technology economics, and the economics of transcontinental transmission. As water resources with better conditions for development in Asia have already been fully developed, the focus has started to shift to the upper and middle reaches of Asia's major river basins. By 2050, the capacity of hydroelectric development is expected to reach approximately 4900 TWh, up 300% from 2010, indicating the exhaustion of hydropower resources of average and inferior quality. Currently, wind, and solar energy resources in Asia are at the initial stage of substantial development. Major growth can be expected with greater government support and improved technology economics. By 2050, wind power

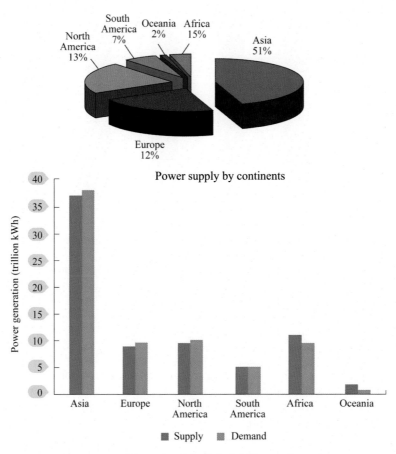

FIGURE 4.61 Global Power Supply, Power Supply and Demand by Continents (Including Arctic and Equatorial Regions) in 2050

generation is expected to reach 10,400 TWh globally (formally the Arctic region), whereas solar energy generation globally (formally the Middle East) is expected to reach 10,100 TWh (including 4600 TWh photovoltaic energy and 5500 TWh photothermal power). In the next 20 years, some developing countries in Asia will continue to develop nuclear energy to satisfy their own higher power demand and the need for low carbon development. Asia's nuclear energy generation is expected to peak at about 1300 TWh in 2040. Given that existing nuclear power units will be decommissioned gradually over the next 30 years, nuclear energy generation is expected to decline to 1250 TWh in 2050. Considering the economic capacity and technological level of China, India and other developing countries in central Asia and south-east Asia, Asia is expected to retain, into 2050, a share of *gas* and *coal* to the extent of about 12% of continent-wide electricity supply. Asia enjoys abundant ocean energy resources and is at the forefront of ocean energy utilization. Taking technology economics into account, ocean energy

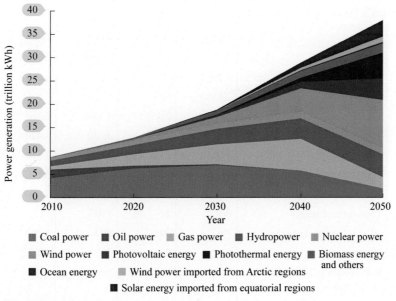

FIGURE 4.62 Asia's Power Supply

generation in Asia is expected to reach 300 TWh in 2050, ranking first in the world. Considering the contribution of the Arctic region in Asia to wind power and the contribution of the Middle East to solar energy, *Asia's continent-wide power supply*[20] is expected to reach 37,000 TWh in 2050, accounting for 51% of the world total. Of this capacity, 34% is attributable to photovoltaic and photothermal power, 31% to wind power, 13% to hydropower, and 3% to nuclear energy. As a vast continent with an enormous population, Asia is rich in distributed generation resources such as microhydropower, wind, solar, and biomass energy resources. In the future, distributed generation will grow more rapidly to meet the demand of remote regions and city centers. It is expected that distributed power generation in Asia will reach about 4500 TWh in 2050, accounting for approximately 12% of the continent's total electricity demand.

Taking an integrated view of the development potential and supply economics of various generation sources, large-scale renewable energy bases, including those in the Arctic and equatorial regions, are expected to reach a supply capacity of approximately 3700 TWh/year in 2050, accounting for 10% of Asia's total demand. Of this supply capacity, wind power of about 1200 TWh/year will be exported from the Bering Strait, the Kara Sea, and Sakhalin Island to northeastern Asia, while solar energy of 2500 TWh/year will be exported from the equatorial regions in the Middle East to South Asia. In addition, 2.6% of Asia's total power demand, at 1000 TWh, will have to be met by imports from Australia. See Fig. 4.62 for details of Asia's power supply and Table 4.4 for the continent's power supply structure.

---

[20]Intracontinental power supply capacity in this section refers to the sum total of generation from the renewable energy bases of the Arctic and Equatorial regions and other generation sources on the continent, excluding power imports. The same definition applies hereinafter.

Table 4.4 Asia's Power Supply Structure (TWh)

	Type	2010	2020	2030	2040	2050
Power generation mix	Total	8,694.1	12,817	18,850	28,900	38,020
	Coal power	4,209.7	6,230	6,960	5,780	1,920
	Oil power	1,835.8	619	170	0	0
	Gas power	697.6	2,453	4,280	6,760	2,480
	Hydropower	1,209.7	1,952	3,330	4,420	4,890
	Nuclear power	584.5	985	1,070	1,310	1,250
	Wind power	72.6	268	1,420	5,120	10,370
	Photovoltaic energy	4.9	50	410	1,370	4,600
	Photothermal energy	0.3	12	110	910	5,540
	Biomass energy and others	79	246	800	1,450	1,970
	Ocean energy	0	2	50	170	300
	Wind power imported from Arctic region	0	0	50	510	1,200
	Solar energy imported from equatorial regions	0	0	200	1,100	3,500
Mode of generation	Total	8,694.1	12,817	18,850	28,900	38,020
	Distributed type	8.7	275	950	2,210	4,470
	Centralized type	8,685.4	12,542	17,650	25,080	28,850
	Arctic and equatorial regions (intracontinental)	0	0	250	1,310	3,700
	Arctic and equatorial regions (extracontinental)	0	0	0	300	1,000

5.3.2 Europe

Europe's power demand is moving towards the saturation point. As a traditional electricity importer, Europe is seeing a huge demand for imported renewable energy resources triggered by electricity substitution. Europe's population has moved into a period of steady decline. According to United Nations forecasts, the continent's population will be 710 million in 2050, down 4% from 2010. Taking into account factors such as the development of electric applications and electricity replacement of its relatively underdeveloped countries, Europe's power demand is expected to reach 9500 TWh in 2050. Electricity consumption per capita will rise to 13,000 kWh/year, up to 90% from 2010.

Under the objective of low-carbon and clean development, all coal and gas-fired generating units in Europe will be decommissioned around 2040, leaving only a few gas-fired generators for peak load regulation. In the future, Europe's clean power supply will come mainly from wind, solar, hydroelectric, biomass and ocean energy. The continent's prospects for hydropower development are limited, as all hydropower resources are expected to be fully developed by around 2030 with annual generation estimated at about 1200 TWh. Given the impact of the nuclear-free policy being pursued by countries like Germany, nuclear safety concerns and the gradual decommissioning of existing nuclear power units, Europe's installed nuclear generation capacity is expected to drop continuously after 2020. Annual nuclear power generation will be about 650 TWh in 2050, down about 50% from now. Looking ahead, Europe will actively develop distributed wind and solar energy at the same time as it will focus on developing North Sea wind power and Southern European solar energy in resource-rich regions. In 2050, wind power generation will reach 2600 TWh in Europe (excluding Greenland, the Norwegian Sea). The Norwegian Sea and the Barents Sea will enter the initial stage of wind power development around 2040. Subject to breakthroughs in UHV undersea cable technology, Greenland will usher in wind power development, which is expected to reach a capacity of 1800 TWh in 2050. In 2050, solar energy generation in Europe will reach 1200 TWh (including photovoltaic energy of 600 TWh and photothermal power of 600 TWh). In particular, distributed wind and solar power generation will reach 860 TWh, or 9% of total power demand. In 2050, Europe's power supply capacity is expected to reach 9000 TWh, of which 49% is wind power attributable to Greenland, the Norwegian Sea and the Barents Sea, 13% to solar energy, and 13% to hydropower.

In 2050, it is expected that Europe will need to import electricity of 1500 TWh from other regions (mainly North Africa) to balance supply and demand. Taking an integrated view of the development potential of different energy sources and supply economics, the wind power bases in the Artic regions of Greenland, the Norwegian Sea, and the Barents Sea are expected to achieve annual generation of 1800 TWh. Of this generation capacity, 800 TWh is exported to meet approximately 8% of Europe's total demand, taking into account transmission distances/paths, economics and other factors. Exports to North America are estimated at 1000 TWh. See Fig. 4.63 for Europe's power supply and Table 4.5 for the continent's power supply structure.

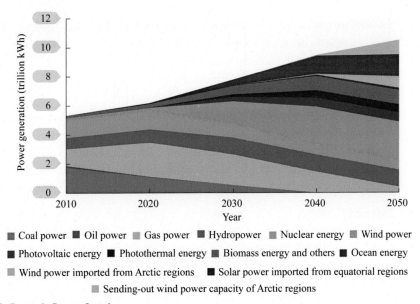

FIGURE 4.63 Europe's Power Supply

Table 4.5 Europe's Power Supply Structure (TWh)

	Type	2010	2020	2030	2040	2050
Power supply structure	Total	5,330.4	6,193	7,790	9,490	10,500
	Coal power	1,755.8	1,088	530	0	0
	Oil power	122.7	36	10	0	0
	Gas power	1,145	2,336	2,100	1,440	400
	Hydropower	789.7	932	1,170	1,180	1,180
	Nuclear power	1,202.5	1,270	930	700	650
Power supply structure	Wind power	152.4	195	1,550	2,600	2,640
	Photovoltaic energy	21.8	29	300	610	640
	Photothermal energy	1.4	7	190	520	550
	Biomass energy and others	139.1	228	570	970	980
	Ocean energy generation	0	1	30	100	150
	Wind power imported from Arctic region	0	0	0	50	800
	Solar energy imported from the equatorial regions	0	71	410	1,270	1,510
	Sending-out capacity of the Pole regions	0	0	0	50	1,000
Mode of Development	Total	5,330.4	6,193	7,790	9,490	10,500
	Distributed type	1.7	53	180	420	860
	Centralized type	5,328.7	6,069	7,200	7,700	6,330
	Imports from Arctic and equatorial regions (intracontinental)	0	0	0	50	800
	Imports from Arctic and equatorial regions (extracontinental)	0	70	410	1,270	1,510
	Sending-out capacity of Arctic and equatorial regions (intracontinental)	0	1	0	50	1,000

5.3.3 North America

North America's power structure is geared towards achieving a self-balance between supply and de-mand, importing only an appropriate amount of electricity from the Arctic region. According to United Nations forecasts, North America's population will rise steadily and reach 450 million in 2050, up 29% from 2010. Taking into account the progress of electricity replacement, the continent's power demand is expected to reach 10200 TWh in 2050. Electricity consumption per capita will grow to 23,000 kWh/year, up 50% from 2010.

North America is rich in energy resources such as wind power, solar energy and natural gas, etc. With the massive development of shale gas, the price of natural gas is likely to remain low. North America's gas-fired generation is expected to peak at 2700 TWh in 2030, then easing to about 1500 TWh in 2050 amid the growing substitution of clean and low-carbon energy for conventional power sources. Despite its rich hydropower resources, North America faces the problems of excessive development of quality resources, with development costs expected to rise significantly in the future. The continent's hydro-power generation is estimated at about 1700 TWh in 2050, with nuclear power generation at about 170 TWh, down 80% from 2010. North America is also rich in wind power and solar energy. The wind and solar energy resources in the Midwest are suited for large-scale, centralized development, while regions with more decentralized, medium-rise residential buildings are ideal for developing distributed solar and wind power systems. In 2050, wind power generation in North America is expected to reach 2300 TWh and solar energy generation, 2500 TWh (including photovoltaic energy of 1300 TWh and photothermal power of 1200 TWh). The distributed generation of photovoltaic energy and wind power is expected to reach 1600 TWh, representing 15% of total demand. Continent-wide, power supply capac-ity is expected to reach 9200 TWh in 2050, with wind and solar energy accounting for a dominant 52% share, followed by gas-fired generation (16%) and hydropower (19%).

In North America, solar resources are mostly concentrated in the southwest and wind power resources, in the west. Load levels are relatively high in the east. Taking an integrated view of a comparison of eco-nomics between developing wind and solar power in the west and delivered to the east region and the importation of Greenland wind power into the eastern regions of North America, North America is expected to import 1000 TWh/year of power from Greenland to achieve a balance between supply and demand. See Fig. 4.64 for North America's power supply and Table 4.6 for the continent's power supply structure.

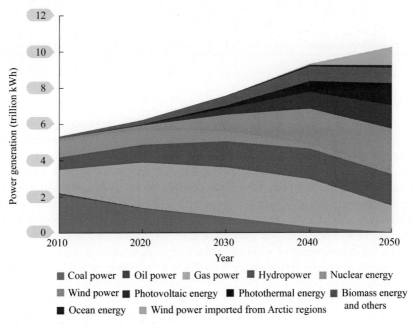

FIGURE 4.64 North America's Power Supply Structure

Table 4.6 North America's Power Supply Structure (TWh)

	Type	2010	2020	2030	2040	2050
Power generation mix	Total	5,333.7	6,239	7,580	9,310	10,240
	Coal power	2,116	1,360	880	320	0
	Oil power	102	33	10	0	0
	Gas power	1,277	2,503	2,730	2,630	1,490
	Hydropower	698	984	1,450	1,690	1,740
	Nuclear power	935.5	837	560	220	170
	Wind power	105.5	222	900	1970	2,340
	Photovoltaic power generation	5.5	50	360	920	1,280
	Photothermal power generation	0.6	12	150	610	1,230
	Biomass energy and others	93.6	237	510	800	840
	Ocean power generation	0	1	30	100	150
	Wind power imported from Arctic regions	0	0	0	50	1,000
Mode of development	Total	5,333.7	6,239	7,580	9,310	10,240
	Distributed type	3.1	97	340	780	1,570
	Centralized type	5,330.6	6,142	7240	8,480	7,670
	Arctic and equatorial regions (intracontinental)	0	0	0	0	0
	Arctic and equatorial regions (extracontinental)	0	0	0	50	1,000

5.3.4 South America

South America's electricity demand is growing relatively fast. With solar energy generation in equatorial regions, the continent can achieve on its own an overall balance between supply and demand in electricity. According to United Nations forecasts, South America's population will rise steadily and reach 780 million in 2050, up 31% from 2010. Taking into account the future development of electricity demand and ongoing electricity replacement, power demand in South America will reach 5100 TWh in 2050. Electricity consumption per capita will grow to 6550 kWh/year, up 2.6 times from 2010.

South America is rich in hydropower resources. In the future, the hydropower resources of the Amazon River Basin will be developed with more vigor to reach a generation level of 1500 TWh in 2050. Currently, the development of wind and solar energy generation is still at its initial stages, but will accelerate to meet growing power demand. In 2050, wind power generation is expected to reach 950 TWh and solar energy generation, 1900 TWh (including photovoltaic energy of 1100 TWh and photothermal power of 800 TWh). South America is also rich in biomass resources such as agricultural and forest crops, as well as geothermal energy. Through steady growth, the continent's power generation is expected to reach approximately 400 TWh in 2050. Continent-wide, a power supply

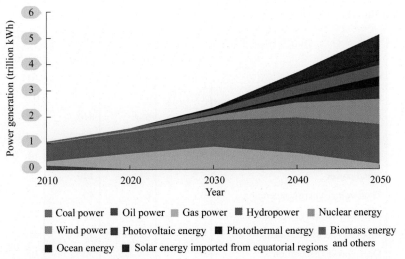

Coal power ■ Oil power ■ Gas power ■ Hydropower ■ Nuclear energy
■ Wind power ■ Photovoltaic energy ■ Photothermal energy ■ Biomass energy
■ Ocean energy ■ Solar energy imported from equatorial regions and others

FIGURE 4.65 South America's Power Supply

capacity of about 5100 TWh is expected to be achieved in 2050. Of this supply capacity, renewable energy is expected to take up a 96% share, with 29% attributable to hydropower and 55% to wind and solar energy.

South America is expected to achieve, on its own, a balance between power supply and demand in 2050. Of the continent's power supply, approximately 1000 TWh will come from the solar power bases in equatorial countries, such as Peru and Chile, mainly for delivery to load centers in the east, like Brazil. See Fig. 4.65 for South America's power supply and Table 4.7 for the continent's power supply structure.

5.3.5 Africa

Africa is an electricity exporter well endowed with renewable resources. According to United Nations forecasts, Africa's population will rise rapidly in the future and reach 2.39 billion in 2050, up 132% from 2010. Given the future development of power demand and electricity replacement, power demand in Africa is projected to reach 9500 TWh in 2050. Per capita electricity consumption will grow to 4000 kWh/year, up 5.3 times from 2010.

Africa has the world's most abundant solar and wind power resources, together with very rich hydropower resources. Africa is currently underdeveloped, with relatively small power demand and per capita electricity consumption. Conventional fossil energy and hydropower play a dominant role in power supply, while the share of renewable energy resources like wind and solar energy is very low. In the future, wind and solar power generation will see rapid growth driven by population expansion and socioeconomic development. In 2050, wind power generation is forecast to reach 2900 TWh and solar energy generation, about 2100 TWh (excluding the export capacity of North Africa and East Africa). Of this level of solar power generation, 600 TWh is attributable to photovoltaic energy and 1500 TWh to photothermal power. In the future, large-scale hydropower development will accelerate in the river basins of the Congo River, the Zambezi River, and the Nile River. In 2050, the continent's hydropower

Table 4.7 South America's Power Supply Structure (TWh)

	Type	2010	2020	2030	2040	2050
Power generation mix	Total	1,062.3	1,563	2,340	3,670	5,130
	Coal power	22	20	20	10	0
	Oil power	137	1	0	0	0
	Gas power	162	584	840	600	200
	Hydropower	671.1	807	1,000	1,350	1,510
	Nuclear power	21.7	37	40	10	10
	Wind power	3.5	24	140	560	950
	Photovoltaic energy generation	0	6	40	180	480
	Photothermal energy generation	0	1	10	110	380
	Biomass energy and others	45	82	160	350	430
	Ocean power generation	0	1	30	100	170
	Solar energy imported from equatorial regions	0	0	60	400	1,000
Mode of development	Total	1,062.3	1,563	2,340	3,670	5,130
	Distributed type	1.6	49.9	170	400	810
	Centralized type	1,060.7	1,513.1	2,110	2,870	3,320
	Arctic and equatorial regions (intracontinental)	0	0	60	400	1,000
	Arctic and equatorial regions (extracontinental)	0	0	0	0	0

generation is projected to reach about 650 TWh. Given Africa's underdeveloped economy compared with the other continents, a certain level of coal and gas-fired generation capacity estimated at about 610 TWh will be retained in 2050. African power supply is expected to reach about 11,000 TWh in 2050, of which 6% is attributable to gas and coal-fired generation and 92% to solar, wind and hydro-power generation.

The solar energy bases in North Africa and East Africa will become important sending-end bases in a globally interconnected energy network. Given the power supply and demand situation in different continents, the solar energy bases in North Africa and East Africa are projected to export 4500 TWh of electricity in 2050, to meet 6% of global power demand. Of this export capacity, 1500 TWh will be designated for Europe and approximately 3000 TWh for the southern, western, and central regions of Africa. See Fig. 4.66 for Africa's power supply and Table 4.8 for the continent's power supply structure.

5.3.6 Oceania

As an electricity exporter, Oceania is rich in resources with relatively small power demand. According to United Nations forecasts, Oceania's population will rise steadily in the future and reach 60 million in 2050, up 55% from 2010. Given the future development of power demand and ongoing electricity

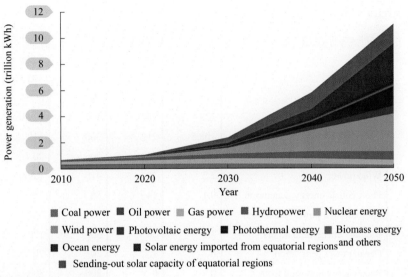

Coal power ■ Oil power ■ Gas power ■ Hydropower ■ Nuclear energy
■ Wind power ■ Photovoltaic energy ■ Photothermal energy ■ Biomass energy
■ Ocean energy ■ Solar energy imported from equatorial regions and others
■ Sending-out solar capacity of equatorial regions

FIGURE 4.66 Africa's Power Supply

substitution, Oceania's power demand is expected to reach 700 TWh in 2050. Per capita electricity consumption will approximate 13,000 kWh/year, up 57% from 2010.

Oceania is endowed with rich solar and wind power resources. The continent's future power supply will come mainly from hydroelectric, wind and solar power generation. In 2050, a supply capacity of 1700 TWh is expected to be reached, of which 12% is attributable to wind power, 71% to solar energy, and 6% to hydropower.

Taking into account the power supply and demand situation and economics across different continents, Australia's power capacity for export to Southeast Asia is projected to reach approximately 1000 TWh in 2050.

See Fig. 4.67 for Oceania's power supply and Table 4.9 for the continent's power supply structure.

5.4 DEVELOPMENT TREND OF GLOBAL ELECTRICITY FLOW

The global landscape of electricity flows is shaped by export-oriented energy bases in the Arctic and equatorial regions, exchanges of electricity across neighboring continents, and balanced consumption at the intracontinental level achieved by large-scale energy bases. It holds the key to the world's energy sustainability in the future.

5.4.1 Power Exports from Arctic and Equatorial Regions

Taking an integrated view of global power demand, renewable resources and their development potential, transmission paths, economic competitiveness, and other factors, priority should be accorded to the development of solar resources in North Africa in building the export capacity of the Arctic and equatorial regions. An export capacity of 920 TWh is initially considered for the Arctic and equatorial regions for 2030, with the equatorial regions accounting for 870 TWh of solar energy and the Arctic

Table 4.8 Africa's Power Supply Structure (TWh)

	Type	2010	2020	2030	2040	2050
Power supply structure	Total	688.2	1,057	2,370	5,800	11,010
	Coal power	263	317	350	320	210
	Oil power	71	36	20	0	0
	Gas power	228	292	400	4,200	400
	Hydropower	108.5	212	360	540	650
	Nuclear power	12.1	14	20	20	20
	Wind power	2.2	75	410	1,600	2,860
	Photovoltaic energy generation	0.1	14	80	280	600
	Photothermal power generation	0	5	50	360	1,500
	Biomass energy and others	3.3	20	50	120	150
	Ocean power generation	0	1	20	70	110
	Solar energy imported from equatorial regions	0	0	200	800	3,000
	Sending-out solar capacity of equatorial regions	0	71	410	1,270	1,510
Mode of development	Total	688.2	1,057	2,370	5,800	11,010
	Distributed type	5.8	184	640	1,480	2,990
	Centralized type	682.4	802	1,120	2,250	3,510
	Arctic and equatorial regions (intracontinental)	0	70	610	2,100	4,500
	Arctic and equatorial regions (extracontinental)	0	0	0	0	0

region, 50 TWh of wind power. In 2040, the Arctic and equatorial regions are expected to have an export capacity of 4200 TWh, including 3600 TWh of solar energy from the equatorial regions and 600 TWh of wind power from the Arctic region. In 2050, the Arctic and equatorial regions are projected to boast an export capacity of 12,000 TWh that accounts for 16% of global power demand. Of this export capacity, 3000 TWh will come from the Arctic region in the form of wind power and 9000 TWh from the equatorial regions in the form of solar energy. See Table 4.10 for the development timetable and export capacity of the Arctic and equatorial regions.

As shaped by the electricity flows from the renewable energy bases in the Arctic and equatorial regions, a pattern of movement will present itself showing power radiating southwards from the Arctic region and radiating northwards and southwards from the equatorial regions. Asian power demand is huge and fast expanding. Europe's total power demand is also massive with an urgent need for renewable energy development due to a lack of resources. Because of their strategic positions, Asia and Europe enjoy a clear geographical advantage in developing wind power in the Arctic region, while

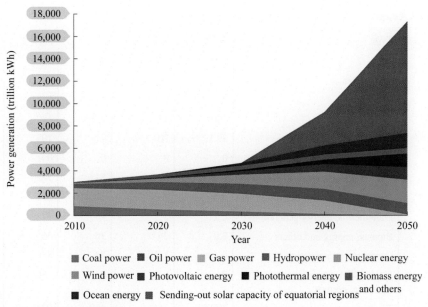

FIGURE 4.67 Oceania's Power Supply

Europe is better placed to develop solar energy in the equatorial regions. By 2050, 12% of Asia's and 24% of Europe's power requirements are expected to be met by imported renewable energy sourced from the Arctic and equatorial regions. Africa and South America are continents that both stretch across the equatorial regions with rich solar energy resources. But as solar resources are unevenly distributed at the intracontinental level, the condition and potential is right for developing transnational transmission across the continent. In 2050, the transnational transmission capacity of solar energy in Africa's equatorial regions will meet 32% of the continent's total power consumption. The comparative figure for South America is approximately 20%.

With its abundant wind, solar and hydropower resources, North America can basically satisfy its future power demand. Transcontinental power transmission can accommodate mutual backup operations across regions in different time zones and the outward power transmission from the Arctic region. The continent is expected to import wind power from the Arctic and equatorial regions to meet approximately 10% of its energy requirements. See Fig. 4.68 for the power export capacity of the Arctic and equatorial regions and the receiving capacity by continent. The export capacity and direction of electricity flows of the Arctic and equatorial regions are shown in Fig. 4.69 and Table 4.11.

5.4.2 The Direction and Capacity of Electricity Flows Across Continents

Before 2030, electricity flows across countries on each continent and also the region between North Africa and Europe, being close to each other, are given priority for development as part of the bigger efforts to build transcontinental interconnections. But the capacity of electricity flows is limited at the initial stage of development. At this stage, the major areas for building transnational electricity flows at the intracontinental level include Mongolia for wind and solar energy, Russia for

Table 4.9 Oceania's Power Supply Structure (TWh)

	Type	2010	2020	2030	2040	2050
Power generation mix	Total	300.5	366	469	926	1,745
	Coal power	84.8	54	35	13	0
	Oil power	0	0	0	0	0
	Gas power	160	175	158	120	10
	Hydropower	39.1	70	91	103	110
	Nuclear power	0	0	0	0	0
	Wind power	7.1	30	81	155	209
	Photovoltaic energy generation	0.5	8	31	67	113
	Photothermal energy generation	0.1	2	13	41	120
	Biomass energy and others	8.9	26	36	46	49
	Ocean power generation	0	1	24	81	134
	Sending-out solar capacity of equatorial regions	0	0	0	300	1,000
Mode of development	Total	300.5	366	469	926	1,745
	Distributed type	0.5	16	60	130	280
	Centralized type	300	350	409	496	465
	Arctic and equatorial regions (intracontinental)	0	0	0	300	1,000
	Arctic and equatorial regions (extracontinental)	0	0	0	0	0

Table 4.10 Development Timetable and Export Capacity of Arctic and Equatorial Regions (TWh)

Type	2030	2040	2050
Wind power in Arctic region	50	600	3,000
Solar energy generation in equatorial regions	870	3,600	9,000
Total	920	4,200	12,000

hydropower, central Asia for renewable power generation destined for load centers in Northeastern Asia (including China), the Arctic regions of the Kara Sea and the Bering Strait for wind power destined for load centers in north-east Asia, the equatorial regions of North Africa and East Africa for solar and wind power destined for western and southern Africa. At this stage, the most important transcontinental electricity flows are mainly those that deliver solar power from solar energy bases in North Africa to Europe, at a transmission level of 100 million kW, and also Eurasian interconnections

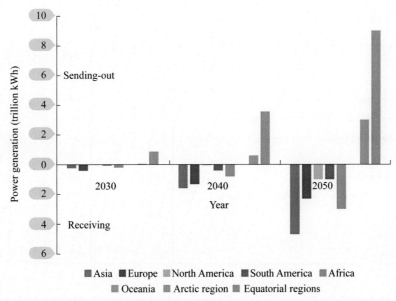

FIGURE 4.68 Export Capacity of Arctic and Equatorial Regions and Receiving Capacity by Continents

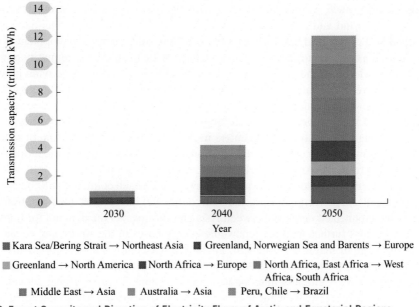

FIGURE 4.69 Export Capacity and Direction of Electricity Flows of Arctic and Equatorial Regions

Table 4.11 Export Capacity and Direction of Electricity Flows of Arctic and Equatorial Regions (TWh)

	2030	2040	2050
Arctic region	50	600	3,000
Kara Sea/Bering Strait → Northeast Asia	50	500	1,200
Greenland, Norwegian Sea and Barents → Europe	0	50	800
Greenland → North America	0	50	1,000
Equatorial regions	870	3,600	9,000
North Africa → Europe	410	1,300	1,500
North Africa, East Africa → West Africa, South Africa	200	800	3,000
Middle East → Asia	200	800	2,500
Australia → Asia	0	300	1,000
Peru, Chile → Brazil	60	400	1,000

for delivering power from Central Asia to Europe where large renewable energy bases are built in Kazakhstan, Xinjiang (China), and Siberia (Russia) to deliver power to load centers in Europe. See Fig. 4.70 for the world's electricity flows across continents and across countries at the intracontinental level in 2030.

From 2030 to 2050, the renewable energy bases in the Arctic and equatorial regions will enter a period of significant growth, with the growing maturity of clean energy and transmission technologies. On each continent, wind and solar energy resources with favorable conditions have already been well developed on a centralized basis. Distributed wind and solar energy development is now mainly focused on densely populated urban areas and on a large scale to address community power demand. As mankind's knowledge of the Arctic deepens, the conditions for large-scale wind power development in the region are falling into place. The focus of global renewable energy development will shift gradually from large-scale clean energy bases on each continent to the Arctic and equatorial regions. The development of wind power in the Arctic region and solar energy in the equatorial regions can satisfy the increasing demand from load centers in Asia, Europe, and North America. At this stage, global electricity flows mainly constitute transnational transmission at the intracontinental level and electricity outflows from the Arctic and equatorial regions. With growing power demand and clean energy replacement, the large-scale renewable energy bases on each continent will focus on satisfying local power demand and the original Eurasian intercontinental interconnections will transform themselves into an interconnected network between Asia and Europe. As the demand for transcontinental electricity flows grows further with the renewable energy bases in the Arctic and equatorial regions moving into the stage of major development, global electricity flows across continents will be centered on the delivery of power from the Arctic and equatorial regions to all continents at a capacity of over 100 million kW. A globally interconnected grid system featuring global allocations of clean energy will come into being, with the focus on clean energy bases in the Arctic and equatorial regions. See Fig. 4.71 for the world's major electricity flows in 2050.

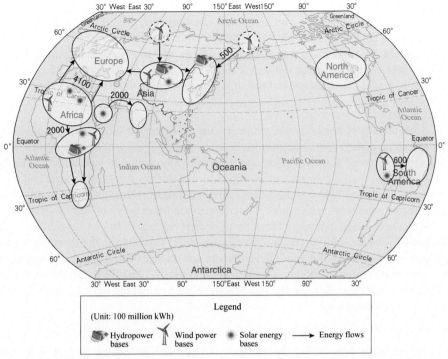

FIGURE 4.70 World's Electricity Flows Across Continents and Across Nations at Intracontinental Level in 2030

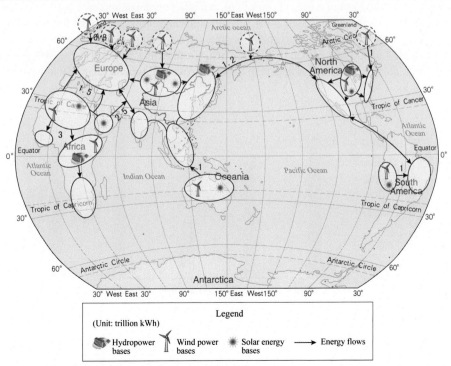

FIGURE 4.71 The World's Major Electricity Flows in 2050

The previous discussion is based on an overall consideration of global electricity flows under a scenario for quicker development of clean energy and global energy interconnection. In the course of actual development, the capacity of future electricity flows will be uncertain due to concerns over the economics of clean energy development on each continent and clean energy technologies, as well as the economics of power development and transmission concerning the Arctic and equatorial regions.

5.4.2.1 The Arctic

Despite its exceptionally rich wind power resources, the Arctic region faces tough conditions in power-related development, construction, operation and maintenance, as well as the problems of long transmission distances. The transmission of wind power from the Arctic region to load centers in Europe, Asia, and North America will largely depend on the economics of wind power development and transmission in the Arctic region, especially in comparison of economic competitiveness with the supply costs of various power sources in the receiving regions. Under a scenario for delivery of 3000 TWh from the Arctic region in 2050, it is assumed that major breakthroughs can be achieved in wind power development and transmission technologies, with greatly improved economics. Considering the uncertainty surrounding the technology economics of wind power development and transmission in the Arctic region, the export capacity of the Arctic region in 2050 may drop to approximately 2000 TWh, and the wind power bases in Greenland, the Norwegian Sea, the Barents Sea, the Kara Sea and the Bering Strait will likewise see a decline in operational capacity.

5.4.2.2 Africa

Judging by the electricity flows across the continent, Africa's demand centers in 2011 were concentrated in the southern and northern regions. The short- to medium-term planning developed by the Southern African Power Pool, the Eastern African Power Pool, and other organizations envisages a flow of electricity to the south from major power bases in the northern and central regions. In the future, with the disappearance of the "without access to electricity" population and rising consumption per capita, the total electricity consumption in West Africa, southern Africa, North Africa, and East Africa will grow to a relatively high level, though power consumption in Central Africa will remain relatively low (see Fig. 4.72).

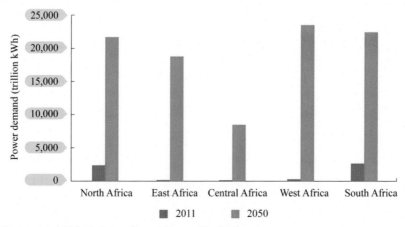

FIGURE 4.72 Forecasts of Africa's Future Power Demand by Region

Given the abundance of water resources in central and eastern Africa with stable output and long utilization hours, together with the world-leading quality solar energy resources in eastern and southern Africa, the central, eastern and southern regions of Africa will see large-scale development in hydropower and solar resources. This will have an impact on the capacity of power transmission from solar power bases in North Africa to central and southern Africa. In view of this, the capacity of electricity flows from North Africa and East Africa to the west and south of the continent may drop to 1000 TWh in 2050, from the 3000 TWh envisaged under a previous scenario.

5.4.2.3 South America

Judging from the development conditions of solar energy bases in South America, the cost of solar power development is relatively high in Peru and Chile where wind and hydropower resources are abundant. In addition, the transmission of solar power from energy bases in South America eastwards to load centers in Brazil and other countries will have to overcome the challenge of traversing the Andes mountains and rainforest reserves, which will affect the development and export capacity of solar energy bases in the equatorial regions of Peru, Chile, and other South American countries.

5.4.3 Research and Forecasts by other Organizations

Systematic research has been conducted by relevant international organizations on future energy development in a bid to address climate change issues and the requirements of low-carbon energy development. Different research objectives and boundary conditions have been set, with the following findings obtained.

The IEA predicts bigger room for growth in global nuclear energy over the next 20 years, from 390 GW in 2013 to 620 GW in 2040. In the New Policies Scenario,[21] conventional fossil fuel-fired generation is expected to increase from 15,000 TWh in 2012 to 22,000 TWh in 2040, with its share of total power generation decreasing from 68% to 55%. Over the same period, nonhydrorenewable energy generation is expected to increase from 1000 TWh to 7000 TWh, with its share of total power generation rising from 5% to 17%.

According to a WEC research report, taking into account the future energy demand of different countries, coal-fired generation is expected to increase from 8700 TWh in 2010 to 20,000 TWh in 2050. The portion of coal-fired generation with carbon capture and storage (CCS) is estimated at 1000 TWh. Over the same period, nuclear power generation is expected to grow from 2800 TWh to 3300 TWh, with installed nuclear capacity rising from 370 GW to 420 GW. Under the Symphony Scenario, nuclear energy generation and installed nuclear capacity will reach 7000 TWh and 880 GW, respectively.

According to the BP Energy Outlook 2035 issued by BP, coal-fired generation will increase by 35% and nuclear energy by 53% in 2012–2035. See Fig. 4.73 for the generation mix by energy type in 2035.

All in all, while a consensus has been reached among research institutions on the determination to tackle climate change issues in the future, views differ over how climate change has progressed.

---

[21]The New Policies Scenario has taken into account the policies and measures adopted by 2014 with implications for the energy market, as well as relevant policy recommendations that need to be implemented through detailed provisions.

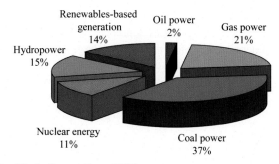

FIGURE 4.73 The Generation Mix by Energy Type, 2035

Source: BP, BP Energy Outlook 2035.

According to an IPCC report, to meet the target of limiting global temperature rise to 2°C, global efforts need to be stepped up to pursue clean energy substitution and electricity substitution. And human society's sustainability targets can only be better accomplished by improving the development of renewable energy and its share in power generation. To achieve this objective, the development of large-scale renewable energy bases in the Arctic and equatorial regions and on each continent should be expedited and expanded to incorporate the world's major renewable energy bases into the sphere of global allocations.

SUMMARY

1. Based on a study of future socioeconomic development, energy resource supply, energy environment constraints, energy technology advancement, and energy policy regulation, global primary energy consumption is expected to rise from 18.8 billion tons of standard coal in 2010 to 30 billion tons of standard coal in 2050, at an annual rate of 1.2%, lower than the average rate of economic growth. Global power demand will rise from 21,000 TWh in 2010 to 73,000 TWh in 2050, at an annual rate of 3.1%, higher than the average rate of economic expansion and energy demand growth. The energy structure will undergo a fundamental change from one oriented towards fossil fuels and supplemented by clean energy to one focusing on clean energy and supplemented by fossil fuels. Asia, Africa, and South America will play a growing role in global power consumption. Despite their falling share in electricity consumption, the developed regions of Europe and America will remain the world's most energy-intensive regions in terms of electricity consumption.

2. Wind power resources in the Arctic region, solar energy resources in the equatorial regions, and renewable energy resources on each continent will be fully developed. Clean energy generation is expected to reach 66,000 TWh by 2050, accounting for a dominant 90% share of total generation. Of this capacity, 26,000 TWh is attributable to solar energy (photovoltaic and photothermal), accounting for a higher share of 35%, and 22,000 T kWh to wind power generation, accounting for 31%. Both solar and wind resources will become the absolute key drivers of clean energy generation in the future.

3. Before 2030, global electricity flows will orient towards transnational transmission at the intracontinental level and transcontinental transmission between two locations close to each other (for instance, North Africa–Europe and Central Asia–Europe). The capacity of transcontinental electricity flows is still at the initial stage of strong growth. After 2030, the centralized development of wind and solar resources with good conditions and easy access will be gradually completed. The focus of development will then slowly shift to the Arctic and equatorial regions. The capacity of electricity outflows from these regions is expected to reach 12,000 TWh in 2050, accounting for 16% of global electricity demand, which will necessitate the development of a globally interconnected energy network to optimize the global allocations of clean energy. In the event of the hydropower and solar resources in central and southern Africa having been fully developed, the capacity of electricity outflows from the Arctic and equatorial regions may decline to 10,000 TWh in 2050, accounting for 14% of global electricity demand.

BUILDING GLOBAL ENERGY INTERCONNECTION

CHAPTER OUTLINE

Global Energy Interconnection. http://dx.doi.org/10.1016/B978-0-12-804405-6.00005-1

1 ROBUST SMART GRID AND GLOBAL ENERGY INTERCONNECTION

Driven by higher energy and electricity demand, the world's power grid system has leapfrogged from conventional grids to modern grids, and from isolated urban grids to transregional and transnational interconnected large grids. It has now entered a new stage of development with robust smart grids as a prominent feature. In response to the new requirements of the two-replacements, robust smart grids will move faster in the direction of global integration for building global energy interconnections to provide more secure, economical, cleaner, and sustainable energy needed for global socioeconomic development.

1.1 ROBUST SMART GRID

The world electric power industry has been growing for more than 130 years since the world's first thermal power plant was completed in Paris in 1875. To construct robust smart grids – a modern grid system capable of allocating transnational and transcontinental electricity and of flexibly adapting to new energy development and diversified service needs – has become a direction and strategic option in the global grid development of the twenty-first century.

1.1.1 The Evolution of Global Grid Development

Driven by greater demand and faster technological progress, the evolution of global grid development is characterized by a shift from low voltage to higher voltage, from small scale to larger scale of interconnection, and from low level to higher level of automation.

1.1.1.1 Voltage Upgrading

As energy loss through grid transmission is directly proportional to the square of line current, by raising grid voltage and reducing line current, electricity can be transmitted more effectively over long distances, in large capacities and with low transmission losses, assuming the same load for delivery. The continued growth in system capacity and power loads has led to higher transmission power requirement, and the voltage levels of transmission lines need to be raised steadily. Typically, introduction of a higher voltage grade is best timed with the point when a fourfold or more increase in a system's peak load power is reached. Traditionally, in a period of continued economic growth, it would take 15–20 years for a new, higher voltage grade to come into being. In 1891, the earliest alternating current (AC) transmission lines built in Germany featured a voltage of 13.8 kV. By 1990, 60 kV HV lines had been built in the United States. In 1952, Sweden completed the world's first 380 kV EHV line. By 1969, EHV lines at 765 kV had been in operation in the United States. China's first 1000 kV commercial UHV AC line was put into service in 2009. Direct current (DC) transmission technology has undergone the two major development stages of mercury arc valve-based conversion and thyristor-based conversion. Currently, flexible DC transmission technology based on gate-controlled switch devices is being developed and applied, with transmission voltage continuously upgraded to meet growing demand. After the completion of the first 1.5 kV DC transmission line in Germany in 1882, a ±100 kV HVDC transmission line was completed in Sweden in 1954. In 1962, the Soviet Union completed a ±400 kV EHV DC transmission line. In 1984, the UHV DC transmission project in Itaipu, Brazil, featured a voltage of ±600 kV. In 2010, China's ±800 kV UHV DC transmission project commenced operations. See Fig. 5.1 and Fig. 5.2 for the development of AC/DC transmission voltage grades in the world.

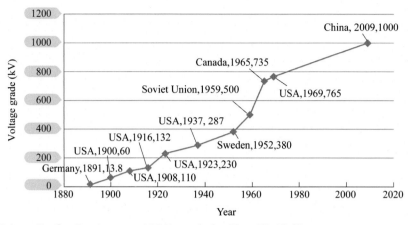

FIGURE 5.1 Voltage Grading Development of AC Transmission Lines Worldwide

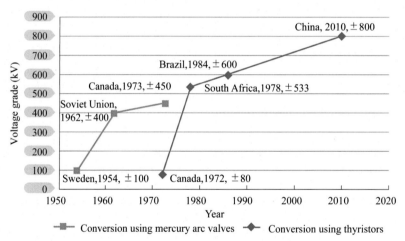

FIGURE 5.2 Voltage Grading Development of DC Transmission Lines Worldwide

1.1.1.2 Greater Grid Interconnection Capacity

From the end of the nineteenth century to the mid-twentieth century, grid development was oriented toward municipal grids, isolated grids, and small grids. They were small in scale and aimed at achieving a balance at the local level. However, the increasing installed capacity of generating units for grid integration has given rise to the need for expanding resource allocation capacity and transmission coverage. Grid development has witnessed a shift to large-scale, interconnected grids characterized by high voltage and strong interconnection capability, gradually forming a large grid operating at the transregional level at EHV of 330, 500, and 750 kV and UHV of 1000 kV. Currently, a number of transnational interconnected grids have been developed, including the interconnected power grid of North America, the synchronous power grid of Europe, and the Russian Baltic Sea power grid. China has also completed the world's first commercial 1000 kV UHV AC transmission project in 2009. UHV transmission marks an important milestone in grid development. The transmission distance of 500–750 kV EHV lines is 700–1000 km, compared with a much longer distance of 2000–5000 km for UHV DC transmission, which provides an extensive geographical reach covering the world's major countries and regions. This has given power grids the capability to allocate energy resources over larger areas, thus laying a foundation for extensive interconnection of the world's power grids at the transnational and transcontinental levels.

1.1.1.3 Higher Automation Levels

Over the past century, with the continued innovation in electronic information technology and the upgrading of automation technology, grid development has demonstrated ever-higher levels of automation and, in electricity generation, the level of informatization, automation, and interaction has also increased. Between the late nineteenth century and the mid-twentieth century, power outages caused by grid faults were frequent and supply reliability was low, reflecting the shortcomings of a grid that relied on simple protection, experienced-based dispatching, an automation system limited to a single automation device, oriented toward safety protection, and process automatic adjustment with a generally lower level of automation. Between the mid- and late-twentieth century, the advent of transregional grid networks called for meeting new requirements in terms of system stability, economical dispatch, and overall automation. Power grids have since achieved more sophisticated protection and dispatching, greater use of automation devices, and an extensive application of tele-mechanical communications technology. Supervisory control and data acquisition systems have also appeared, microcomputers have been gradually integrated into relay protection equipment, and grid automation swiftly enhanced, resulting in significantly improved supply reliability. Since the late-twentieth century, along with the growing capacity and coverage of power grids, advanced technologies in modern control, information communication, and other areas have been widely applied, and information flows processed automatically by power systems have grown ever bigger. In addition, more factors have been taken into account, directly observable and measurable ranges have become increasingly wider, and more targets have been subjected to successful closed-loop control. Meanwhile, the quest for greater security and reliability through smart grid technology, has become a dominant trend. The modern power system has brought together computers, controls, communications, electrical equipment, and power electronics equipment, contributing to greatly improved grid security and stability.

1.1.2 Power Grid Development Enters New Stage of Robust Smart Grids

The world's power grid development can be divided into three stages (Fig. 5.3). *The first stage evidenced the development of small-scale power grids.* From the late nineteenth century to the mid-twentieth century, generating units were small, with limited installed capacity, low voltage grades, small interconnection footprint, and weak interconnection among different power grids. Power grids were predominantly small, isolated grids featuring regional distribution of power at the municipal or local level. *The second stage involved the development of large-scale interconnected grids.* Between the mid- and late-twentieth century, generating units grew in size and installed capacity, with higher grid voltage grades, stronger grid interconnection, and the continued emergence of transregional and transnational grids to drive grid development in the direction of large-scale synchronous power grids with national or transnational capability of power distribution. *The third stage is centered on robust smart grids.* In the twenty-first century, with the rapid growth of renewable energy and advanced technologies in information communication, modern control, and UHV transmission, global grid development has moved into a new era of robust smart grids.

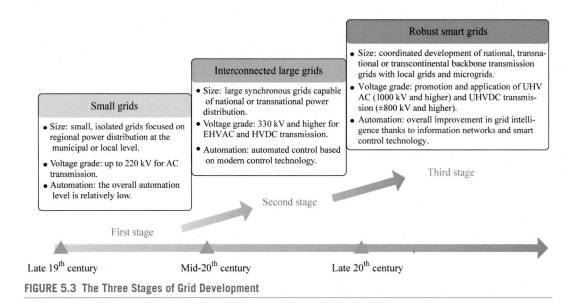

FIGURE 5.3 The Three Stages of Grid Development

One of the earliest developers of smart grids, China has initiated a new concept of robust smart grid development, combining the characteristics of energy resources distribution and the requirements of strong socio-economic growth. It is based on the implementation of a "One Ultra and Four Larges" strategy to accelerate the construction of UHV grids and promote the intensive development of large coal-fired power plants, large hydropower stations, large nuclear capacity, and large renewable energy bases.

Supported by a backbone UHV grid, the robust smart grid is a new, modernized power grid featuring coordinated development of grids at different levels and covering different segments of power operations, including grid access, power transmission, transformation, distribution, utilization, and dispatching. It integrates modern communications information, automatic control, decision support, and advanced power technologies, with all the features of informatization, automation and interaction. It is adaptable to flexible access to and exit from various power sources and powered equipment. It is also capable of friendly interaction with users, with smart response and system self-recovery capabilities to substantially improve the safety, reliability, and operational efficiency of the power system. "Robustness" and "intelligence" are two basic requirements of modern power grid development. "Grid robustness" is the base for assuring the capacity for resource allocation on a large scale for safe and reliable power supply. "Ubiquitous intelligence" is also crucial, meaning that smart technologies are widely applied across the electricity system for greater adaptability, controllability, and safety in all aspects. Modern grid development must put equal emphasis on "robustness" and "intelligence", both being indispensable qualities.

With the continued development of grid technology and extensive integration of smart technology, modern grids are undergoing profound changes in form and function. Functionally, grids are being upgraded from single carriers of electric power to smart platforms with strong capabilities to optimize allocation of energy resources. As the two-substitution policy progresses more rapidly, clean energy is being utilized on an ever-greater scale, the share of electricity in end-user energy demand is growing, and grid allocation of energy resources has become more efficient. This will further facilitate the realization of grid interconnection and clean energy allocation at the global level to form a globally interconnected robust smart grid system.

1.2 GLOBAL ENERGY INTERCONNECTION

Global energy interconnection refers to the development of a globally interconnected, ubiquitous robust smart grid, supported by backbone UHV grids (channels), and dedicated primarily to the transmission of clean energy (Fig. 5.4). Comprising of transnational and transcontinental backbone grids and ubiquitous smart power grids in different countries covering the transmission/distribution of power at different voltage grades, the globally interconnected energy network is connected to large energy bases in the Arctic and equatorial regions, as well as different continents and countries. It can adapt to the need for grid access for distributed power sources with the capability to deliver wind, solar, ocean, and other renewables to different types of end users. Generally speaking, a global energy interconnection is in effect a combination of "UHV grids plus ubiquitous smart grids plus clean energy," forming a green, low-carbon platform for global allocation of energy with extensive coverage, strong allocation capability, and a high level of security and reliability. It can link up the grids on different continents divided by time zones and seasons to remove resource bottlenecks, environmental constraints and spatio-temporal limitations, realizing mutual support and backup between wind and solar generation and across different regions. This will result in greater energy security, improved economic benefits and reduced environmental losses to effectively resolve issues of energy safety, clean development, efficiency improvement and sustainability. This development will turn the world into a

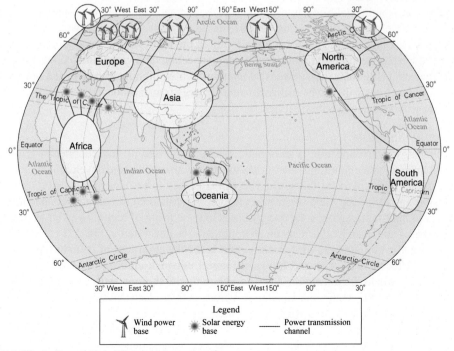

FIGURE 5.4 Illustration of Global Energy Interconnection

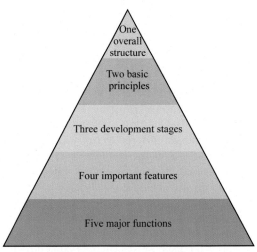

FIGURE 5.5 Development Framework for Global Energy Interconnection

"global village" characterized by abundant resources, greenliness with clear bright skies, and peace and harmony.

The development framework for global energy interconnection can be summarized as "one overall structure, two basic principles, three development stages, four important features, and five major functions" as illustrated in Fig. 5.5. The global energy interconnection will feature an overall structure comprised of transcontinental grids, transnational grids, and national ubiquitous smart grids supported by coordinated development of local grids at different levels. Adhering to the two basic principles for clean energy development and global energy allocation, the network will undergo three stages of intracontinental interconnection, transcontinental interconnection, and global interconnection, with four important features being robustness, extensive interconnection, high intelligence, and open interactivity. Five major functions will also be realized, namely energy transmission, resource allocation, market trading, industrialization, and public service.

1.2.1 Overall Structure

The global energy interconnection is an organic whole, comprised of transcontinental grids, transnational grids, and national ubiquitous smart grids, together with coordinated development of local power grids at different levels. Around the world, the global energy interconnection will rely on advanced UHV transmission technology and smart grid technology, structured to facilitate connections to wind power bases in the Arctic region, solar energy bases in the equatorial regions as well as major renewable energy bases and main load centers on all continents, as illustrated in Fig. 5.6.

At the core of a global energy interconnection is the development of transnational and transcontinental backbone networks and intercontinental network channels to link up clean energy bases around the world, including the Arctic and equatorial regions, with major load centers. The clean energy bases in the Arctic and equatorial regions deliver electricity to the load centers through multiple channels. Specifically, the wind energy bases in the Arctic region deliver electricity to Asia, Europe, and North America; the solar energy bases in North Africa and the Middle East deliver power to Europe and South Asia; and the solar energy bases in Australia deliver electric energy to Southeast Asia. Grid interconnections across different continents mainly include those between Asia and Europe,

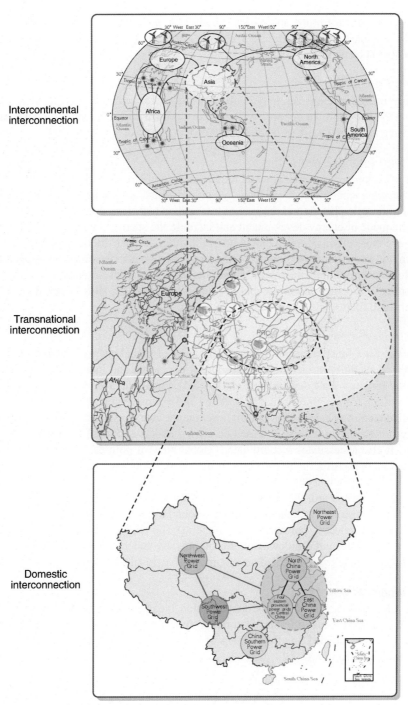

Intercontinental
interconnection

Transnational
interconnection

Domestic
interconnection

FIGURE 5.6 Illustration of Development of Global Energy Interconnection

Asia and North America, Europe and Africa, South Asia, and Africa, and North America and South America. Also included are the transmission channels used by the large energy bases on all continents to supply local load centers.

1.2.2 Basic Principles

Global energy interconnection has important implications for realizing the global energy view and implementing the two-replacement policy. As the most critical core element of global energy development, adherence to the two basic principles as follows is essential.

1.2.2.1 Clean Energy Development

Clean energy development is a fundamental requirement for addressing climate change and mankind's sustainable development. After an extensive, global consensus is reached, all countries in the world should build a strategic plan around the goal of low-carbon and clean energy development, to expedite changes in the mode of energy development and increase the share of clear energy, with concerted efforts to achieve more efficient development and utilization of clean energy globally. Focusing on the goal of clean and low-carbon development of energy globally, efforts should be stepped up to design and build global energy interconnections to promote an efficient development and utilization of various concentrated and distributed clean energy sources and drive the shift in development focus from conventional fossil fuels to clean energy.

1.2.2.2 Global Allocation

Global allocation is a function of the reverse distribution of global energy resources and load centers. By nature, the availability of clean energy resources is random and intermittent. Since clean energy resources good for large-scale development are located typically far away from load centers, consumption issues resulting from massive capacity building and extensive grid access can only be resolved by optimized allocation over larger areas to fully leverage the role of clean energy. The development of a global energy interconnection must be based on a good understanding of the world's energy resource endowments. It also necessitates a coordinated, global view of political, economic, social, and environmental factors to construct a global platform for energy allocation to link energy bases with load centers and facilitate efficient development, optimized allocation, and efficient utilization of energy at the global level. In this regard, the development of UHV transmission technology with large capacity and long-distance transmission capability will lay a sound technological foundation for large-scale and efficient allocation of electric power across continents. The ability to allocate clean energy globally can also transform the resource endowment of economically underdeveloped regions into an economic advantage and contribute to coordinated development of regional economies.

1.2.3 Development Stages

Based on an integrated view of global energy allocation, clean energy development, energy supply and demand, energy transmission and other factors, the future development of a globally interconnected energy network can be divided into three stages of intracontinental, transcontinental, and global interconnection, as illustrated in Fig. 5.7.

> *First stage*: To promote the formation of a consensus before 2020. With reference to the comparative advantages in technology and economics among different continents, the development of large clean energy bases should be initiated by 2030 to improve grid

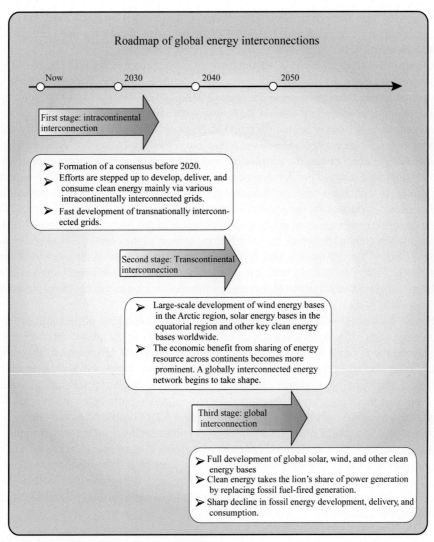

FIGURE 5.7 Roadmap of Global Energy Interconnection

interconnection among countries in each continent. Development of clean energy should be expedited among different continents to deliver and consume clean energy via power grids interconnected at the intracontinental level, such as those in Northeast Asia, North America, Europe (through a synchronous grid system), Latin America, and Central and Southern Africa. While meeting the increasing demand for clean energy, the rapid growth of transnationally interconnected grids enables different countries to support and complement each other in different seasons/periods and across different types of electric energy to improve the efficiency and economics of energy systems. All countries should strengthen the construction and nationwide interconnection of domestic robust grids to better integrate them into a grid interconnected on a continent-wide basis and to receive more effectively clean energy allocated at the intracontinental level.

Second stage: Between 2030 and 2040, the focus of development is, by building on the grid interconnection among the major countries in each continent, to achieve major progress on the development of large energy bases in the Arctic and equatorial regions and grid interconnections across continents, in ascending order of difficulty. With continued improvement in the structure of transcontinentally interconnected grids and large-scale development of wind energy bases in the Arctic region, solar energy bases in the equatorial region, and other key clean energy bases around the world, long-distance transmission, and grid interconnection across continents have become a dominant trend in the development of a global energy interconnection. Supported by the increasingly significant benefits of mutual support and back-up across different continents and time periods, a global energy interconnection has taken shape. The emphasis should be on promoting the construction of power export channels in the Arctic and equatorial regions, the transmission channel between Asia and Europe as well as the corridors of interconnections between Asia and North America, between Europe and Africa, South Asia and Africa, and North America and South America. In addition, an agency should be established to promote and coordinate collaborative efforts in global energy interconnection as the beginning of a cooperative mechanism in this area.

Third stage: Between 2040 and 2050, guided by the strategic thinking of pursuing key breakthroughs and full progress, the construction of a global energy interconnection will be gradually shaped to fulfill the two-replacement goal. With the full development and completion of solar, wind, and other clean energy bases worldwide, clean energy will replace fossil energy to assume a dominant role in power generation, while the development, transmission, and consumption of fossil energy will significant decrease, bringing into being a global platform for optimized allocation of clean energy supported by a fully-completed global energy interconnection. The global energy interconnection will be well-organized with a sound operational mechanism, with the establishment of a global system operation center. The center will form part of a larger control system, operating on a zoned and stratified approach to control, with operation centers in continents and countries. A well-developed worldwide power market will substantially increase the share of transnational and transcontinental electricity trading in total power consumption.

1.2.4 Important Features

As a new global platform for energy allocation, the global energy interconnection comes with four important features: robustness, wide interconnection, high intelligence, and open interactivity.

1.2.4.1 Robustness

As the foundation for global allocation of energy, a robust grid structure is a prerequisite for the construction of a global energy interconnection. Extensive interconnection and large-area allocation of global energy resources is possible only through a robust, reliable grid structure of transnational or transcontinental interconnections. Across the world, only power grids that are scientifically planned, logically structured, safe, reliable, and operationally flexible can meet the requirements of extensive access to and consumption of wind power, solar photovoltaic energy, and distributed energy sources.

1.2.4.2 Extensive Interconnections

A global energy interconnection is basically shaped by extensive interconnection. The interconnection allows efficient development and broad allocation of global energy resources and related public service resources. The coordinated development and seamless connection of intercontinental backbone grids, transnational grids at the continental level, national grids, local grids, distribution grids, and microgrids can form an extensive system for allocation of energy worldwide.

1.2.4.3 High Intelligence

High intelligence provides key support for global energy interconnection, enabling flexible grid access for various power sources and loads, while assuring security and stability for network operations. Through the extensive use of information networks, wide area measurement, high-speed sensing, high-performance computing, smart control, and other technologies, highly intelligent operations can be achieved at different grid levels and sections to automatically make prejudgements and identify most faults and risks, with a fault self-recovery capability. Real-time exchange of information supports the free flow of all elements across the network, thereby achieving efficient allocation of energy resources across different regions.

1.2.4.4 Open Interactivity

Open interactivity is a basic requirement of a global energy interconnection. The construction of this network requires coordination and close collaboration at the international level. Operationally, the network should provide equal and open access to all countries without discrimination. Grids should be allowed to discharge their functions in the market, while an open, unified, competitive, and orderly organizational system should be developed to facilitate broader exchange between users and powered equipment, two-way interaction with power grids and two-way energy flow between users and suppliers to achieve collaboration and interaction among the interested parties in global energy interconnection.

In the final analysis, a global energy interconnection fully manifests the characteristics and concepts of the Internet. First, the network is characterized by an extensive reach of its connections. Like the Internet, the global energy interconnection provides equal access to the user, be it an individual, a piece of equipment, a family, a building, a factory, or an industrial park. Second, consumers and producers can participate in the system on an equal footing. Similar to the situation with the Internet, the positioning of the users of the global energy interconnection have changed radically in the sense that they are now both consumers and producers, resulting in a much higher level of participation and influence. Third, it is the free flow of energy and information. Global energy interconnection enables energy to flow freely worldwide, like the information on the Internet, so that users may conveniently share clean energy delivered from thousands of kilometers away without any more distance and resource constraints. This will free mankind from the shackles that bind production and life down, and fully unleash productivity as a critical element of energy development. Fourth, it is the diversity of services. Building on global energy interconnection, a comprehensive service platform can be established to launch diverse services so as to drive the rapid development of upstream and downstream operations and related industries and to promote the creation of a positive ecosphere around industries.

1.2.5 Major Functions

With the full-scale promotion and application of UHV power transmission, smart technology, and other advanced technologies, a global energy interconnection is far more than a carrier of electric energy in the traditional sense. Rather, it is a powerful platform for resource allocation, market trading, industrialization, and public service. The platform is able to convert primary energy sources (e.g., coal, hydropower, wind energy, solar power, nuclear energy, biomass, and tidal energy) into electric energy. The diversified energy sources can play a complementary role among themselves and support coordinated development and rational utilization. The platform can be connected to large energy bases and load centers to optimize energy allocation on a larger scale by delivering electricity more efficiently over long distances and on a large scale. By integrating with the Internet, the Internet of Things and smart mobile terminals, it can address the diverse needs of users and facilitate the development of smart homes, smart communities, smart transport, and smart cities. It can be said that the global energy interconnection is an important hub of energy and services for the future, providing a base for integrating energy, information and business flows. A global energy interconnection has five major functions as described here.

1.2.5.1 Energy Transmission

Energy transmission is the most basic function of global energy interconnection. Electric energy transmission is an important mode of energy transport, with the flow of electricity generated from coal as well as hydro, nuclear, wind, and solar energy all transmitted through power grids. As the vehicle for optimized allocation of energy resources, a global energy interconnection can convert various types of primary energy into electricity for transmission through power grids. It enables the transmission of energy and electricity at the speed of light. When the two-replacement is well-advanced in the future, renewable energy used for power generation is expected to become a dominant energy source worldwide. Global energy interconnection will also become a core element of a comprehensive energy transmission system, playing a collaborative and complementary role with conventional means of energy transport such as railway, highway, waterway, and pipelines to form a highly modernized, comprehensive global energy transmission system.

1.2.5.2 Resource Allocation

Global energy interconnection forms an important platform for optimized allocation of various energy resources. Through this platform, the network is connected to different power sources and users to realize the intensive development and efficient utilization of different energy types. With the gradually expanding coverage of grid interconnections at the global level, energy resources are distributed more effectively and on a larger scale. The development and construction of large energy bases far away from load centers, including those in the Arctic or equatorial region, contribute significantly to optimize the structure and distribute patterns of energy. In essence, the process of transmitting and distributing electricity is the process of optimizing the allocation of energy resources. As a network hub linking up different power sources and users, the global energy interconnection can optimize the allocation of energy resources, user resources, and also boost the intensive development and efficient utilization of different energy sources, including coal, petroleum, natural gas, hydropower, wind energy, and solar energy.

1.2.5.3 Market Trading

Global energy interconnection provides a physical foundation for global trading of electricity. While electric energy cannot be stored in large scale, a balance between electricity demand and supply must be maintained at all times. This objective reality means that market trading of electricity must be conducted through power grids and that the physical boundary of the electricity market will be determined by the coverage of these grids. Energy interconnection that covers the world are expected to become the conduit for global trading of electricity, playing a critical role in the development of a global electricity market by serving as a platform for electricity trading while assuming the responsibility for grid frequency modulation, system backup, and reactive voltage regulation.

1.2.5.4 Industrialization

The global energy interconnection is the incubator for fostering strategic emerging industries. An important area of technological innovation and also an important vehicle for adopting new technology, it is a strong driving force behind emerging industries, including new energy, new materials, smart equipment, electric vehicles, and information technology. The development of global energy interconnection can lead to the creation of a positive ecosphere and provide an all-round impetus to upstream and downstream industries and improve industry-level standards and speed of development.

1.2.5.5 Public Services

By serving all families and industries at the community level, the global energy interconnection is a public service platform essential for production and life in the future. Along with its deep integration with the Internet of Things and the Internet, the network will evolve into a multifunctional, highly-smart public service platform to provide users with comprehensive services like energy, electricity, and information, meet the needs for diverse and quality services, and drive production and lifestyle changes. The construction of this global energy interconnection as an irreplaceable hub of public services will be integrated fully into the development of economic societies and smart cities.

2 TRANSCONTINENTAL UHV GRID BACKBONE

The transcontinental backbone of the global energy interconnection comprises primarily of the outgoing transmission channels of large renewable energy bases in the Arctic and equatorial regions, together with channels of intercontinental connections. Given the needs for large-scale, long-distance power transmission and intercontinental interconnections, UHV AC and UHV DC technologies will be adopted in the future for outgoing transmission from large renewable energy bases in the Arctic and equatorial regions and for large-capacity exchange of energy through intercontinental channels.

As the top tier of the globally interconnected energy network, transcontinental UHV grid backbones play a functional role in the transmission of power from large renewable energy bases in the Arctic and equatorial regions and power exchange among continents, including primarily the outgoing power transmission channels in the Arctic and equatorial regions and channels of transcontinental interconnections. According to an analysis of the demand for allocation of clean energy worldwide, the volume of electricity delivered by the channels in the Arctic region is expected to reach 3000 TWh/year by 2050, compared with 9000 TWh/year for the channels in the equatorial region, with both regions collectively accounting for 16% of global electricity demand. This is in addition to a certain percentage of power transmission required for transcontinental exchange to facilitate peak staggering, mutual support in resources, and sharing of reserve capacity. Given the development of renewable energy and transmission technology in extremely cold regions and also the uncertainties surrounding the developable capacity of renewable energy in Africa, electricity flows from the Arctic and equatorial regions may weaken to approximately 10,000 TWh in the future.

2.1 OUTGOING WIND POWER TRANSMISSION CHANNELS IN THE ARCTIC REGION

In the Arctic region, the channels for outgoing transmission of wind power meet the transmission requirement of wind energy bases in the region, including Greenland, the Norwegian Sea, the Barents Sea, the Kara Sea, and the Bering Strait. These channels also form a strategic platform for building the northern hemisphere's three transcontinental grids as part of a globally interconnected energy network. See Fig. 5.8 for an illustration of the Arctic's channels for outgoing power transmission.

Power transmission channels in the Arctic and north-east Asia. In terms of transmission distance, the Kara Sea wind power base in the Arctic is approximately 4400 km from North China, whereas the Bering wind power base is about 5000 km from North China, Japan, and South Korea. Both wind power bases are located within the economic distance of ±1100 kV UHV DC transmission channels. In the future, these bases may be considered for a role in delivering power to Northeast Asia (primarily China, Japan, and South Korea). As the potential channels for power transmission to China are all onshore, UHV DC technology with overhead lines may be adopted, while UHV DC submarine cables may be used to support transmission to Japan and South Korea. In particular, if electricity is delivered to the northern parts of Japan through Sakhalin Island, Russia, then the linear distance of cross-sea transmission is approximately 60 km, compared with approximately 210 km for transmission to Fukuoka, South Japan through Busan, South Korea. The Arctic wind power bases are expected to deliver around 1200 TWh/year of power to Northeast Asia by 2050, which demands a transmission channel capacity of 240 GW.

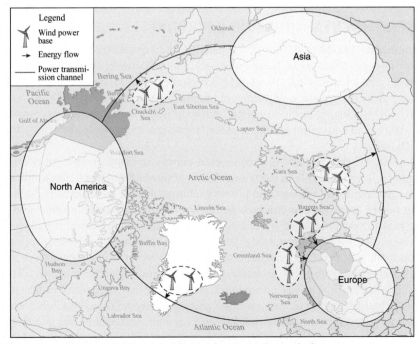

FIGURE 5.8 Illustration of the Arctic's Channels for Outgoing Power Transmission

Power transmission channel between the Arctic and Europe. Following the development of wind power resources in the land of northern Europe and the North Sea in the future, efforts may be stepped up to develop wind energy resources in Greenland, the Norwegian Sea and the Barents Sea for transmission of power to Europe. Wind power generated in Greenland will be delivered over a distance of around 2100 km to northern United Kingdom. The output generated by the wind power bases in southern Greenland can be transmitted via UHV DC submarine lines for up to 800 km to Iceland and, after crossing about 400 km of land into Iceland, will reach northern United Kingdom through UHV DC submarine cables that span approximately 900 km. Offshore electricity in the Norwegian Sea and the Barents Sea may be fed via terrestrial channels to European grids. The wind farms in Greenland, the Norwegian Sea, and the Barents Sea are expected to deliver around 800 TWh/year of power to Europe by 2050, which demands a transmission channel capacity of about 160 GW.

Power transmission channel between the Arctic and North America. At the same time as power from the Bering Strait wind power bases is delivered to north-east Asia, UHV transmission channels may be developed to carry electricity to the load centers on the western coast of North America via the Bering Strait. The Bering Strait wind power bases are approximately 4000 km away from the load centers in the American West in terms of transmission distance. The Bering Strait is about 90 km wide. Using UHV submarine cable technology when it is mature, wind power generated in southern Greenland can be delivered to Canada's eastern coast via UHV DC submarine cables for onward transmission via Ottawa to the load centers in the American East. The southern part of Greenland lies 2000 km or so from Quebec in Canada, covering an undersea distance of around 500 km. Quebec is situated around 1500 km from New York, a distance suitable for using terrestrial ±1100 UHV DC transmission lines. The wind power bases in Greenland are expected to transmit around 1000 TWh/year of power to North America by 2050, which demands a transmission channel capacity of around 200 GW.

The establishment of the above-mentioned transmission channels can not only solve the problems concerning outgoing delivery of Arctic wind power, but is also conducive to building looped interconnections between Asia, Europe, and North America in the northern hemisphere, supported by the key wind power bases in the Bering Strait, Greenland, the Norwegian Sea, and the Barents Sea in the Arctic region to plug the demand and supply gap in north-east Asia, Europe, and North America. The advantages of interconnection accorded by a large grid can also be fully leveraged to achieve significant efficiencies in peak staggering, sharing of reserve capacity and mutual backup at the transcontinental level for the power grids in north-east Asia, Europe, and North America across widely apart time zones. Also, by taking advantage of the time difference between continents, wind power generated in the Arctic can be delivered during appropriate time periods of different continents, to meet the peak load demands in the daytime, and increase the utilization efficiency of wind power from the Arctic.

2.2 OUTGOING SOLAR ENERGY TRANSMISSION CHANNELS IN EQUATORIAL REGIONS

Outgoing transmission channels in the equatorial region are primarily responsible for delivering solar energy from the equatorial energy bases in North Africa, East Africa, the Middle East, Australia, and South America. They are also the major channels connecting the northern hemisphere to the southern hemisphere. See Fig. 5.9 for the outgoing transmission of electricity from solar energy bases in the equatorial region.

Power transmission channel between North Africa and Europe. Plenty of research has been conducted on grid interconnections between North Africa and Europe, such as the Desertec Solar Energy

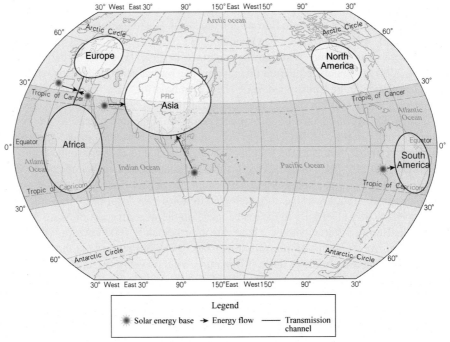

FIGURE 5.9 Illustration of Outgoing Power Transmission from Equatorial Solar Energy Bases

Project. The solar energy bases in some North African countries, for example, Morocco, Algeria, Tunisia, Libya, and Egypt, are just a few dozen kilometers away at the nearest from the power grids in southern Europe via the Strait of Gibraltar, and are no more than 1500 km away at the farthest. The favorable geographical conditions for transcontinental interconnections make power transmissions through grid interconnection possible, by simply using currently available technology. The North African solar energy bases are expected to deliver around 1500 TWh/year of power to Europe by 2050, which demands a transmission capacity of around 300 GW.

Power transmission channel between the Middle East and South Asia. Since the solar energy bases in the Middle Eastern countries, such as Saudi Arabia, Oman, and the United Arab Emirates, lie approximately 4000 km, in terms of transmission distance, away from western India in South Asia, UHV DC submarine cables may be installed via the Strait of Hormuz (at its narrowest point: 38.9 km; average water depth: 70 m) to Iran for onward transmission through land-based UHV DC lines via Pakistan to Bombay, where the load centers of western India are located. The solar energy bases in the Middle East are expected to deliver about 2500 TWh/year of power to South Asia by 2050, which demands a transmission channel capacity of around 500 GW.

Power transmission channel between Australia and south-east Asia. Currently, the infrastructure is relatively weak in Australia. To channel power to Southeast Asia will involve problems as transmission distances will be long, many sections will have to be built underwater, and the technical requirements for interconnection will be challenging. To deliver power from the solar energy bases in northern Australia, UHV submarine cables that are about 500 km long can be built to land on Indonesia, then stretching northwestward onshore to Jakarta before running a short sea distance to end up in Thailand via Singapore and Malaysia. The entire channel involves a distance of around 6000 km, so it is necessary to improve ±1100 kV UHV DC transmission technology and undersea transmission technology. The solar energy bases in northern Australia are expected to deliver around 1000 TWh/year of power to Southeast Asia by 2050, which demands a transmission channel capacity of about 200 GW. Oceania will also be integrated into the global energy interconnection.

The construction of the above-mentioned power transmission channels can not only resolve problems regarding outgoing transmission of power from the solar energy bases in the equatorial region, but also support grid interconnections between continents in the northern hemisphere and continents in the southern hemisphere. Due to little difference in time zones between these continents, solar irradiation intensity is simultaneously aligned with load levels, an example being the synchronization of the sun's overhead period in North Africa with the peak load period in Europe. As a result, solar energy can be utilized more effectively. Moreover, the seasonal difference between the southern hemisphere and the northern hemisphere creates the advantages of complementation in energy allocation. On this basis, a southern channel may be established to link up power grids in Europe and Asia through the energy bases in the Middle East.

2.3 KEY CHANNELS OF TRANSCONTINENTAL INTERCONNECTION AND PROGRESS

Transcontinental UHV grid backbones are not only responsible for global transmission of power generated in large clean energy bases on each continent, but can also increase the efficiency and effectiveness of global energy allocation by capitalizing on the mutual backup and resource sharing made possible by the time difference between the eastern and western hemispheres as well as the seasonal difference between the southern and northern hemispheres. The future of transcontinental grid interconnection is primarily determined by the demand for interconnection among different continents and also the conditions of project implementation. Interconnection projects reflecting the strong aspiration of stakeholders and good technological and economic conditions will be implemented first. Efforts are expected to be devoted to advance grid interconnections between Africa and Europe, between Asia and Europe, and between Asia and Africa around 2030, and grid interconnections between North America and South America, between Oceania and Asia, and between Asia and North America around 2040. Grid interconnection between Europe and North America is also expected to go forward around 2050.

2.3.1 Africa–Europe Grid Interconnection

Africa is not far from Europe with good interconnection conditions, given the prominent economic benefits of interconnection due to the advantages of complementation arising from different climatic conditions and load characteristics. It is technically feasible to build large wind and solar energy bases in northern Africa to transmit electricity to Europe, given a transmission distance of less than 2000 km. Africa has a different generation mix from Europe. In North Africa, power sources are predominantly solar energy, while in northern Europe, the generation mix is dominated by wind and hydroelectric power. Through interconnected grids, wind, solar, and other clean energy sources can be more efficiently utilized to optimize the energy compositions of North Africa and Europe. Europe is relatively short of energy resources, while North Africa is abundant in renewable energy resources with a far lower load level compared with Europe. Therefore, grid interconnection should aim at delivering solar energy from Africa to Europe, supported by the operation of grid interconnection between the two continents for mutual backup. The total capacity of grid interconnection will be around 300 GW. Because the solar energy bases in North Africa are located along the channel of Africa–Europe grid interconnections, we may consider developing an outgoing transmission channel for the solar energy bases in North Africa to allow these solar energy bases to fulfill their role. Currently, for the development of solar energy bases in North Africa and the delivery of clean power to Europe, a development agency – Desertec Industrial Initiative – has been set up, with detailed research and planning already initiated. The outgoing transmission of solar energy from Africa to Europe can not only satisfy the future power demand of Europe, but also contribute to Africa's economic development effectively. The transmission distance between North Africa and Europe is relatively short, covering a limited undersea distance with favorable conditions of project implementation. Grid interconnection is expected to be achieved around 2030, which will lead to improved mutual benefits for both continents.

2.3.2 Asia–Europe Grid Interconnection

Given the significant time difference between the two continents, the load characteristics of Asia and Europe are strongly complementary. In light of the resources and load distribution in Asia and Europe, priority should be given to the development of two UHV transmission channels on the north and the south as part of the grid interconnection between Asia and Europe in the future, as shown in Fig. 5.10. The northern transmission channel will be backed up by the renewable energy bases in Central Asian

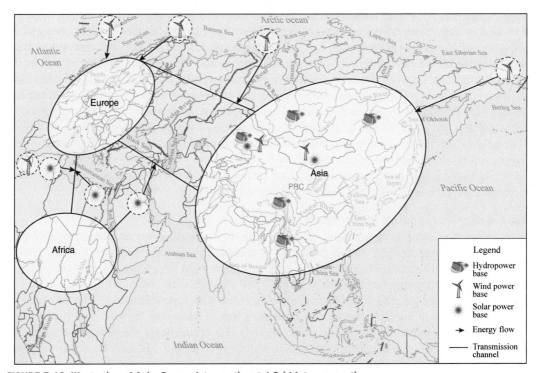

FIGURE 5.10 Illustration of Asia–Europe Intercontinental Grid Interconnections

countries, forming a UHV transmission channel to link up China, Central Asia, and Central Europe. The transmission distance from the Ekibastuz power base in Central Asia to the European load center of Berlin is around 4600 km; the transmission distance from Ekibastuz to the load centers in Eastern and Central China is around 3000 km. The renewable energy bases in Central Asia are strategically located to support the northern transmission channel of Asia–Europe grid interconnections. The State Grid Corporation of China has conducted research on Central Asia–Europe grid interconnections for years. The existing UHVDC transmission technology can meet the requirements for developing terrestrial transmission channels, without any technical problems and with better conditions for implementation. Interconnection is expected to be realized around 2030. **The southern transmission channel will be supported by the solar energy bases in the Middle East and connected to India and Southeast Asia to the east and extended to southern Europe to the northwest to create the second channel of Asia–Europe power grid interconnections, thus achieving the optimal allocation of solar energy resources from the Middle East to Europe, South Asia, and Southeast Asia.** Overseas studies like the Desertec Solar Energy Program and the Mediterranean Solar Plan have covered the subject of Europe–North Africa–Middle East grid interconnections. The southern transmission channel of Asia–Europe grid interconnections will be a terrestrial one expected to be developed along with the progress of the Mediterranean Solar Plan, for scheduled interconnection around 2030. The northern and southern channels of Europe–Asia grid interconnections, together with the Asia–North America–Europe interconnected channels to be constructed via the North Pole, will evolve into a pan-northern hemisphere UHV interconnected system with better resource allocation and stronger mutual backup at the transcontinental level to give fuller play to the role of wind power bases in the Arctic, solar energy bases in North Africa and the Middle East as well as the large renewable energy bases in Central Asia.

2.3.3 Asia–Africa Grid Interconnection

Africa is adjacent to the Middle East geographically, offering marked advantages in terms of grid interconnection to contribute to the optimal allocation of solar energy from North Africa and the Middle East to Europe, Asia, and Africa. The solar and wind power bases in North Africa and East Africa will connect with the southern channel of Europe–Asia grid interconnections through the Middle East to achieve Africa–Asia grid interconnections. The complementary features of the typical daily load curve resulting from the time difference between Europe and Asia will promote the consumption in Europe and Asia of the abundant wind, solar and other renewable energy resources from North Africa, based on a total interconnection capacity of around 500 GW. Asia–Africa grid interconnections will receive an impetus to move gradually forward around 2030, along with the progress of work on the southern channel of Asia–Europe grid interconnections.

2.3.4 North America–South America Grid Interconnection

Due to climate and seasonal differences, the load characteristics of South America and North America are relatively more complementary. The two continents are divided by the Panama Canal, with major climatic, time, and seasonal differences, compounded further by differences in the power generation mix. In South America, power generation is predominantly based on water, wind, and solar energy, while in North America, electricity is generated mainly by wind power, solar energy and water resources and gas. Grid interconnection between the two continents will improve peak-load regulation for hydropower by taking advantage of the characteristics of different power sources for joint operation of power generation by water, wind, solar, and other renewable energy. The load centers in North

America and South America will involve long-distance, mainly land-based interconnection, with the total capacity estimated at 300 GW. The construction of a channel of grid interconnections between South America and North America can be considered in connection with the development of outgoing transmission channels for the solar energy bases in Mexico, Peru, and Chile.

Currently, grid interconnections are existent in North America between Canada and the United States, between the United States and Mexico, and between Mexico and Guatemala, with regional grid interconnections established among Guatemala, Honduras, El Salvador, Nicaragua, Costa Rica, and Panama. Moreover, interconnected lines are already in operation in countries in the northern parts of South America such as Venezuela, Columbia, Ecuador, Peru, and Brazil. Generally speaking, grid interconnection between North America and South America can be initially realized by the completion of grid interconnection between Panama and Columbia. Before 2040, grid interconnection between North America and South America should aim primarily at building connections with adjacent countries for the benefits of peak staggering, mutual support in resources, and sharing of reserve capacity. Taking an integrated view of all factors involved, large-scale intercontinental grid interconnection between North America and South America is expected to be achieved around 2040.

2.3.5 Oceania–Asia Grid Interconnection

In Oceania, the northern parts and central desert areas of Australia abound with solar energy resources, and the northwestern parts are rich in offshore wind energy resources. Conditions in these areas are favorable for transmitting electric power to Southeast Asia. In view of this, some organizations in Australia have conducted research and proposed the concept of grid interconnection between Australia and Asia. With its large population and relatively low electricity consumption per capita, Southeast Asia's electricity demand has room for significant growth. Given their limited energy resources, the island states in Southeast Asia and the countries in the Greater Mekong Subregion have to rely on imported electricity for low-carbon, clean development of energy. As Oceania is relatively far away from Asia, grid interconnection will face greater difficulties caused by the need to traverse the many island chains along the way and the relatively long distance of undersea transmission. Grid interconnection between the two continents is expected to come into being around 2040.

2.3.6 Asia–North America Grid Interconnection

The advantage of time difference between Asia and North America can be leveraged with an interconnected channel from northeastern China and Siberia through the Bering Strait to Alaska in North America before linking up with the load centers in the Pacific West coast of Canada and the United States. North-eastern China is no more than 3500 km from Siberia and a transmission line through the Bering Strait to Alaska will run no more than 2000 km before stretching a further 4500 km to reach the west coast of the United States. As part of this channel, only an undersea cable measuring approximately 90 km is required to be laid across the Bering Strait. Major load centers are located in the eastern parts of Asia and the east and west coasts of North America. Due to a time difference of around 9 h between East Asia and the west coast of the United States and a time difference of around 12 h between East Asia and the east coast of the United States, the typical daily load peaks and valleys of the two continents are strongly complementary. Given the size of some of the loads in Asia and the need for mutual backup to balance peaks and valleys, the capacity of this interconnection project is expected to be more than 500 GW. Since the wind power bases in the Bering Strait are along the channel of interconnections between Asia and North America, the development of outgoing transmission channels for these wind power bases may be considered in connection with plans for building interconnections between the two continents in order to give fuller play to these wind power bases. Given the challenges of the project involving the construction of an interconnection route of close to 9000 km between Asia and North America with the need to traverse extremely cold regions and the Bering Strait, transcontinental interconnection is not expected to be realized until around 2040.

2.3.7 Europe–North America Grid Interconnection

There are significant benefits of peak staggering between the power grids in Europe and North America. In the future, Europe–North America grid interconnections may become achievable with the back-up of the wind power bases in Greenland. In 2050, wind power will be developed on a large scale in Greenland and delivered to Europe and North America. At the same time, an integrated view should be taken to consider the effects of time difference, wind power output curves, the load characteristics of power grids in Europe and North America, and the complementary nature of installed power capacity profiles, with the objective of rationally developing Greenland's wind power bases, consuming power generation therefrom, and jointly operating the power grids in Europe and North America. Because the Greenland wind power bases are located in the Arctic Circle with a long undersea distance, Europe–North America grid interconnections are not expected to be commissioned until around 2050.

3 TRANSNATIONAL INTERCONNECTION IN EACH CONTINENT

Judging by the global distribution of clean energy resources, all continents, including the large renewable energy basis in the Arctic and equatorial regions, are well positioned for building large renewable energy bases, which will be geographically distributed very unevenly as the load centers on each continent. Therefore, building transnationally interconnected power grids on each continent is essential for expediting the development and utilization of renewable energy, and for providing robust support for a continent to import or export electricity. For energy-exporting continents, the focus is on building outgoing power transmission channels as opposed to building robust grids at the receiving end for energy-importing continents to substantially improve the capacity for importing electricity from elsewhere.

For transnational transmission of power through transnational grids and from large renewable energy bases on a continent, UHV AC/DC transmission technology will be employed as the primary technology. For existing power grids, voltage upgrading will be required and grid structures strengthened to cater to the future large-scale development and utilization of renewable energy resources on a continent (for self-use, export, or import) and the import and retransmission of electricity from the Arctic and equatorial regions. The transnational grids on each continent will be constructed in order of increasing difficulty and distance. Priority will be given to those locations with good basic conditions or urgent need for interconnection.

3.1 ASIA INTERCONNECTION

Asia is the world's largest load center with abundant renewable energy resources. In the future, intracontinental interconnected grids will be built with the continent's large renewable energy bases at the sending end, and connected to various major load centers to receive electricity flows from the Arctic and equatorial regions across different continents and countries. To realize low-carbon, sustainable development of energy in Asia, some countries have put forward the concept of "Asian super grid" in which the coal-fired power plants, hydropower stations, wind farms, and solar farms in Northeast Asia will be connected with the load centers in China, Japan, and South Korea. To meet ever-higher targets of low-carbon energy development, Asia must cut down on fossil energy consumption and step up renewable energy development in the future. At the sending end of power through interconnected grids, Asia's major renewable energy bases have undergone more rapid expansion. They include the wind and solar energy bases in Mongolia, the hydropower stations in the Russian far east and Siberia, the wind and solar energy bases in Central Asia, the wind power bases in Northwest China, Northeast China, and North China, the solar energy bases in Northwest China, the wind power bases in the Bering Strait and Sakhalin, and the solar and wind power bases in India. Judging by the distribution of load centers in Asia, Northeast Asia, Southeast Asia, and India will become home to Asia's major load centers, given these regions' large populations and well-developed industries. Judging by the development and interconnection of renewable energy bases, large hydropower bases will be built in Southwest China, Russian Siberia, and the Russian Far East to transmit power to the load centers in China; large wind power bases will be built in Northeast China, Northwest China, North China, Mongolia, and the Russian far east to deliver electricity to the load centers in Northeast Asia. large solar energy bases will be built in Northwest China, Tibet and Mongolia to deliver power to the load centers in Central and Eastern China and Northeastern Asia. Hydro, wind, and solar energy generated in Central Asia (Kazakhstan, Kyrgyzstan, and Tajikistan) will be "bundled," scheduled and transmitted flexibly according to load changes in and time differences between Asia and Europe.

Judging by the consumption of electricity delivered from the Arctic and equatorial regions, the wind power from the Arctic in North Asia will be transmitted primarily to Northeast Asia (East China, Siberia, the Russian far east, Japan, and South Korea) to be utilized and consumed together with the wind, solar, and hydropower generated in Northeast Asia. Solar energy generated in the equatorial regions of the Middle East will be delivered to India for consumption. The solar power generated in the equatorial regions of Australia will be exported to Southeast Asia to be utilized and consumed together with locally-generated hydropower.

On the whole, Asia's grid interconnections will give birth to a number of major regional power grids, including Central Asia, Northeast Asia, Southeast Asia, South Asia, and the Middle East, and further to a network of interconnections linking up the different regions, as illustrated in Fig. 5.11.

3.2 EUROPE INTERCONNECTION

Europe is one of the most important load centers in the world. The continent's interconnected power grids are designed principally to allow access for wind power from the Arctic and the North Sea, and the solar energy from Southern Europe and North Africa. Another purpose is to ensure joint operation of hydropower and other power sources generated in Europe, and continent-wide consumption of these power sources. In 2008, Europe initiated the "European super grid" concept in which a large grid is envisioned for the pan-European region to facilitate continent-wide utilization and consumption of wind power from Northern Europe, solar energy from Southern Europe, and hydropower and other forms of energy generated across the continent. In Europe with a dense population and scarce natural gas resources, calls for denuclearization have been voiced strongly in some countries. To achieve low-carbon, sustainable development of energy, Europe is likely to further reduce the use of fossil energy

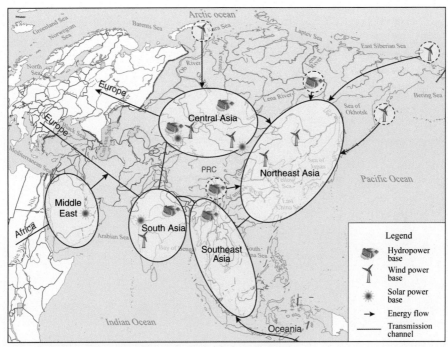

FIGURE 5.11 Illustration of Asia's Transnational Grid Interconnections

and nuclear power while importing more clean electricity, such as wind power from the Arctic and solar energy from the equatorial region by ramping up wind power development in the North Sea and the share of renewables in energy utilization across Europe. Under this scenario, UHV technology will be needed in the future to build a pan-European grid backbone and a robust smart grid to ensure efficient access for and consumption of renewable energy. Power loads in different regions of Europe are comparatively more evenly distributed, with Germany, France, Britain, Italy, and Spain having higher loads. By integrating development and transmission of renewable energy, Europe will see the formation of "three vertical and three horizontal" backbone channels of interconnections. The horizontal channels will include a northern UHV channel for importing wind power from the North Sea, hydropower from Northern Europe and wind power from the Arctic; a central UHV channel connected to the load centers in southern Britain, northern France, Germany, and Poland with capacity for importing renewable power from Central Asia; and a southern UHV channel connected to the solar energy bases in Spain, Italy, and Greece. Among the three vertical channels will be a western UHV channel connected to the wind power bases in Greenland, the Norwegian Sea, and the Barents Sea, the offshore wind power bases in the United Kingdom, the load centers in France and the solar energy bases in Spain and North Africa; a central UHV channel connected to the hydropower bases in Norway, the load centers in Germany, and the solar energy bases in Italy and North Africa; and an eastern channel connected to the wind power bases in the Kara Sea, the load centers in Finland and Poland, and the solar energy bases in Greece and North Africa. In addition, the Europe will be interconnected with Central Asia and North Africa for exchanging electric power at the intercontinental level. Europe's transnational interconnections are illustrated in Fig. 5.12.

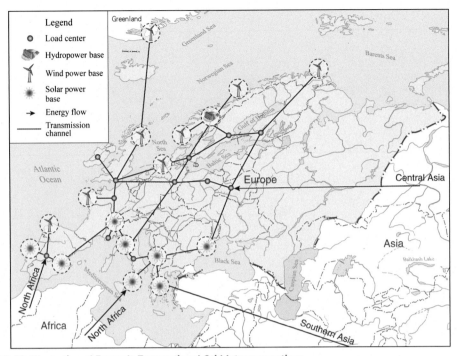

FIGURE 5.12 Illustration of Europe's Transnational Grid Interconnections

3.3 NORTH AMERICA INTERCONNECTION

North America's grid interconnections connect the wind power bases in the central and western parts of the continent, the solar energy bases in the southwestern regions and the hydropower bases in Canada to the load centers in the east and west, importing Arctic wind power from Greenland in the east and with interconnections with Asia's power grids through Alaska in the west, to achieve large-area allocation and efficient consumption of renewable energy at the intracontinental and transcontinental levels. North America is one of the most important load centers in the world, abundant with solar, wind and hydropower resources. In order to promote the development and consumption of renewable energy in North America, the United States Department of Energy has proposed "Grid 2030 Plan" to upgrade its electric power system by building backbone grids in the US and a regional network interconnected with Canada and Mexico. In response to the challenge of climate change and as the two-replacement gains momentum, efforts will be stepped up in North America to develop wind and solar energy bases for joint operation with large river-based hydropower stations in Canada and America to transmit power to the load centers in the eastern and western parts of the continent. As renewable energy bases and load centers are unevenly distributed, electricity flows within the continent will rise significantly, necessitating the development of an interconnected system across the continent with UHV grids as its backbone. The hydropower stations in the Columbia River Basin and the Great Lakes region will deliver power to the load centers in western and eastern parts of North America, forming two vertical backbone channels in the east and west. The wind power bases in midwest America, the solar energy bases in the Southwest and the hydropower stations in the Mississippi River Basin in the south will deliver electricity to the load centers in the eastern and western parts of North America, forming the base for a horizontal channel of

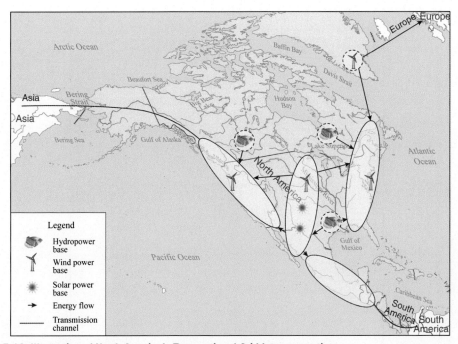

FIGURE 5.13 Illustration of North America's Transnational Grid Interconnections

interconnections across the continent. In addition, the solar energy bases in the American southwest and northwest Mexico will transmit electricity to the load centers in Mexico, creating a grid-covered region in the south as part of the wider grid system of North America and a channel with interconnections with South America. See Fig. 5.13 for North America's transnational grid interconnections.

3.4 SOUTH AMERICA INTERCONNECTION

South America is abundant in energy resources. Grid interconnections in the continent are designed mainly to realize mutual support for power supply between north and south in the western coast of the continent, transmission of electricity from north to south in the eastern regions, and transmission of power from west to east in the central regions. Situated to the west of the Andes with abundant solar and wind energy resources, Chile and Peru have more potential for energy export due to their smaller populations and lower load demand compared with larger countries to the east, such as Brazil and Argentina. Situated in eastern South America, the Amazon, the Tocantins, and the Parana river basins are abundant in hydroelectric resources, enabling them to transmit power to the load centers on the eastern coasts of Brazil and Argentina. The solar and wind power bases in western South America can supply power to load centers in the east through UHV grids. Grid interconnections between east and west will generate better benefits through the combined operation of wind, solar and hydropower bases. The Brazil Belo Monte UHV DC transmission project with a route length of 2092 km transmits hydropower from the Xingu River in the north to Estreito, a load center in the southeast. It is the first ±800 kV UHV DC transmission project on the continent of America.

The future backbone of South America's grid interconnections is designed to realize interconnection among load centers in countries in the north like Ecuador, Columbia, and Venezuela, among load

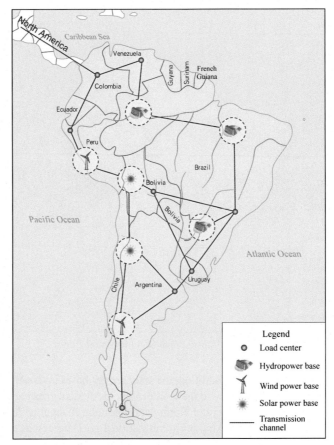

FIGURE 5.14 Illustration of South America's Transnational Grid Interconnections

centers in countries in the east like Brazil and Argentina, among hydropower bases in Brazil, among solar and wind power bases in Peru and Chile, and among various other load centers. See Fig. 5.14 for South America's transnational grid interconnections.

3.5 AFRICA INTERCONNECTION

Grid interconnections in Africa will help achieve the operation of solar and wind power bases in North Africa jointly with the hydropower bases in Central Africa and the solar energy bases in Southern Africa to meet rising power demand in Africa and provide a robust grid at the sending end for solar energy exports from North Africa, shaping a new energy scenario marked by the delivery of electricity from north to south, mutual support for electricity flows between east and west, transmission of power northward to Europe, and interconnections with Asia to the east. To promote hydropower development of the Congo river, the government of South Africa approved in 2012 the draft provisions for developing the Grand Inga Project in partnership with the Democratic Republic of the Congo. Under the project, plans are afoot to employ large-capacity, long-distance power transmission technology to deliver electricity to South Africa, Egypt, Nigeria, and other nations after meeting Congo's power demand.

This project to deliver power to North Africa can reach as far as Southern Europe and the Middle East. Hydropower development of the Congo River will create favorable conditions for the construction of an interconnected power grid in Africa. In the future, solar energy bases in Egypt, Algeria, and other North African countries will deliver electricity to East Africa and West Africa. The transmission channels will then extend southward to interconnect with Southern Africa, forming a horizontal backbone stretching from east to west as part of Africa's wider interconnected system. Hydropower generated in the Nile River Basin will be supplied to Egypt to the north and Tanzania to the south. While supplying power to Central Africa, the hydropower bases in the Congo River Basin and the Zambezi River Basin will jointly operate with the wind and solar energy bases in the north and south to promote continent-wide integration of renewable energy and to supply power to West Africa and Southern Africa.

In the future, North Africa, Central/East Africa, West Africa, and South Africa will evolve into four major power grid regions in Africa, to further develop into a larger, continent-wide grid system through interconnections. North Africa's regional power grid, an important electricity exporter of Africa, covers primarily local solar and wind power bases, and load centers to the north. Central/East Africa's regional power grid, also an important electricity exporter of Africa, comprises hydropower, solar and wind power bases. West Africa's and South Africa's regional power grids carry large domestic loads, making them major power importers in the future. See Fig. 5.15 for Africa's transnational grid interconnections. The backbone structure of the global energy interconnection is illustrated in Fig. 5.16.

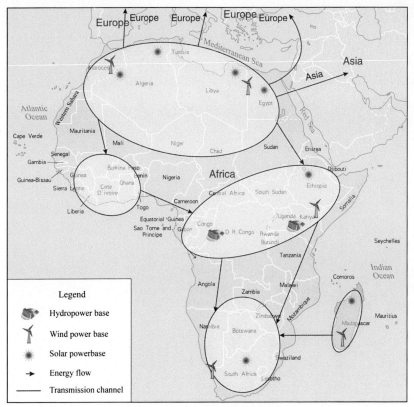

FIGURE 5.15 Illustration of Africa's Transnational Grid Interconnection

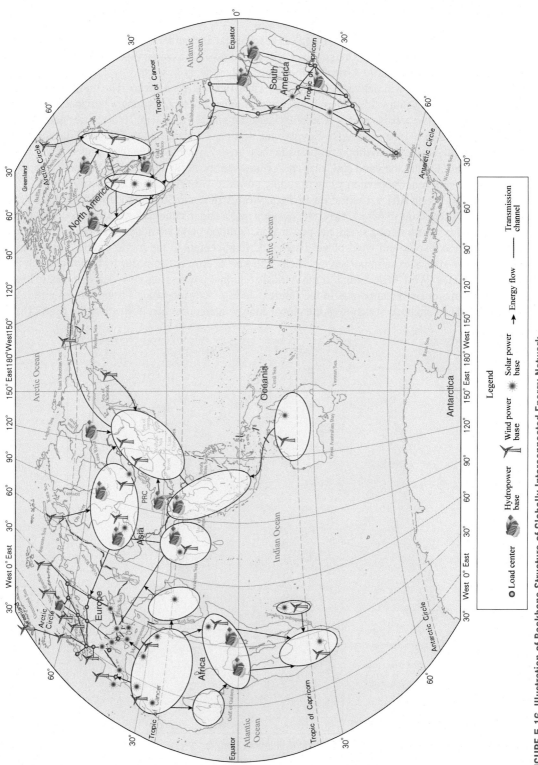

FIGURE 5.16 Illustration of Backbone Structure of Globally Interconnected Energy Network

4 COUNTRY-BASED UBIQUITOUS SMART GRID

Transnational and transcontinental grids form the basic structure of a globally interconnected energy network with capacity for optimizing the allocation of clean energy across the globe. For safe and reliable delivery of electricity to load centers, the construction of ubiquitous smart grids in countries around the world must be accelerated. *Country-based ubiquitous smart grid represents an integral part of a global energy interconnection, connecting extensively to domestic energy bases, distributed generation and load centers, and also interconnecting with energy networks in neighboring countries for allocation of clean energy through the global energy interconnection across countries and continents.* The development of a country-based ubiquitous smart grid should focus on the premises of robustness and intelligence to effectively resolve problems with randomness and intermittence regarding clean energy development by capitalizing on the strength of a large grid and a robust grid structure. The aim is to achieve optimal access for and efficient accommodation of centralized and ubiquitous distributed power sources, thereby safeguarding supply security.

4.1 GRID STRUCTURE

The construction of country-based smart grid structures must be effectively integrated into the backbone structure of a globally interconnected energy network to ensure safe and reliable operation of domestic grids and meet the domestic need for clean energy development and utilization. This should underline coordinated efforts to ensure optimal allocation of energy both at home and abroad by building capacity to import clean energy and export surplus power.

4.1.1 Development Direction

The equal emphasis on robustness and intelligence is the inherent requirement and direction for the development of a country-based ubiquitous smart grid in the future. Surging power demand around the globe and the typical reverse distribution of large clean energy bases and load centers entail the development of robust grid structures in every country, grid capacity and voltage grades befitting the level of socioeconomic growth, and capacity for high-volume, safe, and reliable transmission of electricity. Also required is the ability to realize development of large energy bases and optimal allocation of distribution of energy resources, meet the need for green, low-carbon development of energy, and ensure a supply of safe, reliable, clean, and quality electricity for different user segments. At the same time, the construction of grid structures in individual countries should organically converge with the transnational and transcontinental backbones of a globally interconnected energy network so that UHV/EHV transmission technology can be used effectively and the coverage of grids at different levels can be expanded to achieve large-capacity, long-distance and high-efficiency transmission of electricity.

4.1.2 Development Focus

Country-based ubiquitous smart grids in the future will play a key role in supporting allocation of energy resources across countries and continents through the globally interconnected energy network. They are also an integral part of this global energy interconnection.

First, the focus is on building a robust structure to enhance grid transmission capacity. The structure and layout of grids should be logically planned in line with the development requirements and load distribution of large energy bases and in coordination with the allocation of electricity across nations, with large-scale, long-distance transmission capacity. The grid structure should be able to effectively stand up to natural disasters and untoward incidents, with strong interference resilience and self-recovery capabilities.

Second, equal emphasis should be placed on AC and DC underlined by coordinated development. AC transmission and DC transmission are different in functions and characteristics, as well as the role each plays in grid construction. AC performs the dual function of electricity transmission and grid construction, providing flexibility in grid access, transmission and consumption. The strength of grid structures and transmission capacity are directly proportional to AC voltage grades. DC is used mainly for long-distance, large-capacity transmission, without being able to create a network at this point. To cater to large-scale clean energy generation and large-capacity outward DC transmission, a robust AC grid is required to create a grid system underlined by robust AC/DC operations with mutual backup, support, and fulfillment.

Third, there should be coordinated development of grids at all levels. Coordinated efforts are required to push forward the construction of grids at all levels with logically planned tiers and regions to achieve the organic connection of different voltage grades to form a grid structure that is structurally well-defined, functionally delineated, and logically configured. Given the need to meet different electricity demands in different stages of development in line with the economic growth and energy endowments of different countries, targeted grid planning should be initiated to reasonably expand the coverage of grids at different levels for coordinated development with the economy, society, and the environment.

4.1.3 Functional Requirements

Ensuring safety and reliability. A robust grid structure is the basis for assuring power supply safety and reliability. Building a synchronized UHV/EHV grid can fully leverage the strengths of a large grid, effectively withstand the impact of natural disasters and serious incidents, and substantially enhance grid safety and stability as well as operational stability. The State Grid Corporation of China is accelerating the development of a UHV AC/DC power grid and a transregional power grid to initially form a robust grid architecture embodying a sound structure with coordinated development of AC/DC operations to significantly improve grid safety and stability.

Optimizing allocation of resources. By strengthening the development of a mainframe grid domestically, the problem of uneven distribution of resources can be overcome, the demand for long-distance, large-scale allocation of resources met, and the development and delivery of centralized and distributed power sources ensured. The marked inverse distribution of energy resources and load centers in China has given rise to the urgent need for allocating energy resources in large capacity and over long distances. In the recent years, UHV transmission channels across different regions have been developed to transmit electricity from the coal bases in the east and the hydropower bases in the south-west to the load centers in East and Central China. This has fully leveraged the capability of a large grid to optimize the allocation of resources on a large scale and over a long distance, contributed to the development of energy-rich areas, enabled the intensive development of and reliable electricity delivery by large energy bases, and supported the energy needs arising from the rapid economic growth of east and Central China.

Supporting clean development. The building of domestic transregional power transmission channels is conducive to the effective development and accommodation of clean energy at home. Intensive efforts are underway in many countries to strengthen the construction of power grids to promote clean energy development. The United States is planning to achieve, by 2024, the target of having 30% of its electricity generation from wind power from the American East's interconnected grids, by stepping up investment in interstate capacity building so as to transmit wind power from the central United States and eastern and western coasts. The German Energy Agency is looking to build 850 km and upgrade 400 km of 220/380 kV HV transmission lines to fully accommodate grid connections off the North Sea for transmission of wind power. As with UHV transmission capacity now under development in China, the development and utilization of wind power in the western and northern parts of the country will increase to more than 300 TW from 88 GW.

Achieving mutual backup and access. A robust grid system is conducive to the mutual backup and access for electric power across different regions in China. Hydropower bases across different river basins can complement each other whereas water resource-rich regions and areas relying mainly on thermal power can support each other. These power bases may also form a positive relationship of mutual backup with wind and solar energy bases. State Grid Corporation of China is working on plans to build a synchronized UHV grid and 19 Circuits of UHV DC projects in North China, Central China, and East China by 2020 to facilitate the delivery of electricity from west to east and from north to south, with a trans-regional transmission capacity of 380 TW to realize mutual backup and access for and optimal allocation of energy resources over large areas.

4.2 SMART DEVELOPMENT

High intelligence is an important quality of a globally interconnected energy network and enhancing the level of grid intelligence is an integral part of building such a network. In the future, ubiquitous smart grids built on a robust foundation will become the core network and allocation platform for modern energy, markedly different from conventional grids in terms of development direction, development focus and functional roles.

4.2.1 Development Direction

With the upgrade of information and communication technology, the development of smart control technology, the maturity of grid operation technology, and the application of Internet technology, smart development is covering more areas and exhibiting a move in the following direction:

First, the level of intelligence in the operational control and scheduling of power grids has been increasing. The increasing application in depth of IT and automation technology for operational control and scheduling of grids, together with higher levels of modeling and simulation for large grids, is driving a technology leapfrog from static to dynamic grid observations, from offline to online grid analysis, and from local to overall control. The growing integration of advanced information and communication technology, power electronics technology, optimization, control theories, technologies, as well as new power market theories and technologies, will form the basis for the safe and economical operation of country-based ubiquitous smart grids. This will eventually lead to the formation of a flexible and efficient system for energy supply and allocation, as part of a safe and reliable network of smart grids.

Second, the interactivity among smart grids will be sustainable and deep. The hardware platform for interactive, smart use of electricity has been greatly enhanced, driven by the growing technological development in the Internet and the Internet of Things, and the continued deployment and upgrade of power fiber to the home, smart meters, and other facilities. This will ensure a secure communication base for the diversified, smart, and interactive uses of electricity. Modern information processing technologies like big data analysis and cloud computing have enabled the interactivity among smart uses of electricity to fully unleash the value of mass data, hence to promote the integration and customization of interactive business for the benefit of socio-economic development. Taking distributed generation as an example, users can enjoy greater autonomy and choice in power generation and utilization to gradually become major participants in the operation and interoperation of smart grids.

Third, the smart grid has gradually evolved from being purely a vehicle for power transmission to an integrated infrastructure for carrying smart energy information. The networking advantages of the smart grid and the information and communication resources gathered through power communication networks can be shared in many areas of social life, production, and can facilitate the development of energy and communication facilities into an integrated resource network with pooling and multiplexing capabilities. Grid-related information and data resources can create new value through flexible value-added services and business models. Various types of smart terminals and new powered equipment will be integrated *en masse* into smart grids to form a network covering two-way flows of electricity and information, marking the expansion of the smart grid from being a power grid in essence into a smart power system incorporating energy conversion and equipment utilization.

Fourth, the ubiquity of smart grids will become increasingly prominent. Human society's growing demand for a sufficient supply of reliable, clean and convenient energy necessitates the continued development of smart power grids into ubiquitous networks. Not contented with power supply flexibility, users also look for diversified service offerings that transcend all temporal and spatial confines. Such demand is driving the development of smart grids. Being customer-centered, smart grids are gradually evolving into an infrastructure facility, ubiquitous grid by constantly integrating new networks, and introducing new services, businesses and applications, to not just serve the public, but also provide industry-specific basic applications that contribute to the formation of a value network for integration and optimization of social resources.

4.2.2 Development Focus

Smart development is focused on improving the overall intelligence for power generation, transmission, transformation, distribution, consumption, dispatching, communication, and information.

4.2.2.1 Power Generation

The focus of power generation is on optimizing the mix of power sources, coordinating resource networks, enhancing the capability to accommodate access for various types of clean energy, and facilitating the development and consumption of clean energy in the following three major areas:

1. *Coordination of power transmission and generation*: To perform actual measurement of parameters covering all generators, excitation systems, speed governing systems, and power system stabilizers to increase the precision of smart modeling and simulation; enhance the technical know-how of quick regulation and the capability of peak load regulation of generating units to improve the speed and range of regulation as well as control precision.
2. *Integration and operational control of clean power generation*: To build operation, dispatch and control systems for wind farms and solar power plants to solve problems such as output fluctuations, reactive voltage support, and power quality in the integration and control of intermittent power

sources; develop and use wind and solar energy forecast systems to increase LV and HV ride-through capability and enhance the coordination of clean power plants and grid operation.

3. *Large-scale energy storage*: To perform joint control with the support of energy storage to realize mutual support and backup among different power sources and enable the globally inter-connected energy network to play the role of peak shaving and valley filling; conduct R&D on large-capacity energy storage equipment and large-capacity integrated modular battery systems; set up a large-capacity integrated energy storage management system; execute smooth regulation of intermittent energy power for integrated energy storage; and level out the fluctuations in large-scale clean energy production.

4.2.2.2 Power Transmission

The focus of power transmission is on applying advanced transmission technologies, continuously improve transmission capability and efficiency, ensuring that transmission lines are controllable in all-aspects, and increasing the operational stability of power systems. Covered here are the three major areas:

1. *Application of advanced transmission technologies*: Speed up grid construction by using UHV AC/DC technologies; perform flexible, quick control of AC transmission system parameters, and network structure by using flexible AC transmission technologies; achieve major breakthroughs in ±1100 kV UHV DC transmission technologies, and apply these technologies to engineering projects; promote flexible DC transmission technologies and achieve breakthroughs in multiterminal DC transmission technologies.

2. *Monitoring of transmission lines*: To carry out condition-based monitoring of key transmission equipment, fully promote smart inspection of transmission lines, and extensively conduct condition-based assessment, condition-based maintenance and risk prewarning; execute online preassessment and decision-making, and improve the intelligentization of condition-based assessment of transmission lines.

3. *Management and design of transmission lines*: Perform whole life-cycle management of transmission lines, launch new technological applications like inspections based on helicopters, unmanned aerial vehicles, and smart robots as well as develop interactive and visual inspection devices; promote the integrated application of new technologies, materials and processes to carry out digital surveys, and modular designs.

4.2.2.3 Power Transformation

The focus of power transformation is on improving the intelligentization of substations to support optimal dispatching, operation management of grids, and to improve the asset management and operation of substations. Covered here are the three major areas:

1. *Smart equipment*: To improve the intelligentization of primary equipment for substations; carry out R&D on the application of smart transformers, breakers, and other devices for integrating primary equipment, sensors, and smart components; promote the large-scale application of active electronic-type and passive optical fiber-type current transformers and potential transformers.

2. *Monitoring of substation equipment*: To provide additional real-time monitoring and data acquisition equipment for integrated online monitoring and self-diagnosis of equipment; extract typical characteristic parameters of failure modes of equipment and carry out smart analysis

to generate data on equipment operation status, reliability levels, risks of fault, and service life curves.

3. *Intelligent substations*: To digitalize all the processes for collecting, communicating, processing, and generating substation information; adopt a communication system based on an IEC 61850 standard network; exchange, and share information in and beyond substations based on unified standards and modeling.

4.2.2.4 Power Distribution

The focus of power distribution is on improving power supply reliability, system operational efficiency, and terminal power quality at the distribution level to realize the integration of and coordinate the optimal operation of distributed generation, energy storage, and microgrids, with the objective of achieving efficient and interactive demand side management. Covered here are the three major areas:

1. *Control of distribution networks*: To establish a smart system to promote the integrated control of automation systems and networks for power distribution, in order to expand coverage, execute flexible regulation, and optimal operation of distribution networks; perform smart recovery, adaptive protection to support self-recovery, and network reconfiguration after a network failure.
2. *Coordinated control of distributed generation and microgrids*: To master complementary control technology and coordinated control of source loads of high-penetration and multi-source distributed generation to exercise effective management of random power fluctuations and power quality; meet the demand for controlling distributed generation and micro grids in a complicated work environment marked by grid interconnection, steady-state islanding, and fault islanding operations; leverage the role of distributed generation in improving supply reliability and peak shaving/valley filling.
3. *Operation, maintenance, and management of distribution networks*: To expand new application systems attuned to the future trends of distribution networks and build a system with command, operation, maintenance, and management capabilities for power distribution in order to eliminate information islands and perform effectively integration and interaction of data and functions among different application systems in a distribution network.

4.2.2.5 Power Consumption

The focus of power consumption is on building and improving smart, two-way and interactive service platforms, and related technical support systems for carrying out integration of and interaction with energy flows, information flows, and business flows at the end-use level. Covered here are the three major areas:

1. *Collection and analysis of power consumption information*: To fully promote smart meters, establish a system for collecting information on power usage, and develop and deploy key technologies and equipment related to terminals, communications channels, master stations, and security protection; employ big data or other smart analytical methods to analyze user behavior in support of decision-making.
2. *Diverse, interactive services*: To conduct R&D on key technologies and equipment such as smart home appliances, smart interactive terminals, and smart energy service systems; develop smart communities, buildings, and parks capable of providing consultation and strategic analysis

on energy consumption to promote the use of power in an orderly manner and improve the intelligence level of energy efficiency services to contribute to energy conservation.

3. *New powered equipment*: To establish two-way interaction between user-end distributed generation and energy storage systems on one side and power grids on the other; support flexible access and exit for different smart terminals; construct smart charging and swapping networks for electric vehicles to enable electric vehicle batteries to participate in peak load shifting; promote the application of a variety of new energy-efficient and power-saving equipment.

4.2.2.6 Dispatching

The focus of dispatching is on smart development to build an information-based, automatic and interactive grid dispatching system to maximize the capability for optimal allocation of resources for, and the safe and economical operation of, grid dispatching. Covered here are the three major areas:

1. *Smart dispatching*: To build a technical support system for an integrated smart dispatching system covering monitoring and prewarning, safety audits, and dispatch planning and management; perform full-view monitoring of the dispatching and production processes and realize full integration, data sharing, and multiangle visual presentation of grid operation and analytical results.
2. *Analysis of grid operation*: To establish a platform for analyzing grid operation to meet the need for simulation analysis of demand covering all conditions and times, from online to offline operations and from electromechanical to electromagnetic transients; build a disaster management system for promoting the use of weather information for grid hazard prewarning, disaster prevention, and mitigation as well as dispatch new energy to make grids smarter in coping with natural disasters.
3. *Control of mega grids*: Master the steady-state/transient operations of mega AC/DC grids and the methods for grid fault protection and network reconfiguration; develop simulation technology for mega AC/DC grids well-attuned to the operating environment of a globally interconnected energy network; achieve coordinated control on a multilayered, multizoned basis, and across different countries and continents covering extensive regions to provide technology assurance for the safe and stable operation of the global energy interconnection.

4.2.2.7 Communication and Information

The focus of communication and information is on setting up an information system to support work processes and businesses, together with the development of a platform for building business synergy and interoperability, to facilitate transparent information sharing, and improve modern management. Covered here are the three major areas:

1. *Communication networks*: To construct a power line communication network with high self-recovery capability and adaptability, supported by a large-capacity, high-efficiency transmission backbone that employs a dedicated smart, real-time optical transmission system; adopt an integrated approach to using passive optical networks, power line carriers, and wireless and public communication networks for power distribution and utilization.
2. *Information systems*: To build business systems and information platforms covering multiple segments and levels of smart grids, including a platform for a comprehensive view of grid

integration, an integrated analysis system to support decision-making, and a geographic information system platform; develop a platform for big data management on a historical basis and in quasi real-time to provide normative access to and centralized sharing of application data on different operations.

3. *Application of new technology*: To build an overall framework for the Internet of Things and unified information models and standards; conduct R&D on special sensors and networking technologies; promote the gradual application of quantum communication technology in the electricity industry; apply big data technology to explore the potential value of the huge body of diversified information carried by our future global energy interconnection; apply cloud computing and cloud storage technology to support analysis; and decision-making regarding the operation and management of the global energy interconnection and electricity trading around the world.

4.2.3 Functional Requirements

Through the smart development of grids, the safety, flexibility, adaptability, and interactivity of country-based subiquitous smart grids will play a more important role in assuring energy security, implementing the "two-replacement policy" and providing public services.

4.2.3.1 Making Overall Grid Operations Safe and Efficient

With the rapidly rising share of wind, solar, and other clean energy in total generation capacity, expanding grid capacity and the growing operational complexity of power systems, country-based ubiquitous smart grids need to rely on technological innovations, especially innovations that combine information and electric power technologies, through the employment of advanced smart monitoring, control, operation management, and decision support to achieve reliable and efficient power transmission and distribution, while ensuring safe and reliable grid operations.

Condition-based monitoring of power transmission and transformation equipment. We can ensure that transmission and transformation equipment is in a controllable state and is under control by promoting condition-based monitoring and smart inspections, as well as fully utilizing information technologies to perform online and offline condition-based evaluation, condition-based maintenance, and risk prewarning. Building on a full understanding of transmission and transformation equipment, power and grid resources are maximized through highly-intelligent, optimal dispatch operations to improve line transmission capacity and grid asset utilization, while reducing the energy loss, cost of transmission, and ensuring operational safety and stability for power systems.

Operation control of transmission networks. We can adopt flexible AC transmission technologies to improve the capacity and voltage of transmission lines as well as the flexibility of flow control. With the support of communication, information, and control technologies and by employing satellite positioning, smart monitoring, and advanced inspection technologies, condition-based evaluation, diagnosis, analysis, and decision support can be implemented for grid operations to achieve smart evaluation of transmission conditions. Technologies in condition-based maintenance, whole life-cycle management and smart disaster management can also be utilized to achieve lean management of grid operations.

Smart control and optimal operation of grids. Smart grids must be capable of providing a safe and stable supply of electricity and improved power quality, with highly-intelligent control figuring prominently as the nerve center of a grid system. To ensure higher supply reliability, grids must be able to perform in two areas. One is the ability to avoid extensive blackouts and ensure uninterrupted power supply for key users by withstanding the impact of natural disasters and other external factors and limit damage to a manageable level. The other is the ability to automatically and timely detect any faults that have occurred or are occurring and implement corrective measures to eliminate or minimize the impact on the normal power supply for users. Real-time monitoring and control should be carried out to maintain power quality, while ensuring compliance of the effective values and waveforms of voltages with user requirements, also normal operation of user equipment without shortening its service life. With the support of automation systems and integrated smart control technologies at all levels, flexible control, and optimal operation of grids can be achieved, leading to improved grid reliability and power quality.

Improving grid efficiency and effectiveness. Real-time monitoring of the temperatures, insulation levels, and safety margins of smart grid equipment will increase transmission power and system capacity utilization without affecting safety; line losses can be reduced and operation efficiency improved by optimizing flow distribution; online monitoring and diagnosis of operating equipment can be conducted and condition-based maintenance implemented to lengthen the service life of equipment.

4.2.3.2 Assuring Flexible Access and Operation of Distributed Power Sources

Distributed power supply is an important means of fully utilizing widely-scattered energy resources and also a key approach to the development and utilization of clean energy in the future. Development practices and policy environments in different countries all point to large-scale development of distributed power supply as an emerging trend. It is therefore of utmost importance that smart grids should be able to accommodate and promote access for and safe and economical operation of large-capacity distributed power supply.

Supporting large-scale, high-level access for distributed power supply. When the capacity of distributed power supply in a grid has reached a relatively high level (i.e., high penetration), a conventional grid will find it very difficult to ensure power balance and safe operation as well as supply reliability and power quality. Unlike their conventional counterparts, smart grids do not need to passively restrict access capacity of distributed power supply to ensure operational safety. Rather, they may allow effective access for, and support the plug-and-play capability of, distributed power supply in a way that can facilitate distributed power generation and bring down overall investment costs. By upgrading the protection and control system and standardizing the system interfaces of conventional grids, together with the support of an information and communication platform, smart grids can effect information exchange with distributed power supply and build an open, integrated platform for energy utilization to facilitate equal, convenient, and efficient utilization of distributed power supply.

Supporting safe and economical operation of distributed power supply. Through the data and information platform of a smart grid, data on distributed power supply and grid operation are collected on a real-time basis and highly integrated with offline management data to visualize and control distributed power supply and provide operators with advanced decision support capabilities covering grid operation monitoring, pre-warning and self-recovery. The functions of distributed power supply such as "plug-and-play," two-way measurement, output forecast, and optimized control, will effectively improve the operational characteristics and economics of distributed power supply, and reduce the costs of power system auxiliary service specifically required for distributed power supply.

4.2.3.3 Promoting Electric Energy Substitution

Electric energy substitution is a strategic option to adapt to the trend toward electricity consumption, with significant implications for energy development and social sustainability. It can effectively contribute to energy conservation, emission control, and environmental protection. Accelerating the development of country-based smart grids is the key to effectively promoting electric energy substitution.

Providing a platform for efficient and effective use of electricity: By extending grid infrastructure, smart grids can supply power in a more intelligent and efficient manner to make end consumption more accessible and provide a solid fundamental network platform for electric energy substitution. More extensive and efficient terminal consumption of power will be achieved by building on a grid-based public service platform, with energy efficiency management focused on electricity, and promotion of new smart terminals for using energy.

Promoting substitution of electricity for fossil energy in different fields: By promoting energy efficiency, conservation technology, and equipment in the transport, industrial, business, and residential sectors to substitute electricity for fossil energy, the goal of whole-society energy conservation and environmental protection can be achieved. Meanwhile, continued upgrade and expansion is required for power grids to improve supply stability and reliability for electricity as an energy source of choice for terminal consumption. In the future, the transport sector should be the focus of initiatives to encourage electricity usage, with ongoing efforts to improve the charging infrastructure for electric vehicles and promote transport electrification to replace oil with electricity.

With different levels of socioeconomic growth, different countries have different focuses on electric energy substitution. Less developed countries are focused on accelerating the electrification process for electric energy substitution. Take Africa as an example, annual electricity consumption per capita is estimated at 600 kWh, with many African countries consuming just 100 kWh and over 600 million people still without access to electricity. These countries desperately need to solve the problem of power shortage before they can promote economic development and electrification. As for developing countries, it is essential that terminal energy consumption marked by low efficiency and high pollution be phased out. For instance, backward production capacities in China must be eliminated, coal- and oil-fired furnaces replaced by electric furnaces, and electric heating equipment for industrial purposes utilized. In the commercial sector, efforts need to be stepped up to implement projects to replace coal (gas) with electricity and encourage the use of heat pumps, electric heaters, and electric boilers. In developed countries, programs for electric energy substitution are carried out mainly through the implementation of standards, policies, and meticulous monitoring and management. For example, Japan is known to be a promoter of thermal storage tariffs to reduce heating costs for users. In the European Union, the setting of quality management standards for energy efficiency and energy-consuming equipment has been high on the agenda. Indeed, timely upgrading of the standards for energy efficiency, emission, and new electrical applications plays an active role in promoting electric energy substitution.

4.2.3.4 Ensuring Smart use of Electricity and Fulfillment of Diverse Demands

As power suppliers, conventional grids only provide different types of customers with an electricity supply. However, through smart development, power grids will be able to offer increasingly diverse services to support smart use of electricity to satisfy diverse energy demands. In the future, the Internet concept will be integrated fully into grid-enabled services, bringing profound changes to grid operations in terms of data resource value, business service models, and user experiences. On one hand, users who have become more aware about their participative and interactive roles, will demand the establishment of a new relationship of electricity supply on the basis of an integrated information platform. As a result, their demand will become prominently more personalized and diverse. On the other hand, armed with a fuller and deeper understanding of user behavior, smart grid operators will be able to design more targeted services to help power companies fully unleash the potential of existing resources.

Building a smart-based integrated service platform. Building on the ubiquitous smart grid and its associated information and communication system, network service capabilities can be expanded to promote consolidation and integration with public service resources. Not only conducive to high-level integration and integrated application of energy flows and information flows, this also contributes to the development of a new intelligent, public and interactive integrated service platform based on a smart grid. Through continued efforts to expand into new service areas based on the interactive service platform, value-added offerings like information and communication services can be launched to add more value to power services. By supporting demand-side management and developing smart communities and buildings, the user demand for greater freedom in electricity usage can be met in a timely manner. Advice on smart use of electricity can also be provided and the efficiency of terminal energy consumption improved; customers will be provided with timely and accurate information on tariffs and loads as well as the best solutions and strategies for energy consumption.

Supporting diverse demand for smart use of electricity. By installing advanced metering devices, using smart terminal equipment, building interactive infrastructures and deploying interactive, information-sharing systems, demand for smart use of electricity for diverse purposes can be adequately met to further promote the construction of smart districts, buildings, parks, and communities in support of the development of smart cities. Smart meters will be fully promoted. A system for collecting information on electricity usage will be established. All power consumers and metering gateways will be covered, with online monitoring and real-time collection of important information on consumer loads, consumption levels, and voltages to provide technical support for smart power services. Through the establishment of systems for smart energy services, consumer-side distributed power supply, and energy storage, together with a variety of smart terminals for electricity consumption, real-time interactive response between smart grids and power consumers will be achieved so as to facilitate customer interactions through remote and mobile access to improve a grid's integrated service capabilities. By providing stations to form a logically planned network for smart charging and replacement services for electric cars, two-way energy exchanges between electric automobiles and power grids can be achieved. This will support electric energy substitution and help realize clean energy utilization to meet the demand for low-carbon energy consumption in the transport sector.

Realizing two-way interaction between information and electricity. Smart development promotes the full-scale integration of power grids, information networks, and the Internet to bring information flows and electricity flows together. Consumer-side smart meters capable of capturing, uploading, and distributing information on electricity usage in a timely manner can be used as the access and transit points of mobile networks to upload, via dedicated power grids or the Internet, information on electricity customer behavior, and other external information on heating, gas, and water supply to platforms that provide electricity usage information and data integration and utilization. The storage, processing and utilization of such information and data received can create the functional capability to integrate information with electricity. Business information can then be communicated to consumers,

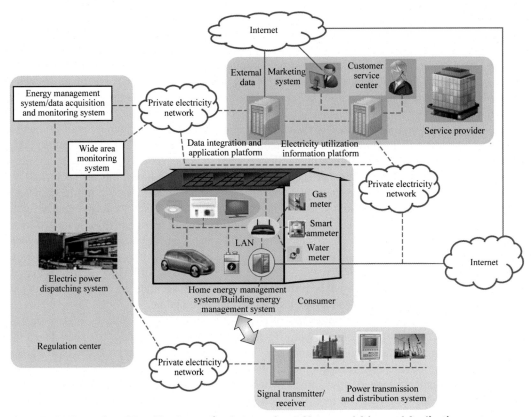

FIGURE 5.17 Illustration of Two-Way Interaction between Smart Meters and Advanced Applications

guide consumer behavior, give timely responses to the changes in consumer demand, and regulate two-way electricity flows via a trading and operation system. See Fig. 5.17 for the two-way interaction between smart meters and advanced applications.

Promoting the integration of the Internet concept into power grids. As smart development embodies the spirit of openness, equality, cooperation, and sharing, we can improve the awareness among consumers of their participative role in interactions by further integrating the Internet concept into grid development, while developing clustering relationships among consumer groups via information integration, as shown in Fig. 5.18. A series of changes will follow: scattered customer-side power resources can be pooled to participate in grid-based interactions. Third-party organizations like smart grid-enabled business service providers will emerge as intermediaries to provide consumers with diverse services. We can also strengthen the management of scattered consumer resources and demands by, for instance, integrating electricity demands from different customer segments in a region with the capability of demand-side management, to centralize two-way interactions with the support of flexible business models and smart power grids. The potential of smart grid infrastructures can be further unlocked and a variety of value-added services, such as energy efficiency management and multigrid integration, can be implemented based on a good understanding of consumer demands.

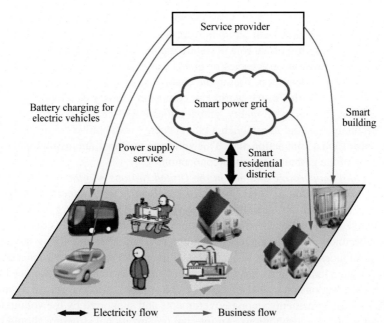

FIGURE 5.18 Application of the Internet Concept to Smart Grid Development

5 COOPERATION MECHANISM FOR GLOBAL ENERGY INTERCONNECTION

The development of a globally interconnected energy network is a major strategy to resolve problems concerning energy sustainability, covering every aspect of world politics, economy, energy, and technology, which requires concerted efforts at the international level to break down all policy barriers and establish an organizational mechanism of interdependence, mutual trust, and mutual benefit. This will bring about broad participation and multiwin cooperation among governments, corporations, communities, and users to pave the way for an efficient operational mechanism as well as a market mechanism to ensure safe and economical operation of the global energy interconnection.

5.1 ORGANIZATIONAL MECHANISM

The development of the global energy interconnection must be secured by a global energy governance mechanism built on all-round cooperation. The construction and development of the global energy interconnection involves large-scale development of renewable energy resources in the Arctic and equatorial regions, and also resource-rich regions of each continent. It also involves transcontinental transmission and intercontinental interconnections of mutual backup, and the upgrading, modification, and intelligentization of transnational transmission networks on each continent and the transmission and distribution grids of each country. This unprecedented example of global energy cooperation requires the establishment of a mechanism and an organizational base of mutual trust and benefit to carry out energy cooperation at the global level. Over the past century, different forms of international energy cooperative organizations have been set up in respect of oil, a core energy resource. Although these organizations have played an important role in stabilizing the oil market as well as supply and demand, no global or integrated cooperative agencies or global governance mechanisms have been developed to facilitate global energy cooperation. Facing the challenges of climatic change, environmental crisis and energy security, no countries in the world can completely satisfy development needs solely with their own resources. As an objective reality, global allocation of energy resources is required. In the long run, it is extremely essential to establish a global energy cooperative organization and to develop a binding mechanism, and a joint action plan. This will form an important institutional base for advancing the construction of the global energy interconnection.

The formation of a cooperative alliance in the United Nations on global energy interconnections will help drive the construction and development of a globally interconnected energy network. The United Nations has more than 20 specialized agencies focusing on the energy issue, but no dedicated energy agency has been set up yet. The focus of future energy development is not only on solving the problems of socioeconomic development of each country, and also on solving the ecological crisis that the whole world faces today. If the construction of the global energy interconnection is at the core of the above-mentioned joint action plan for achieving energy sustainability, we need to establish an alliance in the United Nations on global energy interconnections as a cooperative organization underlined by the framework agreements among governments and supported by the voluntary involvement of relevant departments and corporations. An internal organizational structure should be set up, with well-defined rights and responsibilities at the decision-making, management, and execution levels. At the decision-making level, there is a high-level body founded based on the conferences of energy and environment ministers of all member countries, with the responsibility for procuring a consensus on the direction and goal of building a global energy interconnection. At the management level, there is a specialized management committee supported by a body for day-to-day management, where the work focus is on conducting research, proposing policy recommendations, and carrying out coordination and promotion in the areas of development, economy, technology, and regulation. At the execution level, there is a series of project companies, voluntarily formed by the major corporate members of the above-mentioned alliance.

The cooperative alliance on global energy interconnections will play a leadership role in strategic planning, standard formulation, resource support, and external collaboration. As its goal, the alliance will aim to facilitate interconnected access and global allocation of clean energy; promote technology research, infrastructure development, and safe and efficient utilization concerning global energy interconnections; and encourage equal access to global energy, as well as energy security and reliability. The alliance should play an important role in the following aspects. First, it should conduct research and stipulate development strategies, interconnection planning, economic policy, technical standards, operation rules, and market mechanisms for the global energy interconnection. Second, it should mobilize the necessary technology, funding, and human resources to coordinate and promote the development of clean energy bases around the world and grid interconnection projects across continents to narrow regional differences. Third, it should develop a mechanism for cooperation with relevant international organizations such as the existing electricity alliances of each continent, the International Energy Agency, the International Electrotechnical Commission, and the International Smart Grid Action Network to jointly promote and build a sustainable energy supply system on a global basis. See Fig. 5.19 *for an illustration of the organizational structure of the alliance on global energy interconnection.*

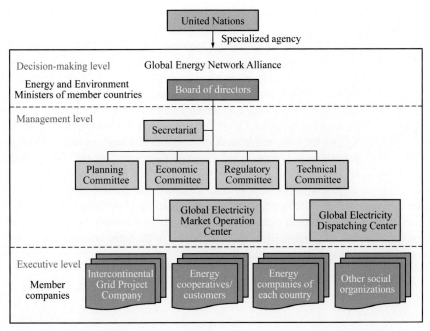

FIGURE 5.19 Illustration of the Organizational Structure of Global Energy Interconnection Alliance

5.2 OPERATIONAL MECHANISM

A globally interconnected energy network requires the support of an efficient collaborative mechanism. The operation management of a power system is a technologically complex systems management discipline requiring seamless organizational skills. The experience in grid development around the world over the past century suggests that an operation mechanism based on unified dispatching rules and collaboration is a basic requirement to safeguard the safe operation of interconnected grids. In the future, this characteristic of coordinated operation will become more prominent, making the formulation of control and operation rules for centralized, grid-wide coordination, and a responsibility system even more important for the operation of a globally interconnected energy network. First, it is because of the size of the network that will evolve into ubiquitous transmission and distribution grids interconnected at the transnational and transcontinental levels, where the grid assets are owned by different companies and scattered in different countries and regions. This situation warrants the establishment of a coordination mechanism to resolve problems and conflicts arising from highly-decentralized property rights and geographical presence and from the need for centralized operations. Second, it is because of the high level of intermittent energy resources involved. In the future, 80% of energy generation worldwide will come from renewables of an intermittent nature. This will require optimizing interconnection operations over larger areas and the support of various energy storage devices. Third, it is because of an increase in the number of mobile devices. With electric vehicles as a typical example, the growing popularity of powered mobile equipment or energy storage devices necessitates efforts to analyze in real time and understand the distribution of loads and available resources in the system, so that energy flows can be balanced according to changes in resources. Fourth, it is because of higher load flexibility where flexible conversion between generation and demand load is made possible for users by distributed energy sources, therefore requiring the support of more timely communication and information services and control measures to make the most of resources.

A dispatch center should be built to ensure the safe and efficient operation of the global energy interconnection. Over the past century, with the gradual development of power grids, in particular the formation of high-voltage large grids, dispatch centers have been built for continental, national and regional grids in different countries in line with the structural and other specific requirements of the power grids concerned. In tandem with the development of the global energy interconnection, a dispatch center should be set up at the global level and subjected to the proposed alliance on global energy cooperation to safeguard the safe, stable, economical, and efficient operation of the global energy interconnection. The global dispatch center shall be jointly governed by the energy dispatch centers of all countries across all continents on the base of global dispatch operation agreements signed by all members. It will form a close-knit dispatch system with different control zones and levels to provide system-wide assurance for the safe operation of the global energy interconnection. Under the centralized guidance of the global dispatch center, dispatch centers at the continental level shall coordinate the operation of individual dispatch centers in each country in the same continent. See Fig. 5.20 for an illustration of the dispatch system of the global energy interconnection.

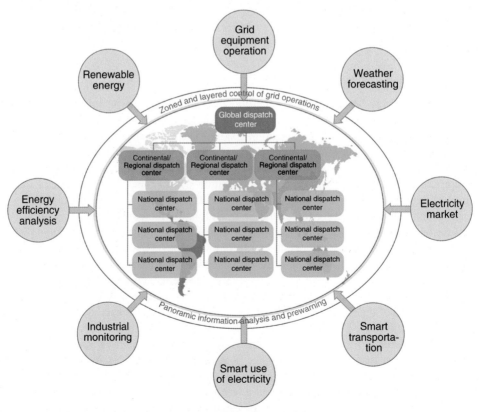

FIGURE 5.20 Illustration of Global Energy Interconnection Dispatch System

The global dispatch center will play an important role in safeguarding global power security and the global allocation of energy sources. The global dispatch center will be the management and coordination center for the construction and operation of the global energy interconnection. Its specific functions will be reflected in the following three aspects. First, it will be responsible for the coordinated operation of power grids worldwide to improve reliability and safeguard supply security globally. Second, it will promote intercontinental connections of power grids, encourage grid investments, and establish a sustainable globally interconnected power network. Third, it will not only create conditions conducive to grid access for renewable energy, but also promote the goal of renewable energy development to the world. The main responsibilities will cover the formulation of rules for grid operation and development; the dispatch operation of intercontinental grid backbones of the global energy interconnection; the coordinated monitoring of power grids across continents for operational safety, and the coordination of backup operations in case of contingencies across continents; and development planning for the global energy interconnection. Moreover, an information and early warning system will be established to provide access for grid operators of all countries to the real-time conditions and data of power grids at the global, continental, and national levels to strengthen collaboration among members for the sake of operational safety.

5.3 MARKET MECHANISM

A globalized market mechanism is an institutional basis for providing impetus to the development of the global energy interconnection. The global energy interconnection will offer great potential benefits in terms of power transmission, system interconnection, and resource-sharing. To maximize these benefits and to reflect the market value of these benefits in the investment returns on the global energy interconnection, a fair, open, and competitive global electricity market mechanism must be set up to guide power companies and users in the direction of full participation. In the past two decades, with the opening of national electricity markets around the world, international electricity trade has witnessed rapid growth, with an ever-growing footprint. A pan-European market covering seven regions has been created in Europe, compared with 10 regional markets in the United States. A national electricity market has also been created in each of Russia, Australia, New Zealand, Argentina, and Brazil. The growing market coverage contributes to the development of interconnections across different regions, such as the pan-European transmission grid under development in Europe. The creation and effective operation of the future global energy interconnection will have to be based on this global electricity market mechanism.

A global electricity market system will be gradually built up. A global electricity market refers to a united market formed on the basis of the markets in different countries and continents. It will start with trading based on transnational/transcontinental multilateral long-term contracts and progress to mutual trade access on unified rules across different countries and continents. This will lead further to the development of short-term, flexible electricity trade, and auxiliary service trade across countries and continents. With improvements in the global energy interconnection as well as transnational and transcontinental electricity trade, some of the functions of the global electricity market and the individual electricity markets in different continents and countries will be gradually integrated, leading to the creation in the long run of a global electricity market open to electricity consumers, corporations, and brokers around the world, for the benefit of free trade. The establishment of a global electricity market is a progressive process from a local level to a global level. Initially, a global electricity market exchange has to be established

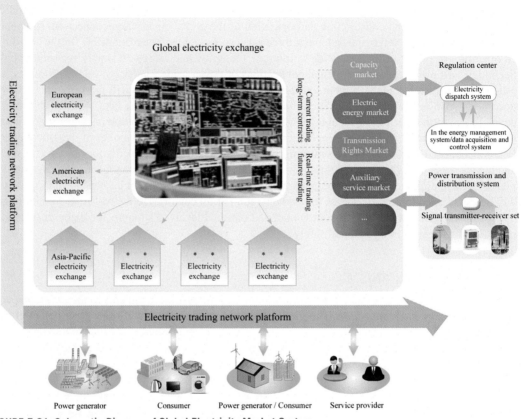

FIGURE 5.21 Schematic Diagram of Global Electricity Market System

through the cooperative alliance on global energy interconnections to formulate, on a centralized basis, marketing plans, trading mechanisms, and operation rules, while spearheading the development of a global electricity market platform. See Fig. 5.21 for an illustration of the global electricity market system.

A robust trading mechanism for transnational and transcontinental electricity markets. In order to ensure safe, stable, and sustainable development of the global electricity market, a market mechanism needs to be designed in such a way as to resolve three problems. First, a flexible trading mechanism needs to be established to promote transcontinental transmission. On one hand, market trade in electricity should focus on medium and long-term contracts in order to provide stable trade volume and revenue expectations for the development of and investment in resources on a transnational basis. On the other hand, to fully leverage the function of the global energy interconnection in resource allocation, it is important to establish a flexible adjustment mechanism covering short-term trades in a monthly, day-ahead, and intraday basis in order to accommodate resource allocation requirements in response to changing renewable energy supply and demand. Second, an investment mechanism to promote the development of transcontinental power grids needs to be established. From the standpoint of investment incentives, a mechanism to ensure stable return on investment is key to attracting investment, given the massive investment required for transcontinental power grids. During the development of global energy interconnections, long-distance, high-capacity transcontinental transmission projects assume a critical role in promoting renewable energy development in remote areas and providing sustainable energy supply to load centers. They also offer significant interconnection benefits where the development and efficient operation of the global energy interconnection is promoted through coordinated design of income based on long-term transmission contracts, regulated income based on grid safety, and trading income based on short-term dispatch operations. Third, a capacity allocation mechanism needs to be set up to enhance the operational efficiency of transcontinental power grids. The roles and functions of power grids must be considered to form a sound basis for designing a mechanism for capacity allocation and cost sharing for transnational transmission facilities, with the objective of maximizing the potential of transmission facilities and attracting investment in grid infrastructure.

A market mechanism and business model for sharing global energy resources will be formed. Global energy interconnection provides an energy infrastructure network serving as a platform for building open interconnections. In the energy sector, an open, interactive, sharing, and win–win business model centering on this global energy interconnection will be developed among market players in the future. Work on the global energy interconnection will be completed jointly by all interested parties. Along with national and regional grid companies, electricity cooperatives will play a role as the major investor and consumer in the global energy interconnection. They will link up users with the same needs and aspirations to establish a network of funding sources, by means of crowd-funding or other financing models, to invest in the global energy interconnection while enjoying the right to use the same. Many professional subcontractors will become intermediaries responsible for the construction, operation and maintenance of the global energy interconnection in accordance with stipulated standards and contracts. All power generators or users will have access to the global energy interconnection via dedicated lines. Network resources are accessible with the payment of a network access fee based on the locations and voltage grades of the connection nodes. Trade in electricity can also be conducted without restriction through an open network platform. Information on electricity trade will be channeled in a timely manner to electricity dispatch and marketing agencies at all levels to facilitate rules-based settlement. Energy network operators, power generators, and service providers will share the trade proceeds upon settlement.

5.4 POLICY SUPPORT

A good policy environment is a key to achieving the goal of developing a global energy interconnection of interconnections. In order to construct this global energy interconnection, a global energy view has to be taken by all countries, with concerted efforts to combat climate change, set low-carbon development goals and strengthen global energy cooperation so as to foster a policy environment based on a global consensus and a win–win approach to cooperation.

First, the world should form a consensus on responding to climate change. A global response to climate change is the driving force behind the development of the global energy interconnection. For this a global consensus on tackling climate change is required. Currently, global efforts to tackle climate change based on United Nations Intergovernmental Cooperation Framework to Address Climate Change are bogged down in the "Cooperation Dilemma," but the world is under increasing pressure to reduce emissions. There have been increasing signs that GHG emissions caused by industrialization are destroying the ecological balance of the Earth, threatening to drag the entire ecological system into a catastrophically unstable state. All countries need to realize that the Earth's biosphere is an indispensable public resource for the survival of mankind, and that we have no alternative but to put aside our arguments and work together to find a solution to problems of human sustainability. The energy sector currently accounts for approximately 80% of the world's total carbon emissions from human activities. According to the roadmap for developing a global energy interconnection of interconnections, efforts to lower energy-related carbon dioxide emissions around the globe to about 11.2 billion tons by 2050 will provide a solid foundation for effective management of climate change. It can be seen that the global energy interconnection is an important vehicle for globally concerted efforts to tackle climate change, where a consensus worldwide has to be reached to realize global emission reduction targets at the lowest possible costs.

Second, energy policies around the world should be advanced on a coordinated basis. Currently, a worldwide consensus on the direction of low-carbon energy development has basically been reached, with important measures proposed for renewable energy development, energy efficiency improvement, and smart grid construction. However, views are still divided over issues concerning energy transition and path selection, such as the clean utilization of fossil energy and the continued development of nuclear energy. In the future, the development of the global energy interconnection will require further coordination at the energy policy level. First, the reliance on fossil energy needs to be further reduced. Carbon capture technology remains a big unknown in terms of technology breakthrough and application and also its subsequent environmental impact. Given the challenges of high costs and unstable performance in this area, all countries need to reach a consensus on a stronger thrust to substitute renewable energy for fossil fuels after 2030. Second, plans for global renewable energy development need to be optimized. In view of a lack of land resources for renewable energy development in densely populated regions and the considerable impact of renewable energy development on the local climate, landscape, production and business activities, all countries should prioritize the development of renewable energy in remote areas blessed with abundant resources. Third, an energy security policy on global cooperation and sharing needs to be developed. Internationally, a more proactive energy policy on international cooperation needs to be established so as to facilitate the development of more liberalized energy markets and also actively participate in international energy resource development for promoting the optimal utilization of global renewable energy.

Third, a geopolitical landscape for win–win cooperation should be created. The energy geopolitics surrounding the quest for fossil energy is driving the world into increasingly turbulent times. Renewable energy development in "the Arctic and equatorial regions" is an important basis for building a new world of energy interconnections. Abundant with fossil energy resources, these regions are currently facing a political game triggered by the competition for resources. The "blue enclosure movement" has been launched in recent years in the Arctic region with massive untapped reserves of oil/gas (accounting for 25% of the global amount) and coal (accounting for about 9% of the global amount). Over the past decade, Africa has recorded annual growth of 58.8% in new oil reserves, accounting for 25% of the new reserves around the world. As a result, countries around the world are clamoring to enter and compete in these resource-abundant regions. However, as sustainable development is the common goal of all human societies, a global energy geopolitical situation capable of engendering win–win cooperation is required before we can start building an energy supply system focusing on renewables. Renewable energy resources are sustainable, instantly gratifying and inexhaustible, but their value can only be realized through development and utilization. For the renewable energy resources in the Arctic and equatorial regions, cooperative development creates more value than simply holding on to these resources. Renewable energy can become valuable resources only through strong cooperation between the private and public sectors in resource-producing nations and their counterparts in resource-consuming countries. An international consensus needs to be reached on moving from competition for resources to working together at the geopolitical level for win–win cooperation in energy development.

6 COMPREHENSIVE BENEFITS OF GLOBAL ENERGY INTERCONNECTION

Building a globally interconnected energy network will generate enormous economic, social, and environmental benefits. First, this global network will boost the development and consumption of renewable energy, sharply reduce the consumption of fossil energy, effectively control GHG emissions and protect the ecological environment. Second, it can link up the power grids across continents to create enormous interconnection benefits, on account of the differences in time zones, climatic conditions, peak/valley load periods, and energy mix as determined by energy resources among different continents. Third, the cost of power supply in electricity-importing regions where power generation is relatively costly can be brought down by importing low-cost electricity from large renewable energy bases. Fourth, it will help drive local economic growth and facilitate the coordinated development of the regional economy by promoting the development and utilization of renewable energy in developing countries.

6.1 ENVIRONMENTAL BENEFITS

The development and utilization of renewable energy resources provides an alternative to the massive consumption of fossil energy and reduces high levels of pollutant discharges and GHG emissions. It can also avoid water consumption and damage to the ecosystem resulting from fossil energy development and utilization.

Amid the quickening pace of development in the global energy interconnection, non-fossil energy generation is forecast to reach 66,000 TWh in 2050, nearly 60,000 TWh more than the level in 2010 and accounting for 90% of total electricity generation. Under the new policy set out in World Energy Outlook 2014 (WEO 2014), clean energy generation will amount to approximately 17,000 TWh by 2040, accounting for nearly 50% of total electricity generation. If the share of clean energy generation given in WEO 2014 is anything to go by, amid the continued development of the global energy interconnection, an increase of 29,000 TWh in clean energy production to replace the equivalent amount of coal-fired generation by 2050 will save 9 billion tons/year of standard coal and 70 billion tons/year of water, while cutting down carbon dioxide, sulfur dioxide, nitrogen oxide, and dust emissions by 25 billion tons/year, 53.7 million tons/year, 56.4 million tons/year, and 9.4 million tons/year, respectively.

Judging by the progress on the development of energy bases, the electricity sent out from the Arctic and equatorial regions is estimated at 900 TWh, 4200 TWh, and 12,000 TWh by 2030, 2040, and 2050, respectively. Based on the equivalent amount of coal-fired generation so replaced, energy consumption will be reduced by 0.3, 1.3, and 3.8 billion tons of standard coal, respectively. Which means, this will reduce annual carbon dioxide emissions by about 0.8, 3.7, and 10.5 billion tons; annual sulfur dioxide emissions by 1.8, 7.9, and 22.3 million tons; annual nitrogen oxide emissions by 1.9, 8.3, and 23.4 million tons; and annual dust emissions by about 0.3, 1.4, and 3.9 million tons. Annual water consumption will also go down by 2, 10, and 29 billion m^3, respectively. Global interconnections will contribute to the development and utilization of clean energy, yielding significant environmental benefits.

On the strength of the emerging global energy interconnection, clean energy is expected to be able to replace fossil energy equivalent to 24 billion tons of standard coal every year by 2050, resulting in 67 billion tons lower carbon dioxide emissions and 0.58 billion tons lower sulfur dioxide emissions. By that year, global carbon emissions is estimated at 11.5 billion tons, accounting for just about 33 and 50% of the emission levels in 2013 and 1990, respectively. An IPCC research report shows that if achieved, the goal of "limiting average global temperature rise to 2°C by 2050" as proposed in the United Nations Framework Convention on Climate Change can fundamentally solve major issues that threaten human survival, such as glacial melting and rising sea levels, and ensure human sustainability.

6.2 ECONOMIC BENEFITS

Assuring energy supply for socioeconomic development. Through the global energy interconnection, widely scattered clean energy resources with great potential can be developed and utilized to assure a long-term and stable supply of energy. Based on average annual growth of 12.4% for wind and solar power generation from now, nonfossil energy will account for 80% of the world's total energy consumption by 2050. At that time, wind and solar energy will become the dominant sources of power; yet even then, we will have only reached no more than five ten-thousandths (5/10,000) of the total developable capacity of these energy resources.

Reducing the cost of energy supply. Through the global energy interconnection, the benefits of large-scale development and outgoing transmission of clean energy can effectively reduce the cost of power supply. Taking intercontinental transmission of power between Asia and Europe as an example, clean energy such as natural gas, wind power and solar energy gathered in the Asian regions at the sending end (Xinjiang in China, Kazakhstan, Russian Siberia, and Mongolia) can be transmitted to Germany by means of ±1100 kV UHVDC transmission technology and up to 5500 h will be usable through the DC channel. Compared with the delivery of offshore wind power into Germany, the delivery of electricity through this intercontinental transmission channel, relayed via St. Petersburg in Russia, is 30.4% cheaper, with the maximum tariff difference being RMB0.3648 per kWh. If electricity is directly transmitted between the two locations, the maximum tariff difference is RMB 0.526 per kWh, 43.8% cheaper compared with the option of importing offshore wind power into Germany. By adopting the option of intercontinental transmission between the two continents, the cost of power supply in German can be effectively reduced, generating remarkable benefits of intercontinental transmission.

Obtaining considerable benefits from interconnections. Given the time differences between different continents and the seasonal differences between the northern and southern hemispheres, the global energy interconnection of grid interconnections across different continents can regulate peak/valley loads across continents and enable the optimal allocation and consumption of global renewable energy by taking advantage of the complementary nature of load characteristic curves in different continents to improve utilization of generating equipment and lower reserve capacity in different continents.

As a result of the Earth's rotation and its revolution around the sun, different continents receive sunlight at different times, giving rise to different time zones. The Earth is divided into 24 time zones, with adjacent time zones being 1 h apart. Asia covers 13 time zones, with China spanning 5 time zones, whereas Europe covers 5 time zones and North America, 8 time zones. As the residents in different continents observe basically the same pattern of daily life, waking up to work when the sun rises and going to bed when the sun sets, the load curves of each continent's countries tend to peak during local daytime and evening (08:00–23:00) and to bottom during the night and in early morning (24:00–07:00).

The daytime is a peak load period for Europe and Africa. Coinciding with this period is the nighttime, an off-peak load period, in East Asia and North America. As more wind power is generated in the nighttime when demand is lower than in the daytime, this power source can be transmitted through global interconnections from East Asia and North America during the nighttime to Europe for consumption. Conversely, wind power can be transmitted to East Asia and North America for consumption across continents from the North Sea in Europe and North Africa when Europe and Africa are in the off-peak period during the night.

Take the interconnections across the northern hemisphere's three major continents – Europe, North America, and Asia – by 2050 as an example. The power grids in North America stretch from Western Zone 4 to Western Zone 10; the power grids in Northeast Asia cover the regions between Eastern Zone 7 and Eastern Zone 9, and the synchronous grids in Europe span the areas from Zone 0 to Eastern Zone 2. After completion of the interconnections globally, the load of global power grids can be optimized by taking advantage of the natural time differences between continents to form relatively smooth load curves to realize the benefits of peak shaving and valley filling. See Fig. 5.22 for a comparison between the load curves before interconnection and the superimposed load curves after interconnection in the three continents of Asia, Europe, and North America. It can be observed that when grids have been fully interconnected at the transcontinental level, mutual backup for load shifting among the three interconnected continents can produce prominent results. The interconnections will lead to a balanced distribution of loads at different times of the day, with the load difference between peak/valley periods in the three continents narrowing from 25–40% to within 10%.

Fuelling global economic growth. To realize the clean low-carbon development of global energy in 2050, global installed generating capacity is forecast to reach 35 billion kW by 2050, an increase of approximately 30 TW from 2020. On a 10-year horizon, installed electricity capacity is expected to grow rapidly in the future, necessitating a correspondingly higher level of investment in capacity building. An estimate of the changes in investment per kW in installed capacity among different sources of power shows the result as depicted in Fig. 5.22.

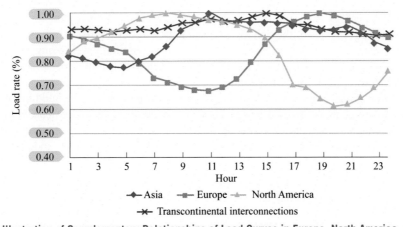

FIGURE 5.22 Illustration of Complementary Relationships of Load Curves in Europe, North America, and Asia

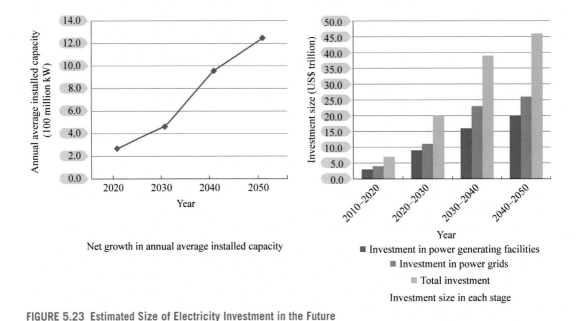

Net growth in annual average installed capacity

■ Investment in power generating facilities
■ Investment in power grids
■ Total investment

Investment size in each stage

FIGURE 5.23 Estimated Size of Electricity Investment in the Future

An investment of approximately US$20 trillion is required for power generation and power grids from 2020 to 2030; approximately US$ 39 trillion from 2030 to 2040; and approximately US$ 46 trillion from 2040 to 2050. These investments will provide a strong impetus to economic development. See Fig. 5.23 for the size of electricity investments in the future.

6.3 SOCIAL BENEFITS

Promoting the shift of focus in developing regions from the advantages of resources to economic strengths. Among others, Africa, Asia, and South America are the major regions where large-scale development of clean energy is absent. By building a global network of energy interconnections, we can help transform the advantages of the rich resources that these regions possess into economic strengths, which will translate into job opportunities for the local population,[1] well-being improvements for people in developing countries, narrower gaps between developing and developed nations, and eventually sustainable development for all mankind.

Driving technological upgrades of the energy and related industries. The building of a global network of energy interconnections will bring about new breakthroughs and extensive applications of technologies, including clean energy generation, UHV transmission, large-capacity energy storage, smart power distribution networks, and microgrids. The traditional materials industries will realize

---

[1]It was proposed in "Re-thinking: 2050 A 100% Renewable Energy Vision for the European Union," a report released by the European Renewable Energy Council in 2000. The report suggested that the installed capacity of renewable energy in Europe would reach 520, 970, and 1960 GW in 2020, 2030, and 2050, respectively, to boost employment in the renewable energy sector to 2.7, 4.4, and 6 million, respectively. In 2009, the employment figures of Europe's renewable energy industry were estimated at 550,000.

technological innovation in areas like nanometers and superconducting. Riding on the crest of technological reforms and innovations in the energy, information, materials, and other sectors, developed countries will gradually emerge from the shadows of economic crisis, while developing countries will see faster speed and improved quality of economic growth to ultimately achieve the goal of shared development among all mankind.

Promoting mankind's harmonious development and utilization of energy. Fossil energy is characterized by its scarcity and also its local and sovereign nature where development and utilization inevitably involves issues of territorial sovereignty and national security. Inexhaustible renewable energy resources can be developed through the global energy interconnection to achieve the peaceful use of energy and essentially improve the relationships of energy production. In the renewable energy system, we have seen mankind shifting from plundering and monopolizing resources to collaborating and sharing. Through synergies and interactions at the international level in the energy development and utilization processes, an energy ecosystem over larger areas will be created. A globally interconnected energy network will fundamentally solve the energy and environmental problems that impact the cultivation of ecological civilization. It will also bring about energy, industry, economic, and lifestyle changes to eventually achieve a harmonious global development.

SUMMARY

1. Driven by voltage upgrading, expanded interconnection capacity and greater automation, grid development around the world has entered a new stage marked by growth in robust smart grids. A globally interconnected energy network marks an advanced phase of robust smart grid development. At the core of this network is a smart grid system focusing on clean energy, supported by UHV grid backbones with extensive interconnections across countries and continents to allocate energy resources on a global scale. As an integral part of the network, powers grids at all levels are well coordinated and flexible access provided for different types of power sources and customers.
2. The global energy interconnection is a major vehicle built on a global energy view for coordinating the development, allocation and utilization of global energy resources. Based on advanced UHV transmission and smart grid technologies, the global energy interconnection will be constructed to link up the wind power bases in the Arctic, the solar energy bases in the equatorial region, and the large-scale renewable energy bases and load centers on different continents for building a global platform, for energy allocation with a strong grid structure, extensive interconnections, high intelligence, and open interactivity, so as to vigorously promote global energy sustainability.
3. The construction of the global energy interconnection mainly covers three phases of interconnection at the intracontinental, intercontinental, and global levels. Efforts are underway to try to bring about a consensus by 2020. By 2030, work on large-scale clean energy bases will start, with a strengthening of intracontinental interconnections. By 2040, the power grids of the major countries on each continent will be interconnected, with major progress made on the development of large-scale energy bases in the Arctic and equatorial regions as well as on transcontinental interconnections. By 2050, the global energy interconnection will nearly be completed to gradually achieve the objective of making clean energy the dominant source of power.

4. Work on the global energy interconnection is carried out in ascending order of difficulty and is guided by an approach to seek major breakthroughs, while maintaining phased progress to accelerate the development of country-based ubiquitous smart grids, intracontinentally-interconnected power grids, and transcontinental UHV grid backbones. Efficient development and utilization will also be promoted for clean energy bases and various distributed power sources in the Arctic and equatorial regions, and on different continents.

5. The construction of the global energy interconnection calls for global efforts to work closely together and break down barriers to establish an organizational mechanism of interdependence, mutual trust, and mutual benefit. An operation mechanism and a market mechanism should also be set up and operated efficiently to encourage extensive participation and win–win cooperation among all governments, corporations, communities, and consumers to ensure the secure and economical operation of the global energy interconnection.

6. The construction of global energy interconnection can generate significant environmental, economic, and social benefits. With the accelerated development of this network, global clean energy is forecast to account for 80% of the world's total electricity generation by 2050. This should be able to ensure a sustainable power supply, effectively control global carbon emissions, reduce power supply costs, and produce other benefits by ramping up transcontinental interconnections and economic growth.

INNOVATION IN GLOBAL ENERGY INTERCONNECTION TECHNOLOGIES

6

CHAPTER OUTLINE

1 DIRECTION AND KEY AREAS

The previous energy revolutions all relied on major breakthroughs in energy technologies in the course of energy development. In the first energy revolution, the invention of the steam engine helped drive a shift from firewood to coal as the dominant energy source. In the second energy revolution, the invention of the internal combustion engine and electric motor brought about a change of the dominant energy source from coal to petroleum and electric power. The currently emerging third energy revolution has spearheaded a transition from the development and utilization of traditional fossil fuels to the massive development of clean energy. This calls for all-out efforts to promote technological innovation in power sources, power grids, energy storage, and information and communication to provide technical support and assurance for accelerating the implementation of the "two replacement" policy and building a global network of energy interconnections.

1.1 ENABLING ROLE

Focusing on cleanliness, low carbon and high efficiency, the application of energy development technology has driven the accelerated growth of clean energy. The efficiency of developing and utilizing traditional fossil fuels has improved significantly after 200 years of growth since the first Industrial Revolution. For example, the overall energy conversion efficiency of advanced supercritical coal-fired thermal power units now amounts to approximately 45%, compared with 40% for automobiles. However, the utilization efficiency of traditional fossil fuels cannot be easily improved further due to the challenge of higher costs and the difficulty of innovating – not to mention the fact that efficiency, even if further improved, will not be able to fundamentally resolve the problems of fossil fuel depletion and ecological deterioration. After over three decades of development, we have seen major breakthroughs in wind and solar power generation technologies, with the generating costs gradually nearing the level that would make mature commercial operation possible. Clean energy will become more competitive in the power generation market if the costs of pollution and carbon emissions from burning fossil fuels are accounted for as part of the total generation cost. In the future, traditional fossil fuels will be gradually replaced by clean energy like wind and solar power. Europe is planning on using renewables as the only energy source by around 2050.

Innovation in transmission technologies has propelled the allocation of electric power toward the direction of building a globally interconnected energy network. With the multiplication of voltage grades, grid interconnection scale and transmission capacity, the longest transmission distance of a single line has exceeded 2000 km with a capacity of over 8 GW. With the development of UHV transmission technologies, power generated by the wind power bases in the Arctic region and the solar power bases in the equatorial regions can be transported over distances of thousands of kilometers to the load centers on different continents to meet urban and rural power demand through the employment of different voltage grades and transmission technologies. By using VSC–HVDC and submarine cable technologies, renewables like wind and solar power can be integrated en masse into the grid for transmission to end-users through UHV lines. Different types of transmission technology can be incorporated to form part of an expansive, globally interconnected energy network that extensively links clean energy bases around the world with load centers across countries and continents.

The integration of information and communications technology (ICT) with electric power technologies has spurred the course of grid intelligentization. The power system is a complex and nonlinear system featuring real-time power balance, and advanced ICT is an important tool to ensure the safe, reliable, and economical operation of power systems. During the initial stages of development of the power grid, the power system was limited in scale and easily operable. As ICT was then at its nascent stage, one could only rely on phone communication and manual control to start/stop generating units, adjust the mode of operation, analyze and calculate the conditions for safe and stable system operations. Load changes could only be forecasted based on experience. With the development and extensive application of ICT in the electricity industry, power systems are gradually moving in the direction of automation and intelligentization. Dispatch centers are now capable of exercising remote control of power generators, while grid safety and reliability calculations can be extended to hundreds of thousands of nodes across different countries and continents. In the future, advanced technologies in optical fiber communication, mobile Internet, the Internet of Things, image identification, cloud computing, and big data will be seamlessly integrated with energy and power technologies to promote the intelligentization of power grids.

1.2 INNOVATION DIRECTION

The transition from traditional fossil fuels to clean energy is posing grave challenges to the innovation of energy and power technologies. A globally interconnected energy network will reshape the future of grid development by extending the coverage of grids from a nation-wide and regional basis to a global level. In this process, challenges have to be overcome in terms of adapting to the large capacity requirement, long-distance transmission, and the often intermittent and volatile interconnected grid operations of large-scale clean energy development, as well as addressing the operational aspects of equipment maintenance and grid construction under inclement weather conditions.

The first challenge is how to improve the controllability of renewable energy to ensure a secure and stable supply of energy. The generation of renewables-based power like wind and solar energy is subject to weather conditions. Output is highly volatile and uncertain, compared with traditional power generation based on coal, petroleum, and natural gas. To satisfy the energy needs of socioeconomic development, climate engineering research should be strengthened to improve the accuracy of wind and solar power projections and the controllability of wind and solar power generation to ensure a sustainable stable supply of energy.

The second challenge is how to reduce the cost of generating clean energy to realize energy sustainability. The energy density of wind and solar power is far lower than that of traditional energy sources like coal, petroleum, and natural gas. Based on equivalent level of output, it is more costly to gather renewable energy than traditional fossil fuels. Currently, although wind and solar photovoltaic power technologies are at a relatively mature stage, the costs of generating wind and solar photovoltaic power remain high at RMB 0.5 per kWh and RMB 0.8 per kWh, respectively, far above the costs of traditional energy sources like thermal, hydro and nuclear power. Adding further to the costs of developing and utilization renewable energy are the relatively short utilizable hours of generation and transmission equipment. Technological innovation is key to lowering the cost of clean power generation by improving the energy conversion efficiency of wind and solar power generation, reducing initial investment, expanding installed capacity, and increasing utilizable hours of equipment. It is also an important foundation for pursuing large-scale development and utilization of clean energy and for implementing the "two-substitution" policy.

The third challenge is how to improve UHV transmission technologies and accelerate the development of large clean energy bases in the Arctic and equatorial regions and on each continent. Through the development of large wind and solar power bases in the Arctic and equatorial regions and on each continent, hundreds of gigawatts of renewables-based electricity will come from the Arctic and equatorial regions thousands of kilometers away. By 2050, the transcontinental electricity flows from the Arctic and equatorial regions are expected to reach over 10,000 TWh and the longest transmission distance, 5,000 km. To accommodate the long distance and high capacity requirements of these electricity flows, research is required into UHV AC/DC transmission technologies that can afford greater capacities and longer transmission distances.

The fourth is how to develop electrical equipment well suited for extreme weather conditions to ensure the operational safety of key equipment and grid construction. The development of wind power in the Arctic will meet with challenging weather conditions like low temperature, high humidity and icy coldness, while the development of solar power in the equatorial regions will face hostile conditions like low humidity, high temperature, and sandstorms. A variety of extreme natural conditions impose higher requirements on wind and solar power generating equipment. For instance, wind turbines must be able to withstand the impact of salt fog corrosion, pollution, storms and low temperatures. Photovoltaic panels have to be able to resist the impact of windstorms, high temperatures, and drought conditions, while power transmission and transformation equipment must demonstrate resilience against new challenges in construction, transportation and installation.

1.3 KEY AREAS

An innovation-based, solution-focused approach is required to addressing issues of feasibility, economics, and safety in the development of a global energy network. The approach calls for major technology breakthroughs in power sources, grids, energy storage, and ICT.

Power source technology. The key areas of innovation in this area include wind, solar, ocean power, distributed generation, and other renewable energy technologies. The direction of wind power technology is toward large-scale development, low wind speed application, resilience against extreme weather conditions, deep sea offshore wind power, accurate projections of wind power intensity, and the construction of grid-friendly wind farms. Solar power generating technology is focused on research into photovoltaic materials with high conversion efficiency, thin-film, and easy production and installation, solar power tracking technology, and improved solar power utilization. The technology for grid-interconnected operation control of photovoltaic power stations is geared toward increasing controllability and intelligence for improving solar-thermal power capacity and reducing costs. Ocean energy remains at the stage of pilot demonstration and future research efforts should be focused on the cost-effective development and utilization of ocean energy. As a vital part of the future global energy network, the development of distributed generation will move toward greater system-friendliness and controllability.

Grid technology. Further research is required into transmission technologies that can afford ultra-long distances and large capacities, with the UHV grid to become the backbone structure of global energy interconnections. The key areas of research include technologies for the operation control of UHV AC, UHV DC, submarine cables, superconducting transmission, and micro and large grids. Also included are the technologies for the configuration, construction, and operation control of future grids, as well as adaptive technologies for grid construction, installation, and maintenance in a physically challenging environment.

Energy storage. Improving the economics and capacity of energy storage equipment is fundamental to storage technology innovation and commercial applications in the future. As the cost of storage equipment is still very high, commercial application has not yet been achieved for power storage, with the exception of electric car batteries. The key to storage technology innovation lies in improving power and energy densities as well as the integration of storage and renewable energy technologies at the operational level.

ICT. Advanced ICT is essential for ensuring the operational safety and efficiency of a globally interconnected network. The global network has higher requirements for technology innovation as

ICT is used to better adapt to the new trends of grid configuration changes and two-way energy and information flows, and also to realize dispatch operation, management, and decision-making, as well as intelligentized trading of electricity.

---

2 GENERATION TECHNOLOGY

The implementation of the "two-replacement" policy to create a new energy landscape oriented toward clean energy and focused on electricity determines the critical role that power source technology will play in the future development of energy. At the center of this development is a continued effort to improve the development efficiency and economics of clean energy, focusing on wind, solar, ocean, and distributed generation technologies. These technological breakthroughs will be the driving force behind the development of global energy interconnections and of vital importance to promoting global development of clean and low-carbon energy.

2.1 WIND POWER

Wind power generation refers to the technology of converting the kinetic energy of the wind into electric power through a wind turbine. The installation produces electricity by collecting and transforming wind power into rotational mechanical energy to drive a generating unit. Wind power generation technology is now relatively mature, with annual generation amounting to 640 TWh, accounting for less than 3% of the world's total energy consumption. Given the more stringent requirements on carbon emission control, the share of wind power in energy generation is expected to increase to 30% by around 2050, with annual generation estimated at 22,000 TWh, indicating great potential for growth. In the coming 40 years, wind power generation technology will see further breakthroughs and, with improving technology, the generating cost is expected to decrease more than 50% to provide more affordable cleaner energy for mankind.

2.1.1 The Latest Technological Progress

Wind power technology has been round in the world for over a century. In the past, due to a lack of economic benefits and supply stability, the scope of application was limited for wind power, which had not been extensively used by the late twentieth century. However, into the twenty-first century, innovations in power electronics, materials, control, and other technologies have led to greatly enhanced installed capacity and efficiency, resulting in constantly expanding scale of commercial application. In 1999, the world's first MW-level wind turbine was commissioned in Denmark. The development of wind turbines from one of 1 MW capacity to the currently largest single-unit capacity of 8 MW only took 10 years, accompanied by a 90% decline in generating costs.

Reflecting its relatively mature stage of development, onshore wind turbine technology at the 3 MW level has been in extensive application. Currently, the world's largest onshore wind farm is located in Alta Wind Energy Center in the United States state of California. With an installed capacity of 1.02 GWh, the facility is under expansion to bring total capacity to 1.55 GWh. Offshore wind power has also moved into the stage of commercial operation, with an offshore wind turbine boasting a single-unit capacity of 8 MW and a blade diameter of 164 m currently in trial operation in Denmark's National Testing Center. Offshore wind turbine installation vessels are a core technology requirement for constructing offshore wind turbines. The world's biggest installation vessel of its kind, the Pacific Orca, has a loading capacity of 8400 tons, capable of carrying and installing 12 × 3.6 MW wind turbines at one go. China is reputed as the world's largest builder of large wind farms, with 16 provincial-level wind power grids each with an interconnection capacity of over 1 GW, together with nearly 200 large wind farms of over 100,000 kW concentrated mainly in the "Three North" region covering Inner Mongolia, Hebei, Gansu, Liaoning, and Xinjiang.

2.1.2 Development Direction and Outlook
2.1.2.1 Wind Turbine Technology
Technology for expanding single-unit capacity. Providing a stable and adequate supply of wind power with least impact on people's lives, offshore regions are one of the focus areas for developing wind power. The large single-unit capacity of wind turbines can increase the windswept areas of the fan impeller to improve the utilization efficiency of offshore wind and the number of utilizable hours per year, resulting in lower generating costs. See Fig. 6.1 for the changes in the leading indicators of wind power generators in 1998–2013. The single-unit capacity of wind turbines is expected to be able to reach a maximum of 20 MW by around 2020.

Low-speed wind turbine technology. A typical double-fed turbine has a start-up wind speed of 4 m/s. However, the wind across areas near cities and some offshore locations has a lower speed. To exploit wind power in these areas requires the development of a technology for low-speed wind turbines. Direct-drive wind turbines can start up at a wind speed of 2 m/s. Compared with double-fed turbines, direct-drive turbines are more costly and larger, where costs need to be brought down further to realize large-scale commercial application. In 2012, China developed the world's first super-low wind speed turbine of 1.5 MW with a super-large rotor of 93 m, which has been connected to the grid to produce

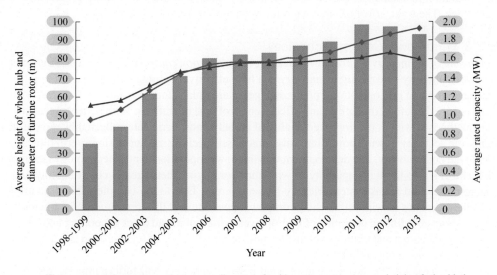

FIGURE 6.1 Changes in Leading Indicators of Wind Power Generators, 1998–2003

Source: Berkeley Library Database.

electricity in the Lai An wind farm in Anhui. Thanks to the research efforts in many countries, the start-up wind speed of double-fed wind turbines is expected to go down to 3 m/s by around 2030. In the future, as long as the site conditions for installation permit, low-speed wind resources in suburban areas can be fully utilized to support rapid development of distributed wind turbines. With the widespread application of low-speed wind turbine technology, the world's utilizable wind resources are expected to treble from the current level.

Wind turbine technology suited for extreme weather conditions. In extremely cold weather, rotor blades can be easily frozen and seriously affect utilization efficiency, with the wind power coefficient shown to be reduced from 0.371 to 0.192, or a 50% decline. When the temperature is below 20°C, the transmission and lubrication mechanisms, storage batteries and controls can be easily damaged, and the tower structures and blades can become brittle, resulting in sharply lower fatigue resistance. Currently, a typical wind turbine will automatically stop running when the temperature drops to −30°C. To accommodate the need for large-scale development of Arctic wind power, research should be focused on developing a technology for turbine insulation, hydrophobic coating on blade surfaces, cold-resistant materials, and so on to address problems regarding turbine resistance against the Arctic region's extreme weather conditions.

2.1.2.2 Wind Farm Technology

The development of large-scale wind farms is shifting toward deep sea offshore locations. The development and utilization of deep-sea offshore wind resources will be supported by wind turbine installation vessels of more than 10,000 tons; floating turbine foundations; large-capacity, long-distance junction stations and submarine cables; and more precise plant design models. By 2030, offshore wind farms of 1 GW capacity are expected to be built extensively, with turbine foundation design, offshore installation, operation maintenance, bus station, submarine cable, and other technologies basically at their mature stage. The wake effect of large wind farms should be assessed accurately.

2.1.2.3 Wind Power Control Technology

Development of technology for precise forecasting and operation control of wind power. High-precision wind power intensity forecasting technology can help effectively mitigate the impact of volatile wind power generation on grid operations, prearrange generation plans for generators, reduce reserve capacity, and ensure the safety, reliability, and effectiveness of grid operations. With better forecasting models and methods as well as the improved quality of global atmospheric data collected by remote sensing systems, the accuracy of wind power intensity forecasts, especially medium to long-term forecasts, will be improved continuously. The development of wind power operation control technology has improved the LV and HV ride-through capability of wind turbines, giving them greater controllability and making large-scale wind farms system-friendly power sources. Due to the complementary nature of the output characteristics of wind, solar and hydropower generation at different times, coordinated control can bring about coordinated operation of hydropower, wind, solar, and other generating units to achieve a more balanced output and improve not just the utilization efficiency of transmission and transformation equipment, but also supply reliability.

Generally speaking, the development of wind power technology will further improve the utilization efficiency of wind energy and reduce costs. With the full commercialization of wind turbines of 10 MW, the cost of onshore and offshore wind power will go down to less than RMB 0.4 per kWh and RMB 0.6 per kWh, respectively. Judging by the progress of current research, wind power technology is expected to fully mature by around 2030 into an important power source technology in support of the development of a globally interconnected energy network. By around 2050, wind turbines with a single-unit capacity of 20 MW can be used to develop and utilize offshore wind farms, with the generating costs falling to below RMB 0.5 per kWh, giving wind power a clear cost advantage over other clean energy alternatives.

2.2 SOLAR POWER

Solar power generation is categorized mainly into photovoltaic and photothermal power generation. Photovoltaic power generation involves the use of solar photovoltaic cells to convert sunlight directly into electric power based on the photovoltaic effect. Solar thermal power generation is a process through which solar power is collected by an array of parabolic dishes and transformed into steam through a heat exchange device to drive a turbine and generate electricity. The most abundant energy source on earth, solar power will become the most promising and fastest growing energy option in the future, with the continued development of solar power generation technology and a globally interconnected energy network. In 2013, solar power generation was estimated at 160 TWh globally, accounting for 0.7% of the world's gross electricity generation. Based on a scenario for accelerated development of clean energy, the capacity of solar power is expected to grow to more than 26,000 TWh around 2050, split equally between photovoltaic and photothermal power generation. Solar energy is expected to account for about 36% of the world's total electricity generation by then.

2.2.1 Photovoltaic Power Generation
2.2.1.1 The Latest Technological Progress
Since the advent of the first photovoltaic cell in 1954, photovoltaic technology has seen considerable progress, having undergone three stages of development. The laboratory stage was in the 1950s and 1960s. In 1954, Bell Laboratory created the world's first applicable single crystalline silicon solar cell with an energy conversion efficiency of 6%. A German by the name of Wei Keer discovered the photovoltaic effect of gallium arsenide and a solar photovoltaic cell by depositing cadmium sulfide films on glass. *The 1970s and 1980s marked the initial stage of commercial application.* In 1973, the United States developed a government-level plan to develop solar power generation with substantially increased funding support for research. A solar energy development bank was also established to promote the commercialization of solar energy products. In 1978, the United States built a 100 kW solar photovoltaic power station. *The 1990s to the twenty-first century marked a period of vibrant growth.* In 1992, the United Nations Conference on Environment and Development was held in Brazil, closely integrating the use of solar energy with environmental protection to promote technological innovation and international cooperation in the solar energy field. The United States proposed the Million Solar Roofs Program in 1997 and the University of New South Wales in Australia created a monocrystalline silicon solar cell with a world record-setting energy conversion efficiency of 25% in 1998. Into the twenty-first century, the solar energy industry has seen rapid growth, with many developed countries providing higher subsidies for new energy generation, and the installed capacity of solar power generation has recorded strong growth. As at the end of 2014, the installed capacity of China's largest solar photovoltaic power stations amounted to 200 MW, and there were three such stations in the country. In 2015, France will build a photovoltaic power station with an installed capacity of 300 MW the largest facility of its kind in the world.

Currently, there are three modes of photovoltaic power generation, namely: silicon-based, thin film-based, and concentrating solar power generation. Comparatively mature, the silicon-based mode has gone into commercial operation, with the highest energy conversion efficiency reaching 20%. The perovskite-type solar cell is a membrane solar cell generating most interest, with its energy conversion efficiency rapidly improved from 3% in 2009 to 16.2% in 2013. It was named one of the Top 10 scientific breakthroughs by Science Magazine in 2013. Concentrating solar power plants use mirrors to concentrate the energy from the sun on photovoltaic materials to improve the light intensity per unit area, with the conversion efficiency raised to more than 40% at a concentration level of 500-fold.

2.2.1.2 Development Direction and Outlook

2.2.1.2.1 Photovoltaic Panels.
Innovative materials can improve photoelectric conversion efficiency. For power stations in commercial operation, a variety of photovoltaic materials hold much promise. These include monocrystalline silicon, polycrystalline silicon, amorphous silicon, microcrystalline silicon, cadmium telluride (CdTe), and copper indium gallium diselenide (CIGS) materials. Silicon-based photovoltaic materials, such as monocrystalline and polycrystalline silicon, have a theoretical photovoltaic energy conversion efficiency of 38%, indicating huge room for future growth if compared with the current conversion efficiency of about 20% in commercial operations. Amorphous silicon, microcrystalline silicon, CdTe, and CIGS materials can be used to produce membrane solar cells, with a cell efficiency as high as 15%, a system efficiency of more than 8%, and a service life of over 15 years. Compared with silicon substrate solar panels, membrane batteries enjoy a clear cost advantage, with the potential for replacing silicon-based solar panels as photoelectric conversion efficiency improves, and for large-scale commercial applications. See Fig. 6.2 for the energy conversion efficiency of different photovoltaic cells.

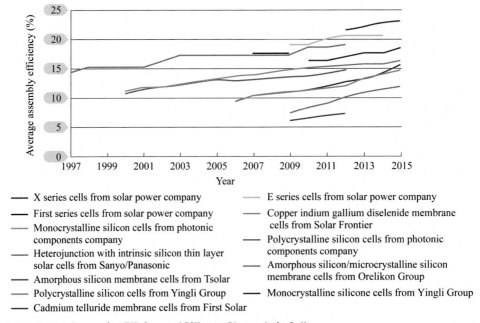

FIGURE 6.2 Energy Conversion Efficiency of Different Photovoltaic Cells

Source: Ref. [104].

The manufacture and installation of photovoltaic panels reflect a trend toward film thinning and streamlining. The cost of photovoltaic power generation is basically a function of material costs, with photovoltaic cells exhibiting a trend toward film thinning. In line with the upgrading of the silicon solar cell manufacturing process, the manufacturing cost of photovoltaic panels has declined by 80% since 2010. On the strength of their manufacturing cost advantages, a variety of membrane cells are gradually expanding their respective market shares to currently account for approximately 15% of the global photovoltaic market. After thin film production is realized, photovoltaic cells can be installed in buildings more easily or simply sprayed on building surfaces, which will save installation costs significantly and expand the rooftop and vertical coverage of solar energy utilization in urban buildings.

2.2.1.2.2 Photovoltaic Power Stations. *Solar tracking technology should be developed and utilization efficiency improved.* As the angles of sunlight change with different seasons and times of the day, a solar tracking system can adjust the angle of photovoltaic panels to gain the highest utilization efficiency by maximizing exposure to the rays of the sun that beat down vertically on the earth. With solar tracking technology, the annual solar irradiation intensity of regions with average solar resources can be improved from 1200 kWh/m^2 to 1500 kWh/m^2. While this technology has been maturely applied in France, the control technology of the tracking system is still complicated and costly. The utilization efficiency and economics of solar energy can be significantly improved when low-cost solar tracking systems become available for commercial application.

Following continued breakthroughs in solar photovoltaic technology and materials, the efficiency and economics of solar power generation will be improved, with promising prospects for large-scale commercial application. The average cost of photovoltaic power generation around the world will decrease from RMB 2 per kWh in 2010 to RMB 0.9 per kWh by 2020, representing as much as a 55% reduction. With the development of a globally interconnected energy network, solar power will become the most important energy source around the world. The costs of centralized and distributed photovoltaic power generation are expected to decline to RMB 0.24 per kWh and RMB 0.27 per kWh respectively around 2050, lower than the current costs of traditional fossil fuel-fired power generation.

2.2.2 Solar Thermal Power Generation
2.2.2.1 The Latest Technological Progress

There are four major types of solar thermal power generation technology, namely: the slot type, the tower type, the linear Fresnel type, and the dish type. The slot type generation technology has been at the stage of large-scale commercial operation. There have been known cases of commercial operation for the tower type, while the linear Fresnel type and the dish type are still at the stage of pilot demonstration. In 1950, the former Soviet Union designed the first tower-type solar thermal power station. In the 1970s and 1980s, many developed countries invested to build a series of experimental solar thermal power stations with government investment, given the higher efficiency and economics of photothermal power generation compared with the then costly solar photovoltaic cells. In 1981–1991, more than 20 solar thermal power stations with a capacity of over 500 kW were built around the world, the maximum capacity being 80 MW. Spain's installed solar thermal power capacity was the highest in the world. Plans are now underway to launch the Desertec project by eight European countries, including Spain, Italy, France and Germany, involving a joint investment of 400 billion Euros to build a super solar thermal power station in the Sahara desert over a period of

40 years. With a capacity of 100 GW, the planned power station in North Africa is expected to meet 15% of Europe's electricity demand.

2.2.2.2 Development Direction and Outlook

Solar thermal power generation technology has been developing in the direction of ever-larger capacity and higher parameters. Currently, solar energy generation can produce a steam temperature as high as 400–500°C, with a generation efficiency of 25%. An ultrasupercritical solar thermal power station capable of producing a steam temperature of over 600°C is under development in Spain. It can improve generation efficiency to over 30% by raising steam temperatures to enhance energy conversion efficiency. The Ivanpah Solar Electric Generating System in the United States, commissioned in February 2014, is the world's largest solar thermal plant with a total installed capacity of 392 MW. In the future, by expanding reflector numbers and installed capacities, the investment and operating costs of solar power generation can be further reduced. The reflectors of solar thermal power stations at the 100 MW level will cover millions of square meters.

Solar thermal power stations will be equipped with heat-storage equipment to provide a more stable output. Molten salt is the medium of thermal storage most commonly used today because of its high specific heat and stable performance under high temperatures. With an installed capacity of 20 MW, Gemasolar thermal power station in Spain can store heat for 15 h to provide an uninterrupted supply of electricity around the clock. In the future, substantially lower heat storage costs brought about by continuous innovation in heat storage technology will lead to a higher proportion of solar thermal power stations equipped with large-capacity heat storage devices. This will eliminate the impact of the day–night cycle on solar energy generation, making it an uninterrupted source of power supply with stable outputs.

The development of air-cooling technology for solar thermal power stations. Solar thermal power plants operate on the same principle as conventional thermal power plants where cooling of steam turbines and power generators is required. The equatorial regions abundant in solar energy resources are typically arid and semiarid desert areas like the Gobi desert with a lack of water resources. Solar thermal power plants equipped with air-cooling technology will be developed to reduce water consumption to adapt to operation in arid climates.

Currently, the cost of solar thermal power generation exceeds RMB 2 per kWh. However, with the progress of technology, it is expected to go down to below RMB 0.5 per kWh in 2050, more competitive than traditional fossil fuel-fired generation. With its characteristic output stability and improvement in energy storage technology, solar thermal power generation can ensure a stable supply of electricity.

2.2.3 Outlook on Forefront Technologies

Solar energy generation is a sunrise industry just beginning to develop. With the widespread application of new materials, solar power generation holds great promise with enormous room for innovation to improve efficiency conversion, reduce generating costs and achieve large-scale commercial application. Many countries hold this innovative technology in high regard, with a greater commitment to materials research.

1. *The technology pathway to higher efficiency conversion lies in new materials research, improved cell structures, and optimized joint operation of solar photovoltaic and photothermal power technologies.*

As a new type of material, perovskites can be used to produce solar cells to improve energy conversion efficiency. Experts from Oxford University in the United Kingdom and Universitat Jaume I in Spain have jointly developed a solar cell with a graphene titanium dioxide compound as its charge collector and perovskites as its light-absorbing material, realizing a photoelectric conversion efficiency of 16.2% in a laboratory environment, with the potential to improve to 50%, way above the comparative efficiency of 8–10% of widely-used photovoltaic materials like cadmium telluride, amorphous silicon, and microcrystalline silicon.

Based on a multi-PN[1] structure, the solar cell can fully leverage the solar spectrum to improve conversion efficiency. The National Renewable Energy Laboratory in the United States has developed two PN junction compounds, which have improved the conversion efficiency of the solar cell to 31.1%. Germany and France have jointly developed four PN junction compounds with a conversion efficiency of 44.7%, a new record in the world.

Optimizing the joint operation of solar photovoltaic and photothermal technologies will significantly improve conversion efficiency. Solar radiation is absorbed by a solar cell in the form of thermal energy, and the heat produced cannot be carried away in a timely manner through natural convection. In summer when the temperature can rise to as high as 80°C, the efficiency of photovoltaic power generation can be severely eroded. By using heat pump technology, we can focus the heat from photovoltaic cells on photothermal generating equipment to realize joint solar photovoltaic and photothermal operations. According to the research findings of Stanford University in the United States, the integrated utilization efficiency of photovoltaic and photothermal operations can reach 46%. A heating demonstration project on the joint photovoltaic, photothermal and heat-storage operations has been established in Gansu Province, China.

2. Cost reduction is achieved mainly through improving and saving materials

Replacement of the organic polymer materials in perovskite cells by inorganic copper iodide materials. According to surveys conducted by the University of Notre Dame in the United States, low-cost, high-performance inorganic copper iodide materials can replace the expensive organic polymer hole-transporting materials in perovskite cells. This will help reduce substantially the cost and improve the stability of perovskite batteries, and lay the foundation for large-scale production of perovskite solar cells.

Two-dimensional batteries can save photovoltaic materials. The two-dimensional solar battery, developed by the Massachusetts Institute of Technology in the United States, consists of one-atom-thick grapheme and molybdenum sulfide. Its thickness is only 1 nm, one of several thousand equal parts of the size of a traditional silicon-based solar cell, but its power density per unit mass is 1000 as much as a traditional solar battery.

Through material and technology breakthroughs covering perovskite batteries, multi-PN structured cells, joint photovoltaic/photothermal operation, copper iodide, and two-dimensional cells, the efficiency of solar energy generation will improve markedly, installed materials saved and costs reduced. If large-scale commercial operation can be realized for these new materials and technologies, currently still in the laboratory stage, the economics and market competitiveness of solar photovoltaic power generation will be significantly enhanced, lending strong support to the development of clean energy alternatives.

---

[1] A PN junction refers to the space charge region in the interface formed by putting the P-type and N-type semiconductors on the same base through diffusion with the adoption of different doping processes.

2.3 OCEAN ENERGY

Ocean energy can be classified into wave power, tidal power, tidal current power, ocean current power, as well as energy from temperature and salinity differences. *Wave energy* is found in ocean waves. *Tidal power and tidal current energy* are produced by the ebb and flow movement induced by the gravitational pull of the sun and moon. The vertical up and down movement of seawater produces tidal ebb and flow, known as potential energy, while the horizontal movement of seawater produces ocean currents, known as kinetic energy. *Ocean current energy* results from density and pressure gradients caused by an uneven distribution of seawater temperatures and salinities. *Temperature gradient energy* is the result of different temperatures between the upper and lower layers of the ocean at low latitudes due to the uneven absorption of solar radiation and the radial transfer of heat through the ocean circulation. *Salinity gradient energy* is the energy created from the difference in salt concentration between seawater and fresh water when a river flows through the estuary into the sea. Different types of ocean energy vary widely in power density. In terms of equivalent water head, tidal energy, wave power, salinity gradient energy, and temperature gradient energy are 10, 2, 240, and 210 m, respectively. As it is more difficult to extract chemical and heat energies than mechanical power, the water head of temperature/salinity gradient energy cannot be compared strictly with that of other energy forms.

2.3.1 The Latest Technological Progress

Of all ocean energies, tidal power is the earliest developed and most mature, having been a research subject for over a century. Currently, a total of 13 countries, including France, the United Kingdom, Russia, Canada, China, India, and South Korea, are conducting planning discussions and design studies regarding tidal power stations. The Lance Tidal Power Station in France, built in 1966, was the world's first facility of its kind in commercial operation. Comprised of 24 units, it boasts an installed capacity of 240 MW and annual generation of 544 GWh. There are currently seven tidal energy stations with a total installed capacity of 520 MW in the world. The Shi Hua Lake Power Station in South Korea is the world's largest facility of its kind, with a total installed capacity of 254 MW and annual generation of over 500 GWh.

Wave energy is a more extensively researched form of ocean energy. The first wave energy generating unit for commercial operation was successfully developed in Japan in the early 1960s. Since the 1980s, the target of wave energy development has shifted from power supply covering near-shore and coastal areas to far-away coastal locations and sea islands, demonstrating the successful practical application and commercialization of this technology on a small to medium scale. There are now over 30 wave energy stations in the world available for demonstration and operational purposes.

Tidal current and ocean current energies generate electricity by using the mechanical power from the horizontal movement of seawater. An experimental generator with a capacity of 2 kW was installed 50 m under the sea in the United States state of Florida in 1976. A 20 kW ocean current generator was developed by United States-based OEK Company in 1985, followed by the undersea installation and operation for a year of a 1.5 m diameter, 3.5 kW ocean current power generator by University of Yam in Japan in 1988. Research on tidal currents was conducted in the Strait of Messina by experts from Italy in 1996 and an experimental power station with a capacity of 120 kW was put into operation in 2004. A 300 kW tidal current generator for experimental purposes was installed on the west coast of the United Kingdom in 2003. In the 1970s, experiments were conducted for the first time on tidal current power generation in Zhoushan, Zhejiang Province. In the mid-1980s, generating equipment with a capacity of 1 kW were built, with the completion of experiments on generating units with a capacity of 70 kW in the late 1990s. China's first tidal current power station, with a capacity of 600 kW was developed in Daishan County, Zhoushan City, Zhejiang Province in 2002.

The utilization of temperature differences to produce energy dates back nearly a century, but substantial progress has only been achieved over the past 40 years. The world's first temperature gradient power generating equipment for practical application was developed in the United States waters of Hawaii in 1979, with a rated power of 50 kW. The maximum power output can reach 53.6 kW when the temperatures of the upper and lower levels of the sea are 28 and 7°C, respectively. Power output was improved to 210 kW in 1993. Over 10 countries, including the United States, Japan, the United Kingdom, France, the Netherlands, South Korea, India, the Philippines, Indonesia, Russia, and Sweden, are conducting research on temperature gradient energy. China has just started to utilize temperature gradient energy, with equipment for practical application yet to be built.

The utilization of salinity gradient energy has a relatively short history. Application technology in this area is basically at the experimental stage. No substantial progress has yet been achieved because of inhibitive costs due to formidable technical challenges.

2.3.2 Development Direction and Outlook

While the development of ocean energy has been around for more than a century, it is still not technologically mature and remains at the research and exploration stage. Compared with other renewable energies, ocean power is technically difficult to develop and is very costly due to constraints on energy density and development conditions. Take the currently most well-developed tidal energy station as an example, the investment cost is as high as RMB 30,000 per kW, three times that of photovoltaic power generation and four times that of wind power generation. The cost of developing wave power, temperature gradient energy, tidal and ocean current generation is even higher, with uncertain prospects for large-scale commercial application.

2.4 DISTRIBUTED GENERATION

Distributed generation technology refers to power generation facilities on the customer side connected to a nearby LV grid or multigeneration systems for integrated gradient utilization (including wind, solar, and other distributed renewable power generation), multigeneration equipment for residual heat, residual pressure and residual gas generation, and small natural gas-fired systems with combined cooling and heating capabilities. In essence, it is a small-capacity generating unit for development, grid connection, and energy consumption based on the proximity principle.

Currently, power from distributed generation covers the following areas of application. First, power is supplied to remote regions like oceanic islands and rural areas to resolve local supply problems. Second, it provides a back-up supply source for customers with high reliability requirements in the event of grid failures. Third, it supports peak shaving by supplying power during peak demand periods to help reduce peak loads. Fourth, it provides diverse energy products through a multigeneration system combining cooling and heating capabilities to meet diversified customer needs and improve integrated utilization efficiency. Fifth, it supports grid voltage readjustment, reduces power loss, and improves power factor. Sixth, distributed generation provides economic benefits that users can enjoy through an investment in grid access. In the future, the development of distributed generation will focus mainly on developing clean energy resources like solar, wind, and small hydropower generation near load centers.

2.4.1 The Latest Technological Progress

Distributed generation is nothing new. When the electricity industry began to develop, installed capacity was limited and voltage grades were low. Power generation was mostly distributional with power sources connected to a small local grid for distribution to local customers. However, with the advancement of power technologies, unit capacity and grid size have been expanding, with ever-higher voltage grades and increasingly conspicuous economies of scale. Contrarily, small capacity generating units have become increasingly less competitive. In the late 1980s, as energy security became a matter of global concerns, the United States and European countries refocused on building distributed power sources and quickly switched to distributed generation. A global transition was started from the traditional mode of centralized supply to a mode of supply based on a combination of centralized and distributed generation. The distributed power sources then were small diesel generators for emergency back-up, early coal-fired captive power plants, and small thermal power stations. However, these were phased out or simply replaced due to their poor technological performance, low efficiency, and environmental impact. In the twenty-first century, with significant efficiency gains and more environmental benefits, small hydro, wind energy, and solar power generation signifies an important direction of future energy development. In terms of single-unit control, distributed generation has achieved, on the customer side, joint operation of gas-fired and electrified equipment combining cooling and heating capabilities, raising the integrated efficiency of energy utilization to over 65%.

2.4.2 Development Direction and Outlook

Innovations in distributed generation technologies will focus on the technology for grid interconnection protection, control and power quality monitoring; the technology for synchronous operation of distributed generation, energy storage and controllable load with transmission grids; the technology for adapting to the demand response to large-scale grid access to distributed generation and the interaction among power sources, grids and loads; the standard technology governing information flows between distributed generation and distribution grids; and the technology for consumption of energy from high-penetration, distributed generation based on a virtual power plant.

Rapid growth of coordination and control technology for future distributed generation. It is expected that by around 2020, a host of technologies will basically become mature to adapt to the control requirements on distributed generation under complicated work conditions of interconnection operation as well as static and fault islanding. These technologies include the simulation technology for high-concentration, multisource coexisting systems at the distributed generation level, multisource complementary control for distributed generation, source charge coordination and control technology, equivalent virtualization technology for distributed generation and microgrids. In around 2030, equivalent simulation technology for centralized access to distributed generation will be mastered to realize flexible, interactive support for distributed generation and main grid power. Coordination and control technology based on flexible load control, energy storage devices and distributed generation will be acquired to exercise effective control over random power fluctuations and power quality in respect of distributed-generation, high-penetration grids. Intelligent recovery control technology and adaptive protection technology can support self-recovery and network reconfiguration after a breakdown of a distributed-generation, high-penetration grid. In around 2050, key information on the global geographical distribution, cluster size and output characteristics of distributed generation will be studied to realize flexible control over the equivalent virtual system for large-scale distributed generation. Based on a high-speed, globally interconnected telecommunication system, coordinated control across zones and layers can be realized over state-level grids, continent-level grids and globally interconnected grids to support efficient backup across different regions and times in response to typical load changes among the large clean energy bases in the Arctic and equatorial regions and across the ocean, relative to a global network of energy interconnections.

3 GRID TECHNOLOGY

A new energy landscape focusing on electricity with global allocations determines the crucial role that power grid technology will play in future energy development. The transmission capacity, allocation capability and economics of power grids need to be continuously enhanced, with the focus on the various components of the power system to accelerate the pace of technological innovation in robust smart grids. The major areas of innovation include UHV transmission technology and equipment, submarine cable technology, superconducting transmission technology, DC grid technology, microgrid technology, and supergrid operation control technology. Breakthroughs in these technologies will lay an important foundation for the development of a global energy network.

3.1 UHV TRANSMISSION AND EQUIPMENT

UHV transmission technology refers to the technology for transmission at an AC voltage of 1000 kV or above and at a DC voltage of ±800 kV or above. In recent years, China's UHV transmission technology has developed rapidly. UHV grids are responsible for transmitting wind and solar power from the northwest and hydropower from the southwest to load centers in the country's eastern coastal areas. The transmission distance has advanced from hundreds of kilometers to thousands of kilometers and single-line transmission capacity to 8 GW. The future global energy interconnection will be supported by backbone UHV grids to achieve large-scale, large-area allocation of clean energy globally.

3.1.1 The Latest Technological Progress

The late 1960s saw the launch of feasibility studies of UHV transmission at 1000 kV (1100 and 1150 kV) and 1500 kV and research and development projects on UHV transmission technology. The 1000 kV UHV AC transmission project commissioned by State Grid Corporation of China (SGCC) in 2009 is the world's first UHV transmission line to go into commercial operation. Since then, three 1000 kV UHV AC projects and four ±800 kV UHV DC projects have achieved commercial operation, the longest transmission distance being over 2000 km and maximum transmission capacity, 8 GW. See Table 6.1 for the major events in the development of UHV transmission projects worldwide.

3.1.2 Development Direction and Outlook

UHV transmission capacity and distance will be further enhanced. On the basis of the current UHV transmission technology, research and development work will be carried out on higher voltage, higher capacity AC transmission technology in the future. UHV transmission technology will be further improved in terms of transmission distance and capacity, with further technological breakthroughs in voltage control technology, insulation and overvoltage technology, electromagnetic, and noise control technology, external insulation interface, and key equipment manufacturing technology.

Research and manufacture of key equipment such as high-reliability converter transformers, converter valves, sleeves, and DC filters. DC transmission technology, key to achieving ultradistance

Table 6.1 Major Events in UHV Transmission Development Worldwide	
Year	Event
1960s	The former Soviet Union, the United States, Japan, and Italy, etc., proposed the development of UHV transmission technology, with work on planning, design and equipment research and development
1974	Work started in the United States on a 1000–1500 kV three-phase test line, which was subsequently put into operation
1978	The former Soviet Union started work on a 1150 kV (270 km) industrial test line stretching from Yitate to Novokuznetsk
1985	The world's first 1150 kV line, between Ekibastuz and Kekeqitafu, commenced on-load (<2 GW) operation under the rated working voltage. Since the early 1990s, the line has been downgraded to 500 kV
1988	Japan started building a 1000 kV UHV transmission line for transmission to Tokyo. The line is 426 km long and operating voltage has been downgraded to 500 kV. In addition, the construction of a UHV test field in Harunamachi was completed
2006	China's first 1000 kV UHV AC project was launched
2007	The Xiangjiaba–Shanghai ±800 kV. UHV DC demonstration project in China was launched
2009	The Southeastern Shanxi–Nanyang–Jingmen 1000 kV UHV AC pilot demonstration project in China was put into commercial operation
2010	The Xiangjiaba–Shanghai ±800 kV UHV DC demonstration project in China was successfully put into operation
2012	The Jinping–South Jiangsu ±800 kV UHV DC project was put into operation
2013	Huainan–North Zhejiang–Shanghai 1000 kV UHV AC project was put into operation
2014	The South Hami–Zhengzhou, Xiloudu–West Zhejiang ±800 kV UHV DC project and the North Zhejiang–Fuzhou 1000 kV UHV AC project were put into operation

and ultralarge capacity transmission, represents one of the major technological means to connect large energy bases with load centers. Recently, breakthroughs have been achieved in the research on the topography of ±1100 kV modular transmission-source converter valves and also in converter valve technology. An all-round breakthrough is expected around 2018 in ±1100 kV UHV DC transmission technology to realize engineering applications to achieve transmission distance of over 5000 km and transmission capacity, 12 GW.

Research and manufacture of UHV transmission equipment suited for extremely hot and cold climates. Currently, UHV projects operate at temperatures as low as −50 to −40°C and as high as 50–60°C. However, the Arctic region's lowest temperature is −68°C and the equatorial region's highest ground temperature is over 80°C, both temperatures beyond the tolerance limits of existing UHV transmission equipment. The insulation performance of electric materials will weaken under extremely high and low temperatures. Therefore, research shall be conducted on the key technologies of equipment well suited for operation in such extremely hot and cold climates. By around 2030, a breakthrough will be achieved in the development of a full set of UHV DC equipment with the properties required. The UHV DC compact converter station concept will go into practical application to meet the export requirements of large clean energy bases in the Arctic and equatorial regions and elsewhere.

Currently, the cost of 1000 kV UHV transmission is only about 72% of that of 500 kV EHV transmission. With the construction of a global energy network, the transmission cost will further decrease subject to successful production of UHV equipment on a large scale.

3.2 SUBMARINE CABLE

To build a globally interconnected energy network, transcontinental interconnections have to be developed across the ocean between each geographical pairing, such as Africa and Europe, Europe and North America, Australia and Asia, and Asia and North America. Submarine cable technology is an indispensable part of the technology set required to achieving transcontinental interconnections as part of a global energy network.

3.2.1 The Latest Technological Progress

Insulation technology is the key point for making a breakthrough in submarine cable technology. Currently, submarine cables feature the following types of insulation: impregnated paper-wrapped cables, applicable to AC transmission lines up to 45 kV and DC transmission lines up to ±400 kV, with a maximum installation depth of 500 m underwater. Self-contained oil-filled cables, applicable to UHV AC or UHV DC transmission lines, can be laid to a depth of 500 m underwater. Restrained by oil-filled pressure, self-contained oil-filled cables provide short transmission distances, making them particularly suitable for transmission over sea channels with a short crossing distance and for export of renewable energy from near-shore locations. Extruded insulation cables, including cross-linked polyethylene insulated cables, are applicable to 200–400 kV AC transmission lines. With its high reliability and long transmission distance, the extruded insulation cable marks an important direction of cable development. Inflatable insulated cables, insulated by impregnated paper, are better suited for longer-distance transmission underwater. However, as installation is required to be performed under deep water at high atmospheric pressure, the design of this cable type and its components is more difficult and installation depth is generally restricted to 300 m underwater.

Since the 1990s, among the world's submarine cable transmission projects, 15 AC transmission projects have been developed, including five at a voltage grade of 500 kV or above. Since 2000, three 500 kV AC submarine cable transmission projects have been put into operation, compared with 67 DC transmission projects, which include 13 HVDC transmission projects under construction or planning. DC submarine cable technology is becoming an important means of building cross-sea grid interconnections and also connections with offshore renewable energy generation. The Norway–Netherlands submarine cable transmission project, featuring HVDC transmission, is scheduled for operation in 2016–2018. The design capacity is 1.4 GW. The 600 km submarine cable across the North Sea is the world's longest submarine cable. Under the North America Power Grid, there are 14 submarine cable transmission projects, with a design transmission capacity of 5.762 GW and a total cable length of 1718 km underwater. The Neptune Project in the United States is based on a ±500 kV DC submarine cable network, with a maximum water depth of 2600 m as the world's deepest submarine cable.

3.2.2 Development Direction and Outlook

High-voltage, long-distance and large-capacity submarine cables represent the major direction of future cable development. Currently, XLPE insulated cables are more popular, with maximum voltage grades at AC 500 kV and DC ±320 kV. The highest voltages of oil-filled cables are AC 765 kV and

DC ±500 kV. The longest submarine cable in China is the Guangdong-Hainan 500 kV AC interconnection project across the Qiongzhou Strait. Featuring oil-filled insulation technology, the 31 km
project can become part of an interconnection across the strait and support the outward transmission
of electricity from offshore wind and ocean power bases in deep-water regions, subject to the successful development of UHV cables at AC 1000 kV and DC ±800 kV with a transmission distance
of over 100 km.

In response to the development requirements of a globally interconnected energy network, research
and manufacture of UHV AC and UHV DC cables must be completed before 2030. After 2030, conditions will be in place for large-scale application of UHV AC/DC cables and submarine cables to better
support the construction of cross-sea projects forming part of the future global energy network.

3.3 SUPERCONDUCTING POWER TRANSMISSION

Superconducting power transmission technology is a transmission technology featuring superconductive
materials of high electric current density. At a superconducting state, the DC resistance of superconductive materials is basically nil with virtually no thermal loss. Since the discovery of the phenomenon of
superconductivity, close to 40 superconductive elements and thousands of superconductive alloys and
compounds have been identified. The transmission capacity of superconducting transmission lines can
reach 3–5 times that of AC lines and 10 times that of DC lines at comparable voltage grades.

3.3.1 The Latest Technological Progress

In 1911, Heike Kamerlingh Onnes of Leiden University in Holland discovered the phenomenon of
superconductivity. When cooled to a super low temperature of −268.98°C, mercury was observed to
have lost all electric resistance. Subsequently, this superconductive property had successively been
found in numerous metals and alloys. Kamerlingh Onnes was awarded the Nobel Prize in physics in
1913 for the discovery. The discovery led to further research into high temperature superconductivity around the world. During 1911~1987, the superconducting temperature increased to 53 K from
the original 4.2 K (0 K = −273.15°C). The discovery of high temperature superconducting makes the
application of superconducting technology physically practicable. A large number of demonstrative
applications had been conducted of superconducting technology at home and abroad. In terms of transmission capacity, the largest one was the demonstrative project on the Long Island AC transmission
line in New York, at 138 kV and rated current 2,400 A with a transmission capacity of 574,000 kW. In
April 2014, a 1,000 m superconducting cable was connected to the grid in Essen City, Germany, then
the longest superconducting line used for an interconnection. The longest superconducting transmission line now under study is in Amsterdam, Netherlands, with a design cable length of 6,000 m. China
has made important breakthroughs in superconducting technology and the superconducting critical
temperature has increased to about −120°C (i.e., 153 K). At the current development stage of superconducting technology, superconducting transmission is still not operable without a low temperature
of below −100°C, and the poor malleability of superconducting ceramic materials can hardly support
long distance transmission.

3.3.2 Development Direction and Outlook

*A high temperature superconductor is generally made of ceramic materials. Due to the poor malleability of
ceramics, high temperature superconductors cannot be used to build cables for long distance transmission.*

To realize long-distance, large-capacity transmission, significant breakthroughs in high temperature super-conducting materials are required. Strict operating temperature constraints on superconducting transmission lines are a key hindrance to the realization of large-capacity transmission. Liquid nitrogen, as the most commonly used and most economical cooling medium for superconducting materials, can reach −196°C. However, the performance of insulation materials will sharply decline under this low temperature. At a maximum voltage grade of 138 kV, superconducting cables are more suited for application in urban distribution networks. Superconducting materials are already costly, at US$300–500 per kA/m. Taking into account other factors such as low temperature requirements for operation and maintenance, the production and operation costs of superconducting transmission lines will be higher. So economics is also an important factor restricting the applicability of high temperature superconducting transmission technology. In 2008, the world's first high temperature superconducting cable operating in a commercial grid was manufactured in the United States at a cost of US$18 million for a total length of 610 m. Judging by the development over the past 30 years, the progress of research into superconducting transmission technology has been slow, with no substantive breakthroughs in sight. For short-distance transmission, superconducting technology can be incorporated into hub substations to reduce energy loss and the space required for large-capacity transmission. For long-distance transmission, the integration of power transmission with the shipment of liquid hydrogen can realize integrated transport of energy to lower costs. But the prospects of large-scale commercial application are hard to predict.

3.4 DC GRID

The DC grid, based on flexible DC transmission technology, is an energy transmission system formed by a massive collection of interconnected DC transmission lines. The objective of DC grid development is to construct a large-capacity power transmission system capable of achieving smooth access for new energy, independent control pf active and reactive power, large-capacity and long-distance transmission, fast and flexible power allocation, as well as power regulation and mutual backup on an overall basis. Compared with a traditional DC transmission system, a DC grid can provide higher supply reliability and equipment redundancy, a more adaptive mode of supply, and flexibility and safety in flow control. Multiterminal DC transmission is the initial stage of DC grid development. It is a transmission system formed by more than two convertor stations linked up in series, parallel or parallel-series connections, featuring multiple sources of supply and multiple points at the receiving end. Compared with multiterminal DC transmission, a DC grid provides better economics and greater safety in terms of large-scale clean energy generation and access for distributed generation, power supply for oceanic islands, clustered output of offshore wind farms, and new city grid construction. The DC grid is one of the important directions for future grid development. With the innovations and breakthroughs made in DC transmission and interconnection technologies, UHV DC grid technologies are expected to come into shape to become one of the key technologies for building backbone structures as part of a global energy network.

3.4.1 The Latest Technological Progress

In 1999, the world's first flexible DC project with access to wind power was commissioned in Gotland, Sweden, at a voltage of ±80 kV and with a transmission capacity of 50,000 kW. Currently, the developed countries of Europe and America are starting to look into the development of a new generation of flexible transmission networks based on DC transmission technologies. In 2008,

Europe announced plans for a "supergrid" of power supplies, with the development of an extensive network of intelligent DC grids based on HVDC transmission (mainly flexible DC transmission). The plan brings together resources, including wind, solar and hydroelectric power, in the North Sea, the Baltic Sea, North Africa, and other regions. Work on different aspects of the proposed program has commenced, covering theoretical fundamentals, key technologies, core equipment and project construction. The plan is now at the stage of phased implementation. UK-based National Grid has developed plans for building, in the East coast region of the United Kingdom and the North Sea regions, a flexible DC transmission network comprised of tens of large-scale offshore wind farms and nearly 50 flexible DC transmission lines. Interconnections with grids in Norway and other countries, are also planned to balance fluctuations over large areas and consume more clean energy. In 2006, China started research on different areas of flexible DC transmission technologies, covering theoretical fundamentals, key technologies, core equipment, experimental capacity building, and system integration. In July 2011, China's first flexible DC transmission demonstration project was commissioned in Nanhui, Shanghai. In July 2014, a ±200 kV multiterminal flexible DC transmission project was inaugurated in Zhoushan, Zhejiang.

3.4.2 Development Direction and Outlook

For future DC grid development, fundamental research has to be conducted on different areas. These include DC grid topology, theoretical fundamentals of grid construction, static characteristics of DC grids and the principle of interaction between DC and AC grids, dynamic characteristics of DC grids and theoretical fundamentals of safety assessment, fault protection for DC grids and grid reconfiguration methodology, and theoretical fundamentals of operational reliability for DC grids and safety evaluation methods. Breakthrough innovations also have to be sought in research into core DC grid equipment as well as equipment research and manufacture. In particular, higher-voltage DC grid technology must be developed to be able to support large-scale, ultralong distance transmission and clean energy consumption.

In around 2020, a basic theoretical framework for DC grids will be initially established, with the completion of the economics analysis of DC grid technology involving interconnections at two voltage grades and associated engineering work. Looking forward to 2030, with the growing maturity of DC grid-related technologies, economics analysis of the technology involving DC grid interconnections at several voltage grades will have been completed, with the relevant projects moving into actual implementation. By around 2050, DC grid technologies will enter a period of active promotion and render strong support to large-scale grid access for renewable energy.

3.5 MICROGRID

Microgrid technology refers to a local-area management technology applied to distributed energy supply systems and loads. Currently, microgrid technology is still at the stage of pilot demonstration at home and abroad, with commercial operation yet to be achieved.

3.5.1 The Latest Technological Progress

The microgrid concept was first proposed by the United States in the 1990s. A number of major power outages broke out in the country in the twentieth century, sparking mounting concerns over the reliability of transmission systems. In 1999, the "microgrid" was first proposed by CERTS to improve supply reliability, with work on technology research and demonstration surrounding microgrids. The European Union and Japan see microgrids as a solution to the grid integration of high-penetration distributed generation. Following the proposal of the smart grid concept, the European Union and Japan have incorporated microgrids as part of the smart grid framework. In China, there are 14 pilot microgrid projects completed or under development. The following features of these projects can be observed. First, the voltage grades of the projects are relatively low, at 380 V or 10 kV. Second, these projects are relatively small, with a capacity of less than 5000 kW. Third, the projects are of different types, including grid-connected microgrids in urban areas, grid-connected microgrids in remote farming and stockbreeding areas, and off-grid microgrids in island regions. Currently, these pilot projects in China are designed mainly for research and validation purposes, with the focus on key microgrid technology.

3.5.2 Development Direction and Outlook

Current research on the operation control of microgrids is centered on simple AC microgrids. In order to boost innovation in microgrid technology and better integrate it into the ubiquitous smart power grids in different countries, in-depth research will be required into complex AC/DC microgrids, combined cooling, heating, and power microgrids, control technology for multimicrogrid parallel operations, and coordinated operation of microgrids and large power grids.

Technology for optimizing coordinated operation of microgrids and large power grids. The wide variety of microgrid equipment gives rise to variance in control methodology and operational characteristics. The complexity of operation control and protection for microgrids calls for research into design standardization in terms of control system structures, communication networks, and control technology, in order to standardize and modularize complete equipment, while improving the universality and expandability of the coordination and control systems for microgrids and large power grids.

Technology for managing microgrid energy optimization. A microgrid incorporates multiple energy sources like solar, wind, and biomass energy, together with a variety of energy conversion devices such as fuel cells and energy storage systems to provide customers with different energy products, like cooling, heating, and electricity. Given the greater uncertainty and time variance involved, more in-depth research is required on the technology for managing microgrid energy optimization so as to optimize microgrid operations and improve overall operational efficiency.

With the development and integration of microgrids and distributed generation technology, plug-and-play distributed power, flexible interactions at the grid level with the demand side as well as coordinated operation with large power grids will be realized and become an integral part of ubiquitous smart grids at the national level.

3.6 LARGE-GRID OPERATION CONTROL

An ultra-large AC/DC power grid ("large power grid" for short) provides an important fundamental platform for aggregation of standard power generation sources, long-distance transcontinental transmission and large-area, flexible allocations. It basically forms the backbone structure of a global energy interconnection, structurally characterized by grid access for diversified energy sources, equipment variety, and wide geographical coverage. Operationally, it is marked by large transmission capacity, frequent flow fluctuations, and complex interferences. The control technology for large-grid operations is key to the development of a global energy network and the assurance of operational safety and stability. Examples of this control technology mainly include control technology for large-grid operations, simulation technology, operation control technology for grid-connected access to large-scale, intermittent energy sources, fault recovery technology, and automatic reconfiguration technology.

3.6.1 The Latest Technological Progress

In the initial stage of electricity development, electric power lines were used just to connect power plants to electricity consumers, falling short of forming a power network. During this period, the focus was on controlling generation output and ensuring the stability of frequencies and voltages, with the dispatch control center usually located in the power plant. In the 1920s, with the rapid growth of electric loads, power networks gradually came into shape to ensure supply continuity and stability through grid distribution and power modulation. At this stage, the scope of control was expanded to cover generating plants, power grids, substations, active and reactive power sources, voltages, power system-level flow and economic dispatch, and power protection devices. After the 1970s, with the growing capacity of generating units and the development of transmission technology, power grids entered a period of UHV development. A large grid features dispatch control usually on a zone-by-zone and layer-by-layer basis, with individual dispatch centers set up to ensure the reliability of automatic control systems and better accommodate system capacity expansion and structural changes.

Into the twenty first century, the method of zone and layer-based control has become relatively mature. Control technology has improved significantly, evidencing a move from "off-line forecasts and real-time matching" on a precontrol basis toward "online forecasts/decision-making and real-time matching" and further to "real-time calculation and real-time control".

3.6.2 Development Direction and Outlook

The security and stability mechanism, characteristics and analytical technology of large grids. For the security and stability mechanism of large grids, one needs to look at research areas such as transient stability analyses, the characteristics of regional interconnected-grid low frequency oscillation and its causes, the mechanism of voltage stability, the mechanism of cascading failure in large grids, and the application of complexity theory to the analysis of the mechanisms of major outages. For the security and stability characteristics of large grids, research areas include the impact on grid security and stability of access for distributed generation, access for major generating units, mixed AC/DC transmission and flexible AC transmission technology applications, as well as new transmission and transformation technologies. For the security analytical technology of large grids, research should cover online analysis and defense systems, online dynamic security assessments and prewarning systems, online monitoring and analysis technology based on wide area measurement systems for power systems, artificial intelligence technology used for online security and stability analysis, online real-time decision-making systems for grid security and stability control, and grid security assessment technology based on risk theory.

Real-time/super real-time simulation and decision-making technology. Offline, online, real-time, and super real-time operations have progressed in chronological order. The growth of grid capacity imposes ever-higher requirements on the timeliness of grid operation analysis and decision-making. In the early days, power system control was based on off-line calculations due to the constraints of computer technology. However, with the progress of information technology, more and more calculations can be performed online or even on a real-time basis. Dynamic security assessments and prewarnings can be conducted online. Through online transient stability analyses, we can not only assess the security level of current systems, but also develop prevention and control strategies to help dispatchers with operational adjustments and improve grid operational safety. Through real-time simulation, we can simultaneously simulate the actual running state of grids. Super real-time simulation is achieved if the simulation results can be obtained before the occurrence of a real incident. In the future, the development of grid-based real-time and super real-time simulation technology will further improve the security control of large grids at the operational level.

Fault diagnosis, recovery and automatic reconfiguration technologies of power grids. On the strength of technology innovations, such as online fault monitoring and diagnosis, new relay protection and wide-area backup protection, fault recovery strategy optimization and smart reconfiguration technology, power grids demonstrate strong security, stability and fault self-recovery capability in different operating environments and in the face of different types of fault. These technologies can greatly improve the defense capability of large grids against cascading faults, extreme weather conditions, and harmful external conditions. With the development of computer technology and control theory, operational control at the grid level is gradually moving in the direction of forecasting, prewarning, and automatic fault recovery. Highly automatic operational control, expected to materialize after 2030, can support day-ahead forecasts for renewables-based power generation within a 5% margin of error and help achieve a global dynamic equilibrium among renewables, traditional energy and demand loads.

4 ENERGY STORAGE TECHNOLOGY

The development of energy storage technology is crucial for ensuring the large-scale development of clean energy as well as the secure and economic operation of power grids. Incorporating electricity storage technology into the power system can add a "flexible" touch to an otherwise "rigid" power system that maintains balance in real time. In particular, this will smoothen out the fluctuations arising from large-scale grid access for clean energy, with improvements in grid-level operational security, economics and flexibility. Energy storage technology generally refers to the technology for thermal energy storage and electric energy storage, with electricity storage to become the dominant application for the future global energy interconnection. The structure of a power grid with energy storage capability is shown in Fig. 6.3.

4.1 LATEST PROGRESS

Electricity storage technology is classified mainly into physical energy storage (pumped-storage, compressed-air energy storage, and flywheel energy storage), electrochemical energy storage (lead–acid cell storage, sodium–sulfur cell storage, flow battery storage, lithium ion battery storage, metal–air cell storage, and hydrogen storage), and electromagnetic energy storage (superconductive electromagnetic energy storage and supercapacitor energy storage).

4.1.1 Physical Energy Storage

As the most mature energy storage technology, *pumped storage* technology is in extensive application at relatively low costs. The world's total installed pumped-storage capacity currently stands at more than 100 GW. The top three countries in terms of pumped storage capacity are Japan, the United States

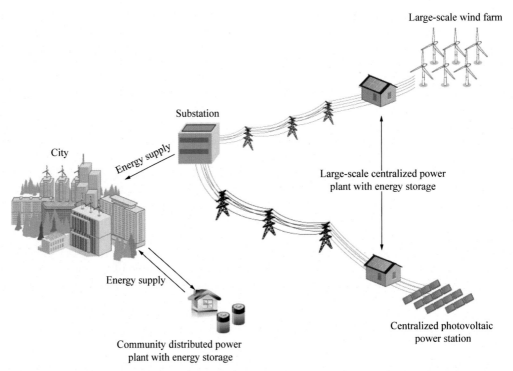

FIGURE 6.3 Structural Diagram of Power Grid With Energy Storage Capability

and China, accounting for 26.27, 22.29, and 21.53 GW, respectively. With a total installed capacity of 2.4 GW, the Guangzhou Pumped Storage Station in China is the world's largest facility of its kind. Given the abundance of hydropower resources in the world, larger-capacity pumped-storage units can be installed by capitalizing on the right terrain to better ensure supply security.

Compressed-air energy storage operates on the principle of transforming electric energy into potential energy storable in compressed air by using surplus electricity produced in low-demand periods to drive the air compressor and press the air into a large-capacity air reservoir. At times of power undersupply, the compressed air is burnt with oil or natural gas to drive the combustion gas turbine for power generation and meet the requirements of system-level peak load regulation. Compressed-air energy storage has the advantages of high capacity, long service life, and good economics. But the generation process requires burning fossil fuels, which results in pollution and carbon emission. Currently, compressed-air stored energy technology is basically at the laboratory modeling or small-scale demonstration stage.

Flywheel energy storage works by accelerating a rotor (flywheel) to a very high speed and storing the kinetic energy generated. The rotor becomes a generator when energy is extracted from the system. Featuring low energy density and instead of large-scale energy storage, this energy storage technology is suitable for short-term energy storage to resolve problems of power quality and impulse-type electricity consumption.

4.1.2 Electrochemical Energy Storage

Electrochemical energy storage is at the forefront of energy storage technology. In recent years, this type of energy storage, like sodium–sulfur cells, flow batteries, and lithium ion batteries, has experienced rapid growth with tremendous potential and broad prospects of application. It is expected to lead the pack in commercial development. To bring down production and operation costs, technology breakthroughs are required in battery materials, manufacturing process, system integration, and operational maintenance in coming years.

With a history of 140 years, *lead–acid cell* storage is the most mature technology of its kind, currently accounting for more than half of the battery market. A low-cost and safe storage option, this technology is used mainly on electric bicycles. But lead–acid cells are not suitable for grid energy storage due to their low energy density, massive size, and also the poisonous materials involved in production.

With a high energy density, the *sodium–sulfur cell* is well-attuned to modular manufacturing and easy on transportation and installation. It is suitable for providing emergency backup to meet special load demands. The sodium–sulfur cell was invented by Ford Motor Company in the United States in 1967. Getting a head start on sodium–sulfur cell research, Japan developed successfully in the mid-1980s, a sodium–sulfur cell with a high energy density of 160 kWh/m^3. In 1992, Japan carried out a demonstrative operation of the world's first sodium–sulfur cell energy storage system. As at the end of 2002, more than 50 sodium–sulfur cell energy storage stations were in demonstrative operation in Japan. In July 2004, the world's largest sodium–sulfur cell energy storage station with a power rating of 9600 kW and a capacity of 57,600 kWh was formally inaugurated in an automatic system plant under Hitachi Limited. In 2009, State Grid Corporation of China completed a medium-sized 2000 kW pilot production line for sodium–sulfur cells, making China the world's second country after Japan to master the core technology of high-capacity, single-cell sodium–sulfur batteries.

Flow cells feature high capacity and a long recycling life, with recyclable electrolytic solutions and separately designable capacity and power. Flow cells are bulky and use vanadium, a toxic compound, as a raw material for production. As early as in 1974, NASA successfully developed the world's earliest flow cell. Currently, the full vanadium flow electricity storage system and the sodium polysulfide/bromine flow electricity storage system are two relatively mature cell types already at the stage of demonstrative operation.

Lithium ion cells use a Li ion-containing compound for the positive pole and a carbon material for the negative pole. The lithium ion battery features large energy density, high average output voltage and low self-discharge. It has no memory effect and operates in a wide temperature range from −20°C to 60°C. Known as a green product, the long-life cell demonstrates superior recycling performance without using any toxic or harmful substance. Currently, the lithium ion battery is used extensively on cell phones, laptops, electric vehicles and other equipment. However, economically it is not well-placed for use on power systems or large-capacity energy storage devices, as a charge–discharge cycle costs more than RMB 1 per kWh.

The metal–air cell is a new fuel cell type that uses metal fuels rather than the hydrogen energy in traditional fuel cells. Nonpoisonous and pollution-free, it offers many advantages, like steady discharge voltage, high energy density, low internal resistance, long service life, relatively low costs, and low requirements on process technology. It is an electricity generating device that incorporates oxygen and metals like zinc and aluminum, with the potential to open up a new generation of green energy storage batteries on account of the low cost, abundance and recyclability of the raw materials used for production. It is also structurally simpler compared with the hydrogen fuel cell.

Hydrogen energy storage technology has commanded widespread international attention as a source of clean, efficient and sustainable carbon-free energy. National plans for hydrogen energy development have been developed in the United States, Japan and some European countries, with a number of large-scale demonstrative applications of hydrogen energy storage already in place. These include an energy storage demonstration system designed for renewable energy integration, a hydrogen refueling station for fuel cell-powered cars, as well as hydrogen-based methane and natural gas–hydrogen blends. The hydrogen cell developed by the National Renewable Energy Laboratory can sustain a driving range of over 1500 km by using 15 kg of liquid hydrogen fuel. The scope of application of hydrogen energy storage technology has expanded from motorized transportation to the energy field. But ion exchange membranes for the fuel cell are still costly. Technology breakthroughs have to be achieved in cost reduction and in hydrogen production and storage before commercial operation can be realized.

4.1.3 Electromagnetic Energy Storage

The supercapacitor developed in the 1970s–1980s is an electrochemical component for storing energy based on polarized electrolytes. As no chemical reaction takes place in the storage process, the energy stored is reversible and the device can be repeatedly charged and discharged hundreds of thousands of times. The supercapacitor features high power density, short charging and discharging times, long recycling life and a wide range of operating temperatures. However, it is not suitable for large-capacity energy storage on the grid, given its low storage capacity.

Utilizing the zero electric resistance of a superconductor, a *superconductive electromagnetic energy storage* device offers the advantages of large instantaneous power, light mass, small volume, zero loss, and quick response. It can be used to improve power system stability and power quality. However, prospects for the future application of this technology are clouded by low energy density, limited capacity, and the fact that it is subject to the development of superconducting materials technology.

4.2 DEVELOPMENT DIRECTION AND OUTLOOK

Large-capacity energy storage can support peak shaving and valley filling in the future global energy interconnection. Large-capacity, long-term storage facilities such as pumped storage devices and compressed-air storage equipment can be used for peak load regulation in support of a large grid. Flow batteries of large storage capacity, multiple circulation times, and long service life can be used to support energy storage devices on a grid. Hydrogen storage can be used to store surplus wind and solar energy to drive fuel cell-powered vehicles.

Large-scale power-type energy storage can be used to smoothen out fluctuations of large-scale clean energy generation. Power-type energy storage devices such as supercapacitors, superconducting magnetic energy storage, flywheel energy storage, and sodium–sulfur batteries, are operated in connection with large renewable energy generation, to quickly respond to the output of wind and photovoltaic power, smoothen out fluctuations of renewable energy generation, and ensure the safety of real-time grid operations.

Small stored energy batteries can be used on electric vehicles. Stored energy devices such as lithium batteries, new lead–acid cells and metal–air batteries, are of higher energy and power intensity but poor homogeneity, which makes them difficult to form large-capacity battery packs. The type of energy storage is used mainly on electric vehicles instead of large power stations. With an extended service life and lower costs, stored energy batteries can meet the requirements of large-scale electric car development.

In the future, stored energy batteries on electric vehicles will be connected to the global energy interconnection to support peak load regulation at well-planned charging times through charging in low-demand periods and discharging in peak-demand periods.

4.3 FOREFRONT TECHNOLOGIES

The key to more advanced stored energy technology lies in making new breakthroughs in materials technologies. Through continuous innovation and development of new stored energy materials, important breakthroughs are expected to be achieved in prolonging the service life of stored energy components, improving energy intensity, shortening charging time, and reducing costs.

4.3.1 Substantial Improvement in Battery Service Life

New lithium–sulfur cells were developed by the Pacific Northwest National Laboratory of the United States Department of Energy, combining the graphite in lithium batteries and the lithium in traditional lithium sulfur batteries to quadruple the battery's service life and increase the number of charging and discharging to 2000 times. The lithium sulfur battery is a solid battery, with high energy intensity and easy application on a small scale. The organic quinine redox flow battery, developed by the University of Southern California, is of the water-soluble type with a life expectancy of 15 years based on 5000 charge cycling tests.

4.3.2 Significant Improvement in Energy Intensity

New solid lithium ion batteries were developed by Toyota Corporation, with an energy intensity of up to 400 Wh/L(watt-hour/liter), twice as much as that of the lithium ion battery now widely used in the market. This battery type has the potential to go into commercial application by 2020. The molten salt air battery is another new type of battery developed by the George Washington University. Based on the chemical characteristics of different molten salts in the lab, the energy intensity of ferric iron, carbanion and vanadium boride-based molten salt batteries can reach 10, 19, and 27 kWh/L, respectively, far above the lithium air battery's 6.2 kWh/L. This battery type features higher stored energy capacity at lower costs.

4.3.3 Significantly Shortened Charging Time

The graphene lithium ion battery is a new battery type developed by Nanotek of Ohio in 2012. It features high charging speed, multiple circulation times, and high specific energy. The negative pole of the lithium battery is made of graphene, on the surface of which lithium ions move around swiftly in large amounts, resulting in greatly improved charging speed. At the end of 2014, the Spanish company of Graphenano, in cooperation with Universidad de Córdoba, completed a prototype design for the polymeric grapheme battery, with a specific energy of over 600 Wh/kg. Currently, electric vehicles equipped with graphene batteries can sustain a driving distance of up to 1000 km, with a full charge completed in 8 min.

4.3.4 Sharply Lower Battery Costs

The rechargeable redox flow battery is a low-cost, membraneless redox flow battery developed by the Massachusetts Institute of Technology. Maximum power intensity can reach up to approximately 0.8 W/cm^2, treble that of the ordinary membraneless battery type and at a production cost of lower than US\$100 per kWh, only one-quarter that of present-day lithium ion batteries.

5 INFORMATION AND COMMUNICATION TECHNOLOGY

Information and communication technology (ICT) provides the very foundation for realizing intelligent and interactive power grids as well as operation control over power grids. ICT covers the information and communication aspects of the technology. Focusing on the coding and decoding of information, information technology is about collecting, identifying, extracting, converting, storing, delivering, disposing, searching, detecting, analyzing and utilizing information. Communication technology, with the focus on information transmission technology, mainly covers transmission access, network switching, mobile communication, wireless communication, optical fiber communication, satellite communication, support management, and private network communication. ICT is regarded as an important driving force behind social development and global economic growth in the twenty-first century. Representing a combination of different technologies involving cross-sector integration of many industries, ICT is bringing about a profound industrial revolution. To cater to the development needs of a globally interconnected energy network, the rapidly growing information and communication contents, and the substantially expanding scope of information and communication, stricter requirements need to be imposed on the safety, real-time and reliability aspects of ICT, which points to the urgent need for greater innovation and breakthrough in the ICT area.

5.1 LATEST PROGRESS

5.1.1 Communication Technology

For more than a century since Samuel Morse's invention of the telegraph in 1835 and Alexander Graham Bell's invention of the telephone in 1876, communication technology has come a long way from analog signals to digital signals, from carrier signaling and microwave communication to optical fiber communication and satellite communication, from wired to wireless communication, and from fixed line communication to mobile communication. Nowadays, optical fiber communication, mobile communication and satellite communication, among other communication technologies, are being constantly upgraded, with transmission capacity growing.

Optical fiber communication technology fulfils the function of communication based on transmission through optical fiber lines. It serves a backbone for information networks, with the advantages of rapid transmission speed, small transmission loss, and high signal quality and safety. Commercial networks currently run at a speed of 400 GB/s. In August 2014, Danish scientists achieved data transmission of 43×10^3 GB/s (5300 GB of data/s) in laboratory conditions using a single optical fiber and a laser generator. This technology makes it possible to download a movie of 1 GB in just 0.2 ms. Experts expect virtually unlimited broadband for users with the continued development of optical fiber over the next 10 years.

Mobile communication technology integrates the newest technology of wired and wireless communications, realizing information transmission between mobile devices as the fastest developing technology of the twenty-first century. Since the advent of 1G (first generation) communication technology in the 1970s, we are now moving into the 4G (fourth generation) era. 4G technology can achieve a downlink peak data rate of 1GB/s and an uplink peak data rate of 500 Mb/s in the 100 MHz broadband. At the same time, 5G technology has become the focus of research around the world. Compared with 4G, 5G can transmit 1000 times more data volume at a data rate, 100 times faster, of 10 GB/s. The number of network-connected devices will also increase by 100 times, and end-to-end delays will shorten by

80%. 5G technology will meet the needs brought on by the fast-increasing intelligent terminals and rapid growth of the mobile Internet.

Satellite communication technology realizes communication between two or more earth stations by using artificial earth satellites as a relay station to retransmit radio waves. Comprised of satellites and earth stations, a satellite communication system is known for its wide coverage and high reliability, not being subject to the restrictions of complex geographical conditions between two communication points or calamities on earth. Satellite communication is good for multipoint transmission and reception, with the capability of providing broadcasts and multiple access communication economically. However, the technology also has its disadvantages, including long time delays, limited communication access in high-latitude regions, and the impact of space radiation on space communications. With technological improvements and lower costs, the launch and application of satellites is now practically possible for corporations.

Apart from improving transmission speed and quality, how to ensure communication security is also an important direction of communication technology research. Based on the principles of quantum mechanics, quantum communication technology can achieve, within extreme physical limits, high performance communication by using quantum effects to physically ensure the absolute security of communication. This brand new technology can resolve problems otherwise unresolvable by other communication technologies, making it the major focus of communication technology research around the world. Currently, quantum communication technology can cover a maximum transmission distance of 300 km and achieve a security rate of 1 Mb/s subject to a communication distance of 50 km. The development of quantum communication technology will ensure information security for important infrastructures, including national defense and military facilities and large grids.

5.1.2 Information Technology

Information technology has grown and progressed rapidly over the past half-century from the advent of the world's first computer in 1947 and the invention of the transistor by Bell Laboratories through the prevalence of the Internet today to the integration of three networks (telecommunications, broadcast TV, and the Internet). We are now at the point of constructing an information highway that demonstrates a trend toward digitalization, intelligentization, personalization, and integration.

The Internet of Things (IoT). An extension of the Internet into the physical world, the IoT is built on an integration of sensor technology, communication technology, and information service technology. It is a dynamic networking infrastructure with self-organizing capability based on standard rules and interoperable communication protocols. Every physical and virtual object in the IoT has its own identification tag, physical attributes and smart interface, enabling seamless integration of the object with the existing information network. IoT technology will bring about full informatization of the real-world physical environment, realize "network ubiquity," and play a significant role in the future development of the global energy network.

The ubiquitous Internet. The Internet, or the worldwide web, was originated in the United States in 1969. With the development of personal computers and smart phones, the Internet has become a household phenomenon and an important ICT platform for work and life. Based on conventional Internet technology, the ubiquitous Internet has extended its reach to facilitate the exchange of information with any people or objects anytime, anywhere. It can provide a variety of information services in response to different demands. Simply put, it is an omnipresent form of the Internet that provides mutual access to information among different objects.

Sensor technology. A detection device and a key technology for gathering information, a sensor converts, based on certain rules, a measured object into electrical signals or other forms of information output as required. Sensor technology started to catch growing attention in the 1980s with the rapid growth of integrated circuit technology and computer technology. It has now evolved into a backbone technology for ICT development. Sensor technology is supported by fundamental research on sensors and research on sensor network systems. Currently, fundamental research on sensors is focused on new materials, with the adoption of micromachining and bionics technologies to improve the sensitivity, accuracy and stability of sensors. The focus of sensor system research is placed on applying near field communication technology to build a stub network with sensors and the information they collect, forming an important foundation for the IoT.

Image recognition technology. Image recognition technology enables computer-based processing, analysis and interpretation of images to identify targets and objects of different shapes and forms. The recognition process includes image preprocessing, image segmentation, feature extraction, and judgment and matching. Simply put, image recognition refers to the computer reading of image contents like the human mind. Image recognition equips machines with the power of sight in place of human-based monitoring of equipment conditions and prewarning of potential danger. This technology provides intelligent navigation for unmanned vehicles, with the capability to judge road conditions in real time based on satellite data.

Cloud computing and cloud storage technology. Cloud computing refers to software sharing, computing and data acquisition on an Internet platform. Cloud storage technology brings together a large collection of storage equipment on the Internet to provide shared data storage and service access capabilities. The development of cloud computing and cloud storage will lead to the formation of a hardware system to accommodate the integrated operation of servers, storage devices and Internet equipment, which will maximize the expandability of software applications. In the area of cloud computing technology, virtualization development and breakthrough will not only enhance the utilization efficiency of computing resources, but also provide dynamic migrations and resource scheduling to allow more efficient management and expansion of cloud computing loads and contribute to the flexibility of cloud computing services.

Big data technology. Big data technology enables the extraction of important information of value from the massive data that is beyond the processing capacity of a traditional database. As big data technology involves enormous volumes of data, the requirements on data processing capacity and transmission rates are far more stringent compared with conventional data processing technology. With the phenomenal growth of the Internet, data and information flows have continued to increase. For the future global energy interconnection, information flows on resources, grid operation and users are expected to extend locally from a single location (a country or region) to the whole world, accompanied by dramatic data growth. This situation will warrant the application of big data technology with important real-life implications for improving the management and operation of the future global energy network.

5.2 DEVELOPMENT DIRECTION AND OUTLOOK

5.2.1 Development of ICT is Marked by Growing Broadbandization, Digitalization, Intelligentization, Personalization, and Integration

The term "broadbandization" refers to the development of a worldwide high-speed broadband communication network supported primarily by fiber optic technology to enable transmission of as much information as possible per unit time. By "digitalization" we mean the complete digitalization of every piece of information in a communication network, be it voice, text, or image, before feeding it once again into the network, without the presence of any analog signals. By "intelligentization" we mean the capability of a communication network to not just transmit and exchange information, but also store, process and control information with flexibility. Information is also processed and transmitted in an optimal manner under all conditions through a software-defined network. By "personalization" we mean the capacity for anyone to carry out, without restriction, any form of communication with any other people anytime and anywhere in the world. Personalization requires the support of large network capacity as well as functional and intelligent flexibility. By "integration" we mean combining all services and networks into one. The integration of three different networks signifies an important trend of development, with the combination of the IoT and the Internet to lead to more integrated services.

5.2.2 ICT Ensures Greater Security, Reliability, and Intelligence for the Development and Operation of the Future Global Energy Interconnection

An ICT network refers to an energy and electricity virtual communication network architecture based on private and public networks, incorporating existing power line and optical cable-based power communication technology, to build a physical and virtual communication network system integrating long-span coherent optical transmission, 4G/5G, satellite communication and public band networking technology so as to provide communication technology support for the future global energy interconnection. To ensure a high level of security for the electricity system, wired communication technology will provide large-capacity, high-reliability communication services with a bandwidth of up to 1000 GB/s to accommodate scheduling and trade communication requirements for the global energy interconnection. Wireless communication technology, such as 4G/5G, satellite communication and public band networking, can be actuated during a system fault to realize, at relatively low costs, long-distance communication not subjected to the restrictions imposed by geographical conditions and natural disasters.

For the IoT, the power communication network will incorporate communication, information, sensor, automation and other technologies, together with sophisticated sensors, to extensively deploy a variety of smart devices with sensing, computing, and actuation capabilities in power generation, transmission, consumption, and management. Through a power sensing network and IoT technology, real-time surveillance can be conducted on grid operation status, intelligent substations, distribution lines, users and power plants to realize panoramic sensing, mutual information access, and intelligent control for the global energy network. The IoT is the foundation for developing the global energy network; only by means of real-time monitoring and sensing can the global energy network operate securely and efficiently based on an established control strategy.

For image recognition technology, surveillance images of power plants and substations, cruise images of aircraft as well as satellite images can be transmitted back to the surveillance and control center of the electricity system to diagnose equipment operation status through image recognition technology and analyze and prewarn damage to insulation materials, corona discharge, short-circuit, icing, and filth. In the icy cold Arctic and scorching equatorial regions, forested and mountainous areas and other locations with unfavorable conditions for work, image recognition technology can recognize equipment conditions intelligently to improve reliability and reduce labor and material resources.

For cloud computing and cloud storage technology, efforts should be devoted to remove bottlenecks in analyzing and processing speed in operation control and power trade management of the global energy interconnection, such that the speed and accuracy of big data analytics can be improved and power scheduling and trading on a global level realized. As the coverage of global energy interconnections expands, power trading technology will mature. With the establishment of a trading system, the complementary nature of resources due to spatial–temporal differences among the electricity markets across different countries and continents will manifest itself more prominently. The evaluation system for power trade through a network of globally interconnected grids and the platform for verifying power trade will be more widely applied, generating strong computing and storage demand. Cloud computing and cloud storage can fully leverage existing global computing resources and supercomputing technology to schedule computing resources according to tasks from time to time to meet the demand for computing and storing global energy data and information.

Big data technology can be employed, given its strong forecasting performance, to conduct fast-time real-time simulation for the electricity system and improve the intelligence level of analytical decision-making. Variables such as temperature, atmospheric pressure, humidity, rainfall, wind direction, wind force, and radiation, are taken into full account to arrive at more accurate forecasts of volatile power sources. Energy and electric power system modeling technology, incorporating big data analytics and numerical weather forecasts, can improve the smoothness of wind and solar photovoltaic power supply. With big data technology, power dispatchers can work out scheduling arrangements ahead of time and contribute to the consumption of more renewable energy.

SUMMARY

1. The global energy interconnection represents a major strategic innovation and an important technological breakthrough. Technological innovation has laid the very foundation for the development of this interconnection. More innovations and breakthroughs will be needed in high-tech fields such as energy and electric power, materials science, information and communication, the Internet and the IoT, in order to constantly improve the economics, controllability, and adaptability of the global energy interconnection.

2. To cater to the needs brought on by the development of the global energy interconnection, it is of crucial importance to improve technological innovation and breakthrough in power generation, grids, energy storage, information, and communication, and other areas, to achieve all-round improvements in the level of technology and equipment.

3. Innovation in power generation technology is key to promoting the "two-replacement" policy and shaping a new energy landscape oriented toward clean energy and centered on electricity. At the core of this development, efforts are required to constantly improve the efficiency and economics of clean energy development, focusing on wind, solar, ocean power, and distributed generation technologies.

4. Innovation in grid technology plays a pivotal role in optimizing the global allocation of energy resources, covering all segments of a robust smart grid that mainly include UHV, submarine cable, superconducting transmission, DC grid, microgrid, large-grid operation control, and other technologies. The focus of innovation is on lengthening transmission distance, expanding capacity, and ensuring the secure, reliable, and economical operation of grids.

5. Energy storage technology is also essential for securing large-scale development of clean energy as well as the safe and economical operation of grids. The focus of innovation is on improving energy and power intensities, lengthening service lives, and lowering costs.

6. Information and communication technology is an important foundation for realizing smart and interactive grids and operation control of large grids. With the objective of providing support for the development of the global energy interconnection, the focus of innovation is on expediting the development and application of fiber optics, mobile communications, satellite communications, and quantum communications, as well as the ubiquitous Internet, the IoT, image recognition, cloud computing, big data, and other advanced information technologies.

R&D ON GLOBAL ENERGY INTERCONNECTION AND PRACTICE

CHAPTER OUTLINE

1 PRACTICE IN CHINA

Since the twenty-first century, China's rapidly developing economy and increasing energy demands have provided a strong impetus to the construction of large-scale power bases of coal-fired, hydro, nuclear, and renewable energy, in particular wind and solar generation. Meanwhile, the geographically reverse distribution of China's energy resources and energy consumption creates the urgent demand for optimized allocations of resources over extensive areas and on a large scale. This calls for leveraging the strengths of robust smart grids in high capacity and long-distance transmission to realize the intensive development of power bases across different energy types, and the export of energy for integration. In recent years, China has vigorously promoted the development of UHV, smart grid technologies, and clean energy, focusing efforts on technological innovation, standard setting, strategic planning, and project construction in respect of robust smart grids, providing technological and other conditions conducive to the development of the global energy network.

1.1 TECHNOLOGICAL INNOVATION

1.1.1 UHV Technologies

UHV technologies cover UHV AC and DC transmission technologies, representing the state-of-the-art level of grid technologies in the world. Since the development of UHV was fully launched in 2004, achievements have been made on the technology and equipment fronts, fulfilling the brand propositions of "Made in China" and "Led by China." As at the end of 2014, SGCC attained a total of 705 patents in UHV transmission technology, including 318 invention patents, 387 utility models, and appearance design patents. According to Klaus Wucherer, former president of the International Electrotechnical Commission (IEC), China now boasts of a world-leading UHV transmission technology, and the country's UHV AC voltage standards (as international standard voltage) will be promoted all over the world.

1.1.1.1 UHV AC Transmission Technologies

As early as 1986, China carried out preliminary discussions and feasibility studies on UHV AC transmission. In 2004, research on key UHV AC transmission technology was conducted, setting a technological framework. In 2006, UHV AC transmission technology was applied in engineering projects. Work on pilot demonstration projects commenced in the same year, and was subsequently commissioned in 2009. Currently, China has fully mastered UHV AC transmission technology.

Key technologies of UHV AC transmission: Through SGCC's proprietary innovations and joint explorations, breakthroughs have been made in the core technologies of system voltage control, secondary arc current suppression, external insulation coordination, as well as electromagnetic environment control, leading China to attain world leadership in this technological area. For system voltage control, measures such as over-voltage amplitude control, protection intertrip, high performance arrester, closing resistor, earth wire optimization, as well as reactive power control have been adopted to attain control of power frequency, operation, lightning, and system operation over-voltage, respectively. For *secondary arc current suppression*, small reactance components are installed on the neutral points of UHV electric reactors, to effectively control secondary arc currents, elevate single-phase reclosing success rates, and ensure system supply reliability. For *external insulation coordination*, measures toward deep suppression of over-voltage level, utilization of the composite insulator and sleeve, and simulation of high altitude insulation characteristics have been pursued to solve problems such as substantially larger external insulation scale, noticeably lower insulation, and voltage withstanding capacity with higher pollution levels and altitudes, to significantly improve system economics, while ensuring security. For electromagnetic environment control, a series of technologies and measures such as complicated multiconductor system power frequency electric field model simulation, wire arrangement optimization, and armor clamp corona control, to effectively reduce the impact of noises, radio disturbance, and also achieve environment friendliness (Fig. 7.1).

Key technologies of UHV AC equipment: Equipment is crucial for UHV development. SGCC, together with electrical equipment manufacturers, has developed a proprietary design of the world's first UHV AC transformer (Fig. 7.2) at a rated voltage of 1000 kV and with a rated capacity of 3 GVA, by successfully mastering the key technologies of transformer insulation structural design, leakage flux control, and local overheating prevention. Gas insulated switchgear has also been developed through breakthroughs in design technologies for UHV AC breaker operating mechanisms and arc-extinguishing chambers (Fig. 7.3). Successful research and development (R&D) has led to the creation of UHV AC shunt reactors (Fig. 7.4) with the world's highest single-phase capacity (320 Mvar) to solve technological problems such as the control of leakage flux, temperature rise, noise and vibration control under high voltage and high capacity conditions. In addition, UHV arresters, capacitor voltage transformers, UHV insulators, and more reliable and stable complete-set secondary equipment of UHV systems have been developed for application in regions of medium and high level pollution.

FIGURE 7.1 UHV AC (1000 kV) Transmission Lines and Towers

FIGURE 7.2 UHV AC (1000 kV) Transformer

FIGURE 7.3 UHV AC (1000 kV) Breaker

FIGURE 7.4 UHV AC (1000 kV) Shunt Reactors

1.1.1.2 UHV DC Transmission Technology

In 2004, China embarked on comprehensive and in-depth research on ±800 kV UHV DC transmission engineering technology, achieving all-round breakthroughs in HV DC, high current, high power transmission technologies, and equipment.

Key technologies of UHV DC transmission: Like UHV AC transmission, UHV DC transmission has encountered unprecedented challenges in over-voltage and insulation coordination, external insulation configuration, electromagnetic environment and noise control, DC system design, and other technologies. Through continuous research, trial and technological innovation, all these problems have now been resolved. For *over-voltage and insulation coordination*, through the calculations, by full

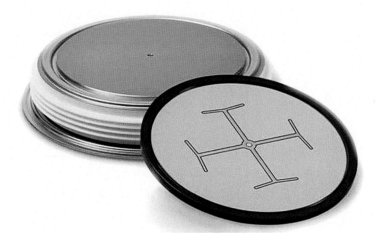

FIGURE 7.5 Thyristor (6 in.)for ±800 kV UHV DC Transmission Lines

simulation, of DC transmission system converter stations and the over-voltage characteristics and levels under different operation and fault conditions along transmission lines, technological solutions to over-voltage problems have been proposed to rationally deploy arresters, optimize arrester parameters, and evenly distribute smoothing reactors. For *external insulation configuration*, through experiments and studies, the long air gap discharge characteristics of transmission lines, the characteristics of real electrode air gap discharges and corresponding altitude correction coefficients, insulator pollution and icing flashover characteristics, and corresponding altitude correction coefficients have been obtained and successfully applied to external insulation configuration in a UHV DC transmission project. For *electromagnetic environment and noise control*, with wire types, minimum wire heights, and minimum corridor widths confirmed through studies, knowledge has been acquired of transverse distribution and the law of change, in various electromagnetic environment factors, leading to proposals for the optimal polar wire arrangement in electromagnetic environments and for optimized configuration of converter stations to effectively reduce the impact of engineering work on the environment. For *DC system design*, after confirmation of UHV DC system rated voltages, rated currents, transmission power, and other key parameters through studies, complete design plans have been proposed for the first time, and world-class standards successfully achieved for proprietary system integration in UHV DC transmission.

Key technologies of UHV DC equipment: In terms of components, the major parameters of 6-in. thyristors have been determined through studies. The development of special monocrystalline silicon pretreatment processes has resolved key technological problems of impurity diffusion uniformity, chip thickness uniformity, chip parameters uniformity, as well as chip deformation control, caused by the expansion of thyristor chip size to 6 in. This has led to the successful development of 6-in. thyristors (Fig. 7.5) and the fulfilment of the goal of creating high current and high blocking voltage thyristors with 6-in. monocrystalline silicon materials. For *equipment*, a ±800 kV UHV DC transmission converter transformer (Fig. 7.6), with the world's highest voltage and capacity levels, has been successfully developed, where problems such as local discharge control, leakage flux, harmonic wave, and temperature rise control have been resolved by adopting advanced regulating winding connection methods.

FIGURE 7.6 UHV DC (±800 kV) Transmission Converter Transformer

UHV DC transmission converter valves (Fig. 7.7) with the highest capacity have been produced, after solving design problems of insulation structures and optimizing valve-side sleeves and outgoing line devices, with the development of DC wall bushing and disconnectors with the world's highest flow capability. In addition, ±800 kV smoothing reactors, ±800 kV DC and AC field equipment, ±800 kV control and protection equipment, wires with 1000 mm^2 and higher sections, ±800 kV DC transmission line insulators, and complete sets of ±800 kV DC wire armor clamps (Fig. 7.8) have been developed.

FIGURE 7.7 UHV DC (±800 kV) Transmission Converter Valve

FIGURE 7.8 UHV DC (±800 kV) Transmission Lines and Towers

1.1.2 Smart Grid Technology

Technological innovation in smart grids has contributed to the improvement of safety, adaptability, economics, and interactivity in power system operations. China has realized all-round breakthroughs in smart-grid equipment monitoring, system operation, smart interactivity, information and communication, and other technologies. As at the end of 2014, SGCC owned 11,312 patents related to smart grids, including 1,622 invention patents and 9,690 utility model, and appearance design patents.

Equipment monitoring: Monitoring and control of key equipment of all components of power system can provide real-time and complete information on equipment operation conditions, offering support to the dynamic optimization of equipment operations and efficiency improvement. Master station and terminal equipment for monitoring transmission and transformation equipment conditions have been developed (Fig. 7.9), with work started on equipment condition evaluation, condition-based monitoring and risk prewarning to achieve controllability of key transmission and transformation equipment. Breakthroughs have also been made in robot and UAV smart inspection technologies (Fig. 7.10), resulting in improved operational safety of equipment. Successful development of integrated monitoring systems at the substation level, auxiliary function-level control, and fully digitalized protection monitoring systems, has realized integrated online monitoring and self-diagnosis for substation equipment, testing of monitoring and automatic devices, as well as automatic monitoring of operation environments. Extensive application of unified support systems for smart power distribution terminals and online risk evaluation and prewarning systems at the distribution level has been realized.

System operation: To achieve smart operation and dispatch by gaining a full understanding of grid operation conditions and improving grid operational safety and stability through online analysis, safety evaluation, forecasting and prewarning, as well as dispatch control. For *super-grid* operation, technological breakthroughs have been made to achieve online, sophisticated, integrated, and practical smart dispatch operations. By "*online*" we mean online tracking based on the three dimensions of time, space, and business to realize panoramic monitoring, smart warning and active security and defense of all

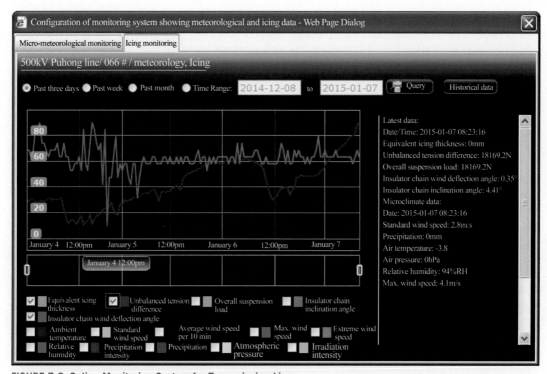

FIGURE 7.9 Online Monitoring System for Transmission Lines

FIGURE 7.10 Robot on Patrol Duty in Smart Substation

aspects of dispatch and production. By "*sophisticated*" we mean the automatic and optimized formulation of large-scale, multitarget and multiple-time dispatch programs for safe, economical, and integrated operation to meet complicated constraints, as well as all round safety and stability checks covering stable, dynamic, and transient conditions. By "*integrated*" we mean the realization of upper and lower linkages, and coordinated operation across different levels of dispatch to form a distributed, integrated support system for smart dispatch technology and effectively support the integrated operation of UHV large grids. By "*practical*" we mean efforts to perfect fundamental data, improve online application in terms of system operation, and promote the application level of day-ahead dispatch plans and safety checks. For *distribution grid operation*, important breakthroughs have been made in self-recovery control, smart distribution terminals, distributed power access, and other areas. Open distribution automation systems and smart distribution terminals have also been developed to realize information sharing and application integration of distribution automation and related systems. R&D work on the control technologies for distributed power generation/energy storage and microgrid access and coordination has played a vital role in improving power supply reliability and peak shaving/valley filling.

Smart interactivity: Multiple technological measures have been adopted to realize information sharing and interaction between systems and users. Breakthroughs have been made in smart meters, metering system monitoring, electric car charging, smart buildings, smart communities, and smart parks, leading to the development of a smart, interactive fundamental platform to realize noticeable improvements in smart power services. Smart meters have been promoted on a full scale and a system for collecting information on electricity consumption established, covering all electricity users and metering points. Online monitoring has also been achieved, with important information on customer-side power load, consumption and voltage collected in real time. Smart power service systems for smart communities (buildings) and demonstration projects on smart power consumption have been accomplished, realizing real-time and interactive response between smart grids and users. Chargers, charging piles, battery swapping, and other equipment have been developed, with a platform for monitoring electric car operations built to form a standard operation management system for electric car–smart recharging/swapping networks. As a result, an exchange of information has been achieved between battery charging/swapping stations and vehicles.

Communication and information: A communication and information platform is a basic foundation for achieving and applying smart technologies on all fronts. China has made breakthroughs in different aspects of power communication and power information technologies. In power communication technology, a high capacity backbone optical transmission network, a demonstrative project on the ice resistance of optical fiber composite overhead ground wire, and a demonstrative project on the electricity-based Internet of Things have been set up. Through the development of special electricity communication chips, breakthroughs have been made in packet transport network, Ethernet passive optical network, industrial Ethernet, power line carrier communication, wireless private and public network communication, as well as special optical fiber composite cable technologies (Fig. 7.11). A technology system for next-generation flip networks has been proposed, with a management system for power telecommunication networks developed and widely applied. A platform for power telecommunication network simulation has been developed and quantum communication technology introduced into the electricity industry for the first time. On the power information technology front, all-round breakthroughs have been made in R&D and application of information technologies in different segments of a smart grid. A host of smart-grid information systems covering transmission, transformation condition monitoring, and mobile operations have been put online. A unitary video monitoring system and

FIGURE 7.11 Section of Optical Fiber Composite LV Cable

a geographic information service system have been developed. The application of information systems covers planning, construction, operation, maintenance, and marketing. An integrated information and dispatch operation system has been built, covering the headquarters, failover centers, and provincial companies. Smart-grid information security protection plans have been announced, with the grade protection evaluation of secondary-system information security protection and management information systems completed for grids higher than Grade III.

1.1.3 Clean Energy Technology

Clean energy technology is an important tool to ensure clean energy substitution. China has developed a host of innovations and applications in clean power generation and operation technologies, giving a strong impetus to large-scale exploration and integration of clean energy.

Key technologies of large-scale wind power dispatch: Breakthroughs have been made in key technologies such as wind power forecasts, dispatch technology for optimal coordination of wind power and other power, and the application of wind power in peak load shifting. This has resulted in the achievement of online wind-power capacity assessment, wind power day-ahead, and intraday scheduling adjustments, and the evaluation of wind power grid integration performance. In the course of operation, wind power is incorporated into start-up arrangements and also the rolling optimization of thermal power plant start-ups, to realize maximum accommodation of fluctuating wind power within its operation range and ensure system operational safety.

Large-scale photovoltaic power grid integration and operation technologies: Through R&D of key technologies for photovoltaic power generation, including output forecasting, operation monitoring, and operation control, all-weather predictions, across different times and spaces based on numerical weather forecasts have been achieved. Closed-loop control of the power of large photovoltaic power stations has been realized. Rolling optimization of day-ahead and intraday scheduling of high penetration photovoltaic power grids has been made possible. Problems such as momentary output fluctuations have been resolved, a hierarchical photovoltaic power generation control system established, and the automatic and smooth regulation of the active and reactive output of photovoltaic power generation realized.

New energy-based generation output forecasting and operation monitoring: A technology system for forecasting and monitoring new energy-based generation output has been established. For *new energy-based generation output forecasting,* technology research and system development has been launched for projecting wind and photovoltaic power output based on numerical weather forecasts, and SGCC operation center for numerical weather forecasts established. The technology for short-term and super-short-term forecasting of wind power output has been acquired, and an output forecasting system that embodies independent intellectual property rights has been developed. A wind power generation output forecasting system, with an accuracy of 88%, has been deployed at provincial-level dispatch centers. Technology research has been carried out on short-term and super-short-term forecasting, with an accuracy of 90%, of photovoltaic power output based on numerical weather forecasts and satellite and foundation cloud pictures. This technology is now in demonstration application in a number of provincial-level grids. A grid-based forecasting technology for distributed photovoltaic power generation has been developed and applied in many parts of China. *For new energy operation monitoring,* advanced coordination of new energy power generation and transmission has been achieved and breakthroughs in wind power operation online monitoring made (Fig. 7.12). Through efforts in building wind power operation monitoring and automation systems, wide-area panoramic presentation and monitoring has been realized of wind farm weather information as well as wind-turbine information on active power, reactive power, voltage, current, and operation conditions of generating units. Real-time information monitoring and instant data collection have been realized for photovoltaic power plant operations, facilitating real-time monitoring of efficiency, power, generation, power quality, and operation conditions to record a variety of static and transient incidents as the base for operation control.

Technologies for energy storage system operation: Technological breakthroughs have been made in promoting the application of intermittent access for energy storage systems. Quantitative models with energy storage technology incorporated to improve wind power access have been established, and control technologies for optimizing wide-area coordination for stored energy stations based on multivariable coordination and multirule switching have been put forward. Through the development of demonstrative projects on wind and photovoltaic power storage and transmission (Fig. 7.13), an understanding has been established of the characteristics of joint operation of wind, photovoltaic power storage and transmission, as well as the optimized deployment of energy storage systems, which has helped to promote the construction of an online monitoring platform for large energy storage stations and to enhance the coordinated operation of energy storage and intermittent power generation.

FIGURE 7.12 Wind Power Operation Monitoring System

FIGURE 7.13 Control Center for Joint Operation of State Wind and Photovoltaic Power Storage and Transmission Demonstration Project in Zhangbei, Hebei

1.1.4 Experimental Systems

In the course of technology R&D and engineering practice on robust smart grids, China has built complete UHV and smart grid research experimental systems, covering a UHV AC experimental base, a UHV DC experimental base, a high altitude experimental base, a UHV tower experimental base, a complete-set design and R&D (experimental) center for UHV DC transmission, a State Grid simulation center, an R&D (experimental) center for large-grid integration systems for wind power, and an R&D (experimental) center for solar energy generation, to provide world-class grid-related experimental research capabilities. These advanced experimental conditions and methods have laid a preresearch platform for key technology R&D and project construction related to the future global energy network.

UHV AC experimental base (Fig. 7.14): Located in Wuhan, Hubei, the base is equipped with a 1 km single and double-circuit UHV experimental line, complete with test facilities including an electromagnetic environment testing laboratory, a UHV AC corona cage, an environment and climate laboratory, a 7500 kV outdoor impact testing ground, a long-term live testing ground for UHV equipment. It provides a favorable platform for impact research on UHV electromagnetic environments, research on the characteristics of external insulation under high altitude, heavy pollution, serious icing, and other special natural conditions, long-term full-voltage live tests on electrical equipment, and research on operation and maintenance of power transmission and transformation equipment.

UHV DC experimental base (Fig. 7.15): Located in Changping District, Beijing, with an outdoor experiment field, an experiment hall, a pollution and environment laboratory, an insulator laboratory, an arrester laboratory, UHV DC experimental lines, a corona cage, an electromagnetic environment simulation, and testing field, as well as all-round experimental and research capabilities covering UHV DC electromagnetic environments, external insulation, system operational safety, equipment and testing technologies, and operational characteristics.

UHV tower experimental base (Fig. 7.16): Located in Bazhou, Hebei, the base meets the need for 1000 kV UHV double-circuit iron tower experiments, dimension and design load experiments on

FIGURE 7.14 UHV AC Experimental Base in Wuhan, Hubei

FIGURE 7.15 UHV DC Experimental Base in Beijing

FIGURE 7.16 UHV Tower Experimental Base in Bazhou, Hebei

1000 kV double-circuit, ±800 single-circuit and ±800 kV double-circuit towers, components experiments as part of the research on new tower structures, as well as overall experimental requirements.

High altitude experimental base in Tibet (Fig. 7.17): Located at Lhasa, the Tibet Autonomous Region, the base boasts an outdoor testing ground, a pollution laboratory, and experimental lines. It is the world's highest electricity experimental base, offering favorable conditions for research on transmission and transformation lines at an altitude of 4000 m and higher, equipment insulation and electromagnetic environment characteristics. It also supports technological innovation in functional design, equipment development, control and test technologies, and engineering application.

FIGURE 7.17 High Altitude Experimental Base in Tibet

FIGURE 7.18 State Grid Simulation Center in Beijing

State Grid simulation center (Fig. 7.18): Located in Haidian District, Beijing, the center is equipped with basic test and research facilities, covering electric power system digital simulation, dynamic simulation, digital-analog hybrid simulation, and operation and security monitoring, with the focus on planning, design, construction and operation technologies for large UHV AC/DC hybrid grids. It forms a well-structured, fully functional and technologically advanced electricity system simulation and research system to achieve complete, multilevel, and multiangle simulation of power systems.

Complete design and R&D (experimental) center of UHV DC transmission (Fig. 7.19): Located in Changping District, Beijing, the center is well equipped to concurrently conduct joint test and experimental research on one UHV DC transmission project and the secondary systems of three conventional DC transmission projects. It can provide technical support, resource sharing and management platforms for key technology research, system design, valve hall design, complete-set equipment design, equipment procurement standards, equipment production supervision, system commissioning, and on-site commissioning for EHV/UHV DC transmission projects.

Large-scale wind power grid integration system R&D (experimental) center (Fig. 7.20): Located in Zhangjiakou, Hebei, the center has fundamental research capability in wind power generation, test capability in mobile wind turbines, experimental capability in wind power, and full test capabilities on all characteristics of wind power generating units. With the most internationally advanced test methods for wind turbine electrical equipment, the center offers model certification and grid connection testing for newly produced wind turbine products.

FIGURE 7.19 Complete-Set Design, Research and Development (Experimental) Center for UHV DC Transmission in Beijing

FIGURE 7.20 Large Wind Power Grid Integration System R&D (Experimental) Center in Zhangjiakou, Hebei

Solar power generation R&D (experimental) center (Fig. 7.21): Located in Nanjing, Jiangsu, the center is equipped with fundamental research capability in solar power generation, experimental and test capability in grid-connected photovoltaic system operation, and mobile test capability in grid-connected photovoltaic power stations. It can conduct performance tests on solar power generation components as well as tests on grid adaptability and resilience against interference. The center can meet the requirements for conducting research and tests on key technology for grid access, key system and equipment technology, as well as planning and design technology.

1.1.5 Research on Global Energy Network

Preliminary research on the development of a globally interconnected energy network has been carried out in China, covering Asia/Europe power transmission, wind power development in the Arctic region,

FIGURE 7.21 Solar Power Generation R&D (Experimental) Center in Nanjing, Jiangsu

and outlook on the global energy network and related key technology and equipment, to lay a solid foundation for the development of global energy interconnections.

Research on Asia/Europe power transmission: Work in this area covers in-depth analysis of the European Union's market demand for power, research on the energy resources and power export potential of Asian regions at the sending end, and analysis of the technology economics of transcontinental transmission between Asia and Europe. Studies have identified some market potential for Europe against a scenario of nuclear energy abandonment and fossil fuel generation decommissioning. With abundant energy resources, Asian regions at the sending end are more capable of ensuring a sustainable supply of electricity. Being economically competitive, power transmission between Asia and Europe may provide a feasible option for Europe's clean energy development. Taking an integrated view of incremental load in the future, fossil fuel and nuclear generation decommissioning, expanding installed capacity of hydropower, new energy development and other factors, the demand for electricity in Europe is expected to amount to 140 GW during 2011–2020, of which 14.6 GW is attributable to Germany. The on-grid tariffs of Asia–Europe transcontinental power transmission will be about RMB 0.3–0.5 per kWh lower than that of wind power from the North Sea region. With growing utilized hours on the transmission channel, on-grid tariffs will fall further and add to the competitiveness of this transcontinental transmission project.

Research on Arctic wind power development and outlook on global energy network: Work in this area includes a series of topical research, covering assessment of the Arctic region's environmental characteristics and wind power resources, evaluation of global renewable energy resources and research on transcontinental grid interconnections centered on the Arctic region, technology research on the Arctic region's wind power development, technology research on transmission and transformation in high cold, high humidity, and high wind regions, and research on plans to export wind power from the Arctic region, and electricity from Siberia in the Russian far east. Comprehensive studies have been carried out on the status of resources, development prospects, and interconnection strategies of the Arctic region and Siberia against the backdrop of a global energy network. Based on preliminary research findings, the Arctic is a region endowed with oil and gas, as well as wind, and other renewable resources and there is a real possibility of meeting the energy requirements of low carbon, sustainable development by developing wind power in the Arctic together with natural gas, hydro, coal, combustible ice, and other resources in Siberia and the Russian far east, followed by bundling and optimized allocation on a transnational basis through UHV grids to supply to China, Japan, South Korea, and North America.

Research on key technology and equipment for global energy network: Work includes research on the key aspects of development of the future global energy network, covering resource analysis, evaluation of development potential, strategic planning, adaptive transmission and transformation of technology and equipment, operation and control technology, and new technological applications. The focus is on construction technology for transmission and transformation projects, optimized engineering design technology, DC grid technology, key technologies related to high voltage, high current electrical and electronic components, key technology for energy storage equipment, operation and dispatch structure for global power trade, simulation modeling, protection and stable control, and information support. All this will render technological support to the development of the global energy network.

1.2 STANDARD FORMULATION

Standard setting is an important part of the push for a global network of energy interconnections. A move to standardize mature technologies and equipment into unified interfaces and specifications will not only facilitate subsequent promotion and application, but also create conditions for building international-level interconnections among grids and related equipment as part of the future global energy network. Building on a foundation of continued innovation in UHV, smart grid, and clean energy technologies, China has actively promoted standard setting to establish a sound standardization system. Since 2005, China has developed 83 national standards related to robust smart grids and 204 industry standards, while actively promoting the internationalization of UHV and smart grid technologies. As at the end of 2014, State Grid was involved in the development of 21 international standards with 611 enterprise standards announced, as shown in Table 7.1.

System of UHV AC transmission technology standards: Riding on the results of UHV AC technology research and engineering development, State Grid has put forward a system of UHV AC transmission technology standards, covering six major areas of planning and design, equipment and materials, project construction, test and measurement, operation and maintenance, and environmental protection

Table 7.1 UHV and Smart Grid Standards Developed by State Grid During the End of 2014		
Fields	**Categories**	**Number of Standards**
UHV	International standards (application pending)	12
	National standards (codeveloped)	46
	Industry standards (codeveloped)	61
	Enterprise standards (developed)	171
Smart grid	International standards (application pending)	9
	National standards (codeveloped)	37
	Industry standards (codeveloped)	143
	Enterprise standards (developed)	440

Table 7.2 System of UHV AC Transmission Technology Standards

Category	National Standards	Industry Standards	Enterprise Standards (Set)
Planning and design	5	1	14
Equipment and materials	19	7	34
Project construction	6	7	31
Test and measurement	–	5	4
Operation and maintenance	3	18	15
Environmental protection and safety	–	3	2

and safety. Of these standards, 100 enterprise standards, 33 national standards, and 41-electricity industry standards have been promulgated. The system of UHV AC transmission technology standards is shown in Table 7.2.

System of UHV DC transmission technology standards: Riding on the results of UHV DC technology research and engineering development, State Grid has developed a system of complete ±800 kV UHV DC transmission technology standards, covering all major areas of UHV DC transmission, including planning and design, equipment and materials, project construction, test and measurement, operation and maintenance, and environmental protection and safety. Of these standards, 71, 13, and 20 industry standards have been promulgated. The system of UHV DC transmission technology standards is shown in Table 7.3.

System of smart grid technology standards: State Grid has formulated and announced *System Planning for Smart Grid Technology Standards*, incorporating eight professional disciplines, 26 technology areas and 92 standards series. A smart grid technology research system has come into being, covering technology development, equipment manufacture, tests and experiments, engineering application and standard setting, which is led by grid operators and involving joint participation by research institutions, equipment manufacturers, and electricity users. So far, State Grid has announced 417 enterprise standards related to smart grids, with involvement in developing 37 national standards and 143 industry standards, which have been promulgated. State Grid's system of smart grid technology standards is shown in Fig. 7.22.

Table 7.3 System of UHV DC Transmission Technology Standards

Categories	National Standards	Industry Standards	Enterprise Standards (Set)
Planning and design	2	1	5
Equipment and materials	3	2	22
Project construction	5	12	31
Test and measurement	1	1	5
Operation and maintenance	2	2	6
Environmental protection and safety	–	2	2

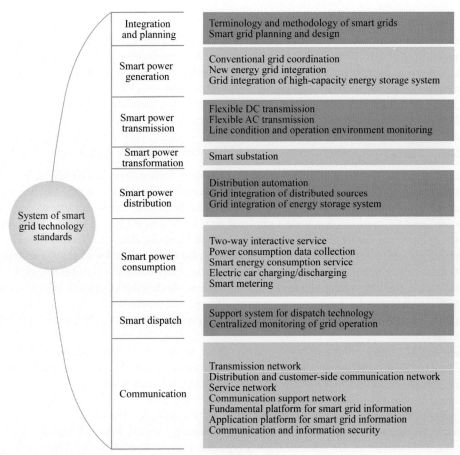

FIGURE 7.22 State Grid System of Smart Grid Technology Standards

Active participation in developing international standards: China's UHV AC voltage standard has been accepted by the IEC. As proposed by China, the IEC has set up three technical committees to deal with HVDC transmission, smart-grid user interfaces, and grid access for renewable energy. The IEC has four secretariats based in State Grid. At the International Council on Large Electric Systems (CIGRE), a number of working groups have also been formed, led by China and covering UHV substation equipment (A3.22), UHV substation system (B3.22), UHV insulation coordination (C4.306), the breaking characteristics and experimental requirements of EHV/UHV AC switchgear (A3.28), onsite test technology applied during the construction, and operation of UHV AC substations (B3.29). At the Institute of Electrical and Electronics Engineers (IEEE), China has led the development of six standards, including three related to UHV AC, one related to energy storage, and two related to international UHV standards. China has also proposed the formulation of three UHV AC technical standards, covering insulation coordination (IEEE P1862), onsite tests (IEEE P1862), and reactive voltage (IEEE P1860).

1.3 STRATEGIC PLANNING

In advancing the development of UHV and smart grids, China has developed a full-fledged strategy with supporting plans. In 2009, SGCC formulated the *Planning Outline of Robust Smart Grid Development*, pioneering the construction of a robust smart grid system supported by a UHV backbone and coordinated development of grids at all levels, with a strong grid structure and smart technology covering all components of an electric power system. In recent years, based on continued improvements in technology R&D and engineering practices, SGCC has carried out rolling grid-planning studies, with a sound planning system already in place.

For SGCC, the development of robust smart grids can be divided into three phases.

- *Phase I (2005–2010): Preliminary stage of development.* Research on key technologies for robust smart grids, together with equipment R&D, was conducted. Technology and management standards were developed and pilot projects in all areas launched. Demonstrative projects on UHV AC and DC transmission were completed, leading to a new era of UHV AC/DC hybrid transmission. Grid interconnections nation-wide were basically achieved. Technological applications were piloted in all segments of a smart grid and a collection of smart substations have been built. Smart dispatch facilities have been basically completed to improve grid coordination, building up enough grid capacity to meet the development needs of renewable energy. Smart meters have essentially covered all large and medium-sized cities in China. R&D work was carried out on smart equipment, and pilot projects on two-way interactive services were launched.
- *Phase II (2011–2020): Full construction stage.* Plans are underway to accelerate the construction progress of UHV grids, urban, and rural power distribution networks, promote significant breakthroughs in and widespread applications of key technologies and equipment, and basically complete the construction of a state-level robust grid supported by a UHV backbone with coordinated development of power grids at all levels. By 2020, a series of UHV AC projects covering Huainan–Nanjing–Shanghai, Ximeng–Shandong, Western Inner Mongolia–southern Tianjin, and other areas, are scheduled for completion to build four vertically aligned and seven horizontally aligned UHV AC grid structures in North China, East China, and Central China, with the completion of nine UHV DC transmission projects to meet the needs of large-scale energy bases to export hydropower from the southwest and coal-fired, wind, and solar energy generation from the western and northern regions. Across the country, five synchronous power grids will be built, including the one in North China/East China/Central China, and the rest in the north-eastern, north-western, south-western, and southern regions. The power grids in the north-eastern, north-western, south-western, and southern regions shall be asynchronously interconnected with the power grid in North China/East China/Central China through UHV or EHV DC lines, to create a new, improved transmission landscape marked by delivery of electricity from west to east and from north to south, with mutual backup and support between hydroelectric and thermal generation sources and between wind and solar power generation sources. By then, SGCC's capability to optimize resource allocations across different regions will be greatly improved to ensure the outward transmission of electricity for consumption from large-scale coal-fired, hydroelectric, nuclear, and renewable power bases. Core smart substations will achieve a penetration rate of approximately 50%, with the full promotion of intelligent scheduling and two-way interactive services in large and medium-sized cities. The electric car charging/swapping

market will be completely opened up, with the development of quick charging networks along expressways. Grid power quality will reach international advanced levels.

- *Phase III (2021–2025) improvement stage.* Construction of robust smart grids will be fully completed, reaching international advanced level in technology and equipment terms. By 2025, resource allocation capability, safety levels, and operating efficiency at the grid level will be significantly improved. From smart meters to user information collection systems, from smart equipment to smart substations, and from fiber-to-the-home service to comprehensive service systems based on "multinetwork integration," all types of smart equipment will be promoted and widely applied in the power system. The level of grid intelligence will be improved markedly to accommodate the development needs of centralized and distributed clean energy sources as well as smart buildings, smart communities, and smart cities. China's power grids will be interconnected with power grids in neighboring countries in a friendly manner to set an example for the development of a global energy network.

After years of hard work, SGCC has achieved the phased targets planned for smart grid development. Looking ahead, SGCC will carry out full-scale construction of smart grids supported by a UHV backbone and strengthen transmission and grid connection with neighboring countries. SGCC will further consolidate and improve its grid structures to form stronger receiving-end grids in North China/East China/Central China, while the capacity in regions outside North China/East China/Central China for receiving electricity will continue to improve to ensure the reception and consumption of power delivered across regions through UHV DC transmission from renewable energy bases in the west and north, and hydropower bases in the southwest. The intelligence level of power grids will be further improved to form an integrated public service platform that seamlessly links up large power bases, distributed generation, charging/swapping facilities, and smart terminal equipment. Smart distribution networks with a more logical structure will be built in urban and rural areas to remarkably improve supply capacity and reliability. Smart terminal equipment shall be widely used to realize full sharing, remote, and automatic control of two-way information. Friendly connections among distributed generation, microgrids and super-grids shall be achieved. Smart load-end dispatch shall be applied to electric vehicles and other smart terminal equipment as well as virtual power plants to effectively realize peak load operations. The construction progress of China's grid interconnections with Russia, the Republic of Mongolia, Kazakhstan, Burma, Thailand, Vietnam, South Korea, North Korea, Japan, and surrounding countries shall be accelerated. China will also be actively involved in research on Asia–Europe interconnections as well as the development of power bases in the Arctic and equatorial regions and power transmission across different regions to lay the foundation for the development of the global energy network.

China highly values the development of UHV and smart grids, making it the strategic focus of national energy development by incorporating it into the government's plan for national economic and social development and its plan and related special plans for energy development, which has given a strong boost to innovation in grid development. China's national plans for UHV and smart grid development are shown in Table 7.4.

1.4 PROJECT ENGINEERING

Guided by technological innovation and development plans, a collection of major UHV and smart grid projects have been completed in China, bringing about significant economic, and social benefits.

Table 7.4 China's National Plans for UHV and Smart Grid Development

Year	Plans
2005	Outline of national medium and long-term planning for scientific and technological development (2006–2020)
2006	Opinions of the state council on accelerating the development of equipment manufacturing industry
2006	The eleventh 5-year planning outline for national, economic, and social development of the PRC
2007	National planning for coping with climate change
2011	The eleventh 5-year planning outline for national, economic, and social development of the PRC
2011	The twelfth 5-year plan for scientific and technological development
2012	The twelfth 5-year plan for development of strategic emerging industries
2012	The twelfth 5-year special plan for smart grid development as a major industrialization project on science and technology
2013	The twelfth 5-year plan for construction of national major innovation bases
2013	The twelfth 5-year plan for energy development
2013	Action plan for atmospheric pollution control

1.4.1 UHV Grid Projects

In recent years, a collection of UHV AC and DC transmission projects have been developed in China, fully confirming the safety, economics, and environment-friendliness of UHV transmission. The successful construction and operation of these projects clearly demonstrate the feasibility of building a global energy network on UHV technology. As at the end of 2014, three UHV AC lines and six UHV DC lines were completed and commissioned in China. In June 2014, the National Energy Administration issued the *Notice on Accelerating the Construction of 12 Major Transmission Channels under the Action Plan for Atmospheric Pollution Control*, including four UHV AC and five UHV DC projects. SGCC is responsible for constructing eight of these projects. In November 2014, work commenced on two UHV AC projects and one UHV DC project. As at the end of 2014, SGCC's UHV transmission lines in operation and under development measured a total length of nearly 16,000 km, with a transformer (converter) capacity of approximately 160 GVA(GW). SGCC's UHV projects in operation and under development are shown in Fig. 7.23 and Table 7.5.

1.4.1.1 UHV AC Transmission Projects

UHV AC transmission projects lay the foundation for building main robust grids, realizing power transmission through regional interconnections, and forming a strong grid at the receiving end to provide structural support for UHV DC, high capacity, and multicircuit transmission.

South-eastern Shanxi–Nanyang-Jingmen 1000 kV UHV AC transmission experimental and demonstration project: Phase I of this 640 km project was commissioned in January 2009 as the world's first UHV AC transmission line to go into commercial operation, as shown in Fig. 7.24. The project involves the development of three substations and two transmission lines. Two substations are located in south-eastern Shanxi (Changzhi) Station in Shanxi and Jingmen Station in Hubei, each featuring a 3 × 1000 MVA HV transformer, and the other one, a switching station, is sited in Henan–Nanyang

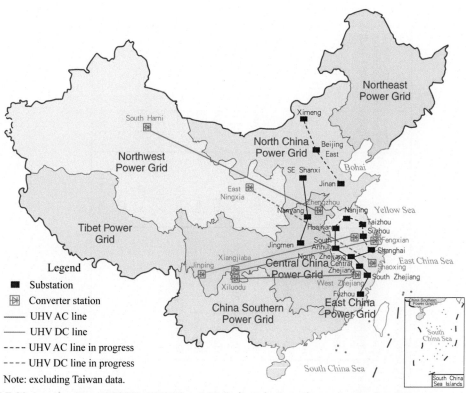

FIGURE 7.23 Location Map of SGCC's UHV DC and AC Projects in Operation and under Development

Station. Phase II was put into operation in September 2011, with the addition of one 3×1000 MVA HV transformer to each of south-eastern Shanxi Station and Jingmen Station. Two 3×1000 MVA HV transformers have also been added to Nanyang Station, with the installation of supporting switchgear and other equipment. With a transmission capacity of 5000 MW, the project can transmit 25 TWh of electricity every year. By linking up the North China Power Grid and the Central China Power Grid, the project has become an important energy transmission channel running in north–south direction. It produces significant economic and social benefits by bringing electricity from coal-rich North China to the south in winter and delivers surplus electricity from hydropower-abundant Central China in summer to the north, while providing backup in the event of a contingency. At an international meeting on UHV transmission technology held on May 21, 2009, Secretary General Koval of CIGRE described this UHV AC experimental and demonstration project as a great technological achievement and a milestone in the history of electric power development.

Huainan-Northern Zhejiang-Shanghai 1000 kV same-tower double-circuit UHV AC transmission project: Featuring a length of 2×648.7 km, the project was commissioned in September 2013 as the world's first UHV AC transmission project with double circuits on the same tower to go into commercial operation. The project involves the development of four substations and three transmission lines, starting from Huainan Substation in Anhui, through Wuhu Substation in southern Anhui, and Anji

Table 7.5 SGCC's UHV Projects in Operation and Under Development

Items	Project Names	Voltage Grades (kV)	Commissioning or Target Commissioning Date	Line Length (km)	Transformer/Converter Capacity (10,000 kVA/kW)
Projects in operation	South-eastern Shanxi–Nanyang–Jingmen	1,000	January 2009	640	600
	Xiangjiaba–Shanghai	±800	July 2010	1,891	1,280
	Extension of South-eastern Shanxi–Nanyang–Jingmen	1,000	November 2011	0	1,200
	Jinping–Southern Jiangsu	±800	December 2012	2,059	1,440
	Huainan–Northern Zhejiang–Shanghai	1,000	September 2013	2 × 648.7	2,100
	Southern Hami–Zhengzhou	±800	January 2014	2,191	1,600
	Xiluodu–Western Zhejiang	±800	July 2014	1,669	1,600
	Northern Zhejiang–Fuzhou	1,000	November 2014	2 × 603	1,800
Projects under development	Huainan–Nanjing-Shanghai	1,000	2016	2 × 780	1,200
	Ximeng–Shandong	1,000	2016	2 × 730	1,500
	Eastern Ningxia–Zhejiang	±800	2016	1,720	1,600
Total				15,693.4	15,920

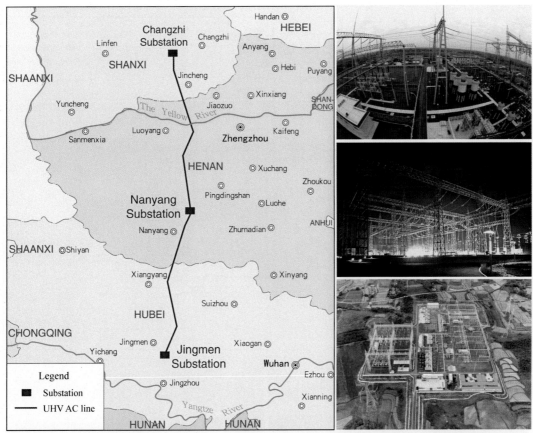

FIGURE 7.24 South-Eastern Shanxi–Nanyang–Jingmen 1000 kV UHV AC Transmission Experimental and Demonstration Project

Substation in northern Zhejiang, to terminate at Liantang Substation in western Shanghai (Fig. 7.25). The project has a transformer capacity of 21 GVA and a long-term transmission capacity of 10 GW. By connecting coal-fired energy bases in Huainan and Huaibei in Anhui to load centers in East China, the project contributes to the construction of a robust grid at the receiving end of the East China Power Grid to better receive high capacity DC transmission from the Xiangjiaba–Shanghai and Jinping–southern Suzhou lines, significantly improving the East China Power Grid's capacity for receiving power imports and also grid operational safety and stability, while promoting coordinated economic and social development in the region.

Northern Zhejiang-Fuzhou 1000 kV same-tower double-circuit UHV AC transmission project: With a length of 2×603 km, the project was put into operation in December 2014. With newly built Central Zhejiang (Lanjiang) Substation, southern Zhejiang (Liandu) Substation, and Fuzhou (Rongcheng) Substation as well as expanded Northern Zhejiang (Anji) Substation, the project has a transformer capacity of 18 GVA, as shown in Fig. 7.26. In the initial stage of operation, the project will have a transmission capacity of 6800 MW and a higher long-term transmission capacity of more than 10.5 GW. It links up

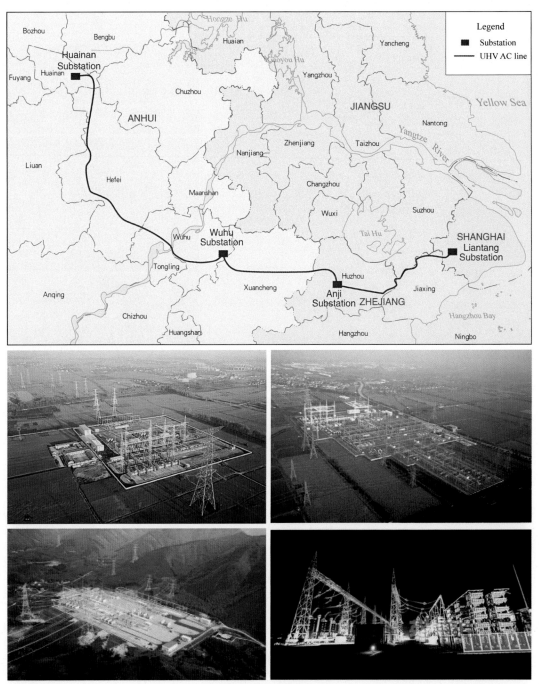

FIGURE 7.25 Huainan–Northern Zhejiang–Shanghai 1000 kV UHV AC Transmission Project with Double Circuits on the Same Tower

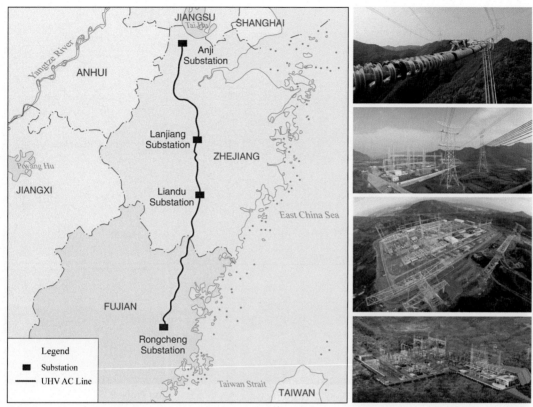

FIGURE 7.26 Northern Zhejiang–Fuzhou 1000 kV UHV AC Transmission Project with Double Circuits on the Same Tower

Zhejiang Province and Fujian Province, forming an integral part of East China's main UHV grid and contributing to the region's capacity for receiving electricity. By further enhancing the ability to optimize allocation of energy resources through a UHV grid, the project plays an important role in ensuring supply security for Zhejiang and Fujian and servicing East China's socioeconomic growth.

1.4.1.2 UHV DC Transmission Projects

Through a UHV DC transmission project, long-distance, high capacity and point-to-point transmission of electric power from large energy bases to load centers can be achieved. SGCC has completed and commissioned four UHV DC transmission lines, effectively transmitting electric energy from coal, hydro, wind, and solar energy bases in the western, northern and south-western regions to load centers in support of eastern and central China's socioeconomic development.

Xiangjiaba–Shanghai ±800 kV UHV DC transmission demonstration project: Stretching 1891 km, the project was put into operation in July 2010. Starting at Fulong Converter Station in Sichuan, the project runs through a total of eight provinces (municipalities), including Sichuan, Chongqing, Hunan, Hubei, Anhui, Zhejiang, Jiangsu, and Shanghai, to end at Fengxian Converter Station in Shanghai

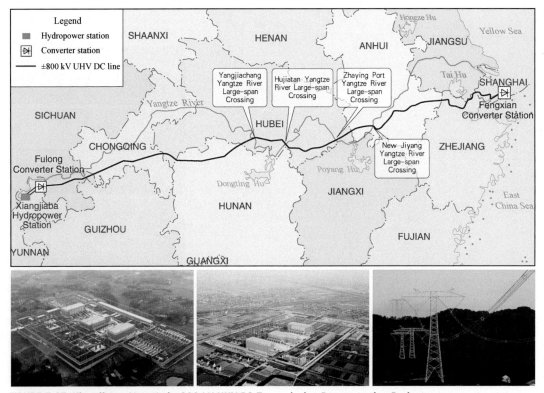

FIGURE 7.27 Xiangjiaba–Shanghai ±800 kV UHV DC Transmission Demonstration Project

(Fig. 7.27). It has adopted the world's first successfully developed complete set of ±800 kV UHV DC equipment, including 6-in. thyristors, 800 kV 321 MVA converter transformers, 800 kV 1750 MW converter valve units, and 800 kV 4500 A dry-type smoothing reactors. At a rated voltage of ±800 kV and rated power of 6400 MW, the project can transmit electric power of 32.5 TWh per year with a maximum continuous transmission power of 7000 MW. It plays an important supportive role in meeting East China's energy requirements through transmission of surplus hydropower from Xiangjiaba, Xiluodu, and Sichuan.

Jinping-southern Suzhou ±800 kV UHV DC transmission project: Stretching 2059 km, the project was put into operation in December 2012. Starting at Jinping Converter Station in Sichuan, it runs through a total of eight provinces (municipalities), including Sichuan, Yunnan, Chongqing, Hunan, Hubei, Zhejiang, Anhui, and Jiangsu, to end at Suzhou Converter Station in Jiangsu with a transmission distance exceeding 2000 km for the first time (Fig. 7.28). The converter transformers, converter valves, smoothing reactors, and DC sleeves employed in the project were then the highest voltage, flow capacity, and highest capacity of any DC equipment in the world. The transmission line has adopted six-splitting 900 mm^2-large cross-section conductors for the first time, to successfully resolve design and construction problems of UHV DC lines in regions of high altitudes, heavy icing, and serious pollution. At a rated voltage of ±800 kV and rated power of 7.2 GW, the project can deliver electric power

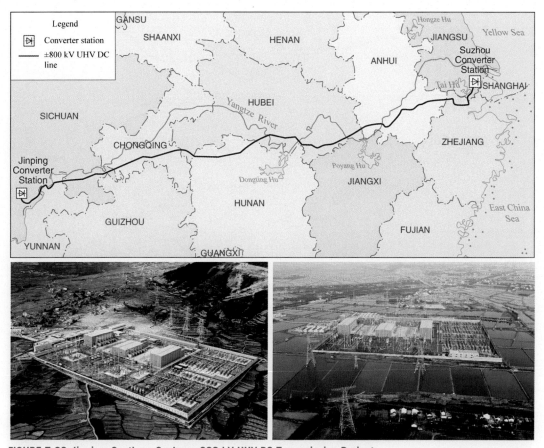

FIGURE 7.28 Jinping–Southern Suzhou ±800 kV UHV DC Transmission Project

of 36 TWh per year, serving as yet another important green energy channel for power transmission from west to east. By ensuring the timely and smooth outward transmission of surplus hydropower from Guandi and Jinping Hydropower Stations, and from Sichuan in flood season, the project effectively relieves the power shortage of East China during peak demand periods in summer and eases the mounting environmental pressure in East China's economically developed regions.

Southern Hami–Zhengzhou ±800 kV UHV DC transmission project: Stretching 2191 km, the project was put into operation in January 2014. Starting at Tianshan Converter Station in Xinjiang, it runs through a total of six provinces (autonomous regions), including Xinjiang, Gansu, Ningxia, Shaanxi, Shanxi, and Henan, to end at Zhongzhou Converter Station in Henan (Fig. 7.29). It is currently the world's longest UHV DC transmission line and the first UHV DC project to bundle and deliver electric power from large-scale thermal and wind power bases. At a rated voltage of ±800 kV, the project leverages the flow capacity of converter valves of 6-in. large cross-section thyristors, capable of transmitting 50 TWh of electricity per year with a transmission capacity of 8 GW. The southern Hami–Zhengzhou project is the first UHV channel for outward transmission of electricity from Xinjiang, contributing to

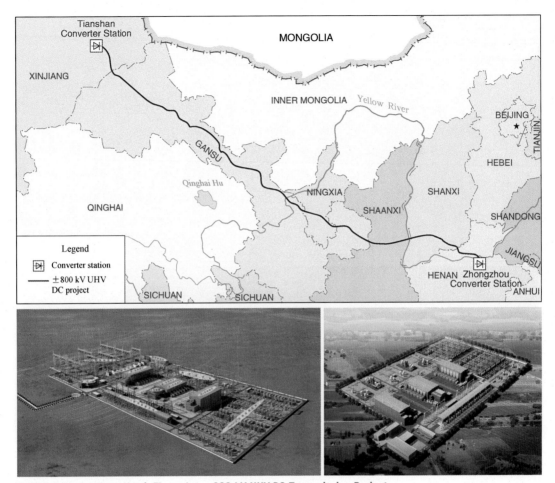

FIGURE 7.29 Southern Hami–Zhengzhou ±800 kV UHV DC Transmission Project

the intensive development of coal-fired, wind, and solar energy generation in north-western China and realizing optimized allocation of electric energy resources across the country to become an "electric power silk road" that connects western border areas to Central China.

Xiluodu–Western Zhejiang ±800 kV UHV DC transmission project: Stretching 1669 km, the project was commissioned in July 2014. Starting at Yibin Converter Station in Sichuan, it runs through a total of five provinces, including Sichuan, Guizhou, Hunan, Jiangxi, and Zhejiang, to end at Jinhua Converter Station in Zhejiang (Fig. 7.30). The project is the world's first single circuit DC project to achieve full-load transmission operation at 8 GW, setting a new record in ultra-large capacity DC transmission. It can transmit approximately 40 TWh of clean hydropower each year to Zhejiang from south-western China. Connecting hydropower bases in the southwest to load centers in the east, it is another clean energy channel after the Xiangjiaba–Shanghai and Jinping–South Suzhou UHV DC transmission projects. The project makes a great difference in terms of ensuring transmission and

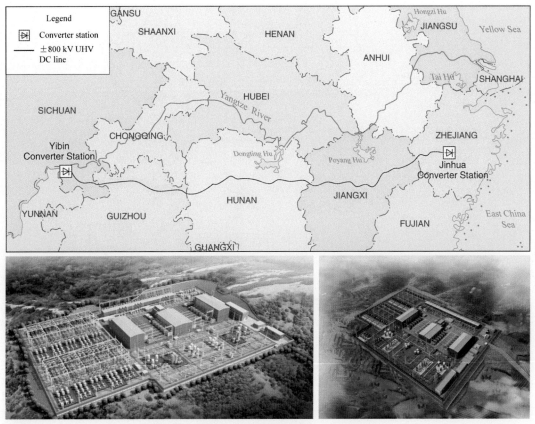

FIGURE 7.30 Xiluodu–Western Zhejiang ±800 kV UHV DC Transmission Project

consumption of surplus hydropower from large hydropower stations in the Jinsha River Basin in wet season, transforming resource advantages into an economic competitive edge, and optimizing the energy structure.

1.4.2 Smart Grid Projects

Marking an important direction of global grid development, the smart grid offers clear advantages in supporting the large-scale development of clean energy, meeting diverse user requirements, realizing fault self-recovery, and improving operational economics to lay a foundation for the development of a smart-based global energy interconnection. China's smart grid development covers power generation, transmission, transformation, distribution, consumption, and scheduling. As at the end of 2014, 358 smart grid pilot projects across 38 categories were arranged by SGCC, with 305 pilot projects across 32 categories already completed. See Fig. 7.31 for the distribution of these projects. The completion and efficient operation of these demonstration projects has set an example for incorporating elements of intelligence into the development of a globally interconnected energy network.

Note: numbers on the map indicate project numbers, excluding Taiwan data.

FIGURE 7.31 Distribution of SGCC's Smart Grid Pilot Projects

1.4.2.1 Power Generation

Meeting the requirement for large-scale grid integration of clean energy generation is an important objective of smart grid development. The key lies in solving the problems of grid integration and consumption caused by the random and intermittent nature of wind, photovoltaic energy and other renewables through major measures such as forecasting wind and solar-based generation power as well as coordinated operational control of diverse clean energy sources. To forecast clean energy generation power, SGCC deployed as at the end of 2014 systems for projecting wind generation power in 20 provinces (autonomous regions and municipalities) and systems for forecasting solar energy generation power in Qinghai, Xinjiang, and Ningxia. This network of forecasting systems plays a vital role in supporting the operation, control and consumption of clean energy generation. In terms of operational control, SGCC has mastered the technology for joint operation of clean energy generation and storage through building a national demonstration project on wind and solar power storage and transmission in Zhangbei (Fig. 7.32). The demonstration project has a total capacity of 670 MW. Phase I involves development of wind power capacity of 100 MW, photovoltaic energy capacity of 40 MW, and energy storage capacity of 20 MW. The comparative figures for Phase II are 400, 60, and 50 MW, respectively. Through in-depth explorations of multiconfiguration, multifunction, adjustable, and schedulable methods of joint operations governing wind and photovoltaic power generation and storage, SGCC has learned about different types of wind turbines and the running characteristics of large-scale power-adjustable photovoltaic generation equipment and chemical energy storage devices of different types and sizes, to realize the joint operation of wind and photovoltaic power generation, storage, and transmission as an innovation model for improving grid integration and consumption of new energy.

FIGURE 7.32 National Wind/PV/Energy Storage and Smart Grid Demonstration Project in Zhangbei, Hebei

1.4.2.2 Power Transmission

Online condition monitoring and real-time diagnosis for transmission equipment are important for intelligent transmission. The major smart equipment is shown in Fig. 7.33. As at the end of 2014, SGCC completed installation of condition monitoring master stations for transmission and transformation equipment in 26 provinces (autonomous regions and municipalities), in which it operates, realizing condition monitoring of 4263 transmission lines as well as transmission and transformation equipment in 3597 substations. Comprehensive monitoring of UHV lines, UHV substations, converter stations, and substations on important cross-regional grids has also been achieved. Through the online condition monitoring and real-time diagnosis system, real-time recognition, monitoring and early warning, analytical diagnosis, and assessment and prediction of the operation conditions of key transmission and transformation equipment has been realized. This provides online monitoring information for production management, giving a full view of the operation conditions of transmission and transformation equipment to significantly improve the level of lean management for transmission and

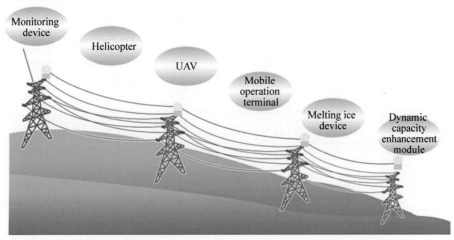

FIGURE 7.33 Major Smart Equipment for Power Transmission

transformation services. Through smart inspection equipment based on helicopters and UAVs, smart surveillance has been conducted to further improve the controllability of real-time observation and monitoring of equipment operations and operational safety.

As an important channel for outward transmission of electricity from Xinjiang, the condition monitoring system of southern Hami–Zhengzhou ±800 kV UHV DC transmission line is composed of transmission-tower front-end monitoring devices and transmission line background monitoring centers. By using wireless mobile communication, data, and images are transferred, images scrutinized, and warning messages received. A total of 17 sets of online monitoring equipment have been installed, including signal measuring, breeze, vibration, video systems, and auxiliary equipment. The project provides more stable power supply and network communication by relying on multiple power sources, including solar panels and wind power, and cable splicing and 3G networks for communication. By using the transmission line condition monitoring system, operation and maintenance staff can obtain accurate real-time technical data of air pressure, humidity, thunder, and lightning, and other natural environmental conditions in the operating environment of the transmission lines, offering a brand new technology for effectively improving operation and maintenance.

1.4.2.3 Power Transformation

As the intelligent development of substations is key to improve the intelligence level of the whole electric power system, it is of paramount importance to realize digitalization, equipment integration, service integration, and design compactness. See Fig. 7.34 for the structure of a smart substation. Full digitalization refers to the complete digitalization of all the signals, equipment, and controls in

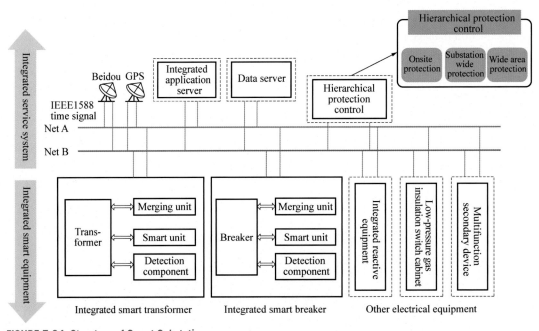

FIGURE 7.34 Structure of Smart Substation

FIGURE 7.35 Smart Substation (750 kV) in Yan'an

a substation to form a digitalized substation model as a platform for complete smart control and efficient management. Equipment integration means putting sensors and smart components together to improve equipment performance, control equipment dimensions, and enhance reliability by applying new technologies, new materials, new processes, optimized transformers, breakers, and other key devices. Service integration refers to the combination of protection control, automation and communication systems, with an integration of online monitoring, onsite inspection, operation, and maintenance operations, to establish an integrated service system that reduces crisscrossing and duplicate efforts for better coordinated control and improved overall efficiency. Design compactness means an integrated design based on different voltage grades and the characteristics of different substation types, to optimize the main wiring connection and station layout for space economy and lower investment costs. As at the end of 2014, SGCC operated 1527 smart substations at 110 (66)–750 kV, including 1135 at 110 (66), 344 at 220, 12 at 330, 29 at 500, and 7 at 750 kV, respectively.

The 750 kV Smart Substation in Yan'an is the first unmanned substation (Fig. 7.35), completely realizing condition visualization, operation sequencing, condition-based maintenance, and intelligent operation, with innovative breakthroughs made in different areas. Among the intelligent primary electrical equipment used are breaker-mounted mechanical, gas, partial discharge condition monitoring devices and smart terminals; oil chromatography, partial discharge, other sensors, and smart terminals embedded in the main transformer, and a smart ventilation system that results in 15% lower energy consumption. Electronic transducers are employed station-wide, including the Rogowski coil electronic current transducers for 750 kV equipment and the Rogowski coil and all-fiber current transducers for 330 kV equipment, which make maintenance and overhaul easier, improves the electromagnetic property of transducers, and enhances protection and monitoring equipment performance. Also employed is a unified condition monitoring platform operating on combined off-line/on-line modes, with a standard platform for data acquisition and diagnostic analysis for condition monitoring equipment to bring about a transition in monitoring parameters from "singular diagnosis" to "integrated diagnosis." The substation also features "one-key" switching operation by combining advanced applications like sequence control, smart alarming, fault reasoning, and analytical decision-making to realize operation automation, and change the traditional mode of switching operations. Such a facility offers the benefits of unmanning by using an integrated panoramic information platform to optimize station-wide data integration, improve substation operation, and realize advanced applications such as primary equipment visualization, condition-based maintenance, and smart alarming.

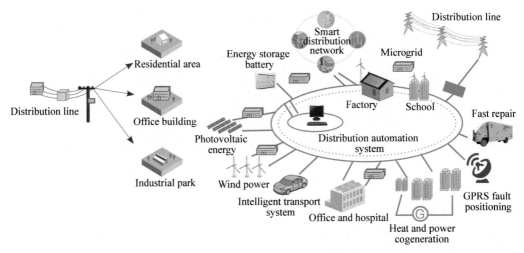

Distribution line

Smart distribution network

Energy storage battery

Microgrid

Residential area

Distribution line

Office building

Photovoltaic energy

Factory

School

Fast repair

Distribution automation system

Industrial park

Wind power

Intelligent transport system

Office and hospital

Heat and power cogeneration

GPRS fault positioning

FIGURE 7.36 Differences Between Smart Distribution Network and Traditional Distribution Network

1.4.2.4 Power Distribution

The development of intelligent distribution is reflected in the major breakthroughs in grid self-recovery control, distribution terminal intelligentization, and access for distributed generation. Please refer to Fig. 7.36 for the differences between a smart distribution network and a tradition power distribution network. As at the end of 2014, SGCC had developed intelligent distribution networks covering the core centers of 78 cities, operating distribution automation systems for over 10,000 10 kV lines. Based on an integration of six systems for smart grid self-recovery control, unified support for smart power distribution terminals, and grid-based online risk assessment and prewarning, the distribution automation systems are further supported by more than 20 applications of smart distribution terminals, interfaces for grid integration of distributed generation, microgrid protection and other smart equipment to improve the intensity of grid power distribution, and the controllability of distribution operation in the energy production process. This has resulted in substantially reduced unplanned outage time, limited scope of fault impact, and improved supply reliability.

The Chengdu power distribution automation project is the largest distribution automation pilot project implemented by SGCC with the largest number of terminals. It has completely realized distribution network operation monitoring, feeder automation, distribution network model management, and distribution network application analysis, with the establishment of an integrated shared distribution communication system as well as the operation and maintenance management systems for the integratedly controlled Chengdu distribution network and the city's urban grids. Covering all the 10 kV feeder lines in the main urban areas, the project features 1190 sets of power distribution equipment, including switching terminals, feeder line terminals, distribution transformer terminals, switchgear stations, and boundary disconnect switches, to provide operational stability and reliability with positive effects. An interactive control model is offered to help dispatchers locate and troubleshoot a fault in one minute, realize fault isolation in 30 s, restore supply through network reconfiguration, narrow the scope of outage, and significantly improve supply reliability.

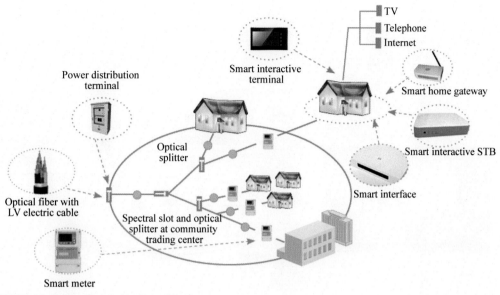

FIGURE 7.37 Power Fiber-to-the-Home System

1.4.2.5 Power Consumption

Smart power consumption concerns every household. China has started work on a series of engineering practice covering smart meters, power consumption information acquisition, interactive marketing, demand side management, user-side distributed power, electric vehicle charging/swapping facilities, power quality monitoring, and power fiber-to-the-home. By the end of 2014, SGCC had established a power consumption information acquisition system with 240 million smart meters, realizing remote automatic meter reading, self-service recharging, real-time power consumption monitoring, line loss monitoring, and orderly power consumption management. Power optical fibers with integrated fiber optics and power lines have been introduced into more than 470,000 households (Fig. 7.37) to provide terminal users with internet, telecom, radio, and TV signal transmission and other value-added services, while distributing electric power, shaping a new mode of grid operation, and providing more convenient, diversified, and efficient services to the public. In Beijing, Shanghai and other locations, 28 smart communities have been developed (Fig. 7.38), with a service platform covering 287,000 households. Construction of electric vehicle battery charging/swapping networks has been accelerated, with 618 charging stations and 24,000 charging piles already installed. In 2014, 133 quick charging stations and 532 quick charging piles for electric automobiles were built along the vertically aligned Beijing–Shanghai and Beijing–Hong Kong–Macau (Beijing-Xianning) Expressways, and the horizontally aligned Qingdao–Yinchuan (Qingdao–Shijiazhuang) Expressway, covering 34 cities with the capability to sustain a driving range of 2900 km. By 2020, quick charging networks will be completed for four vertically aligned expressways serving Shenyang–Hainan, Beijing–Shanghai, Beijing–Taiwan, and Beijing–Hong Kong–Macau, as well as four horizontally aligned expressways serving Qingdao–Yinchuan, Lianyungang–Khorgos, Shanghai–Chengdu, and Shanghai–Kunming, covering 135 cities with the capability to sustain a driving range of 19,000 km (Fig. 7.39).

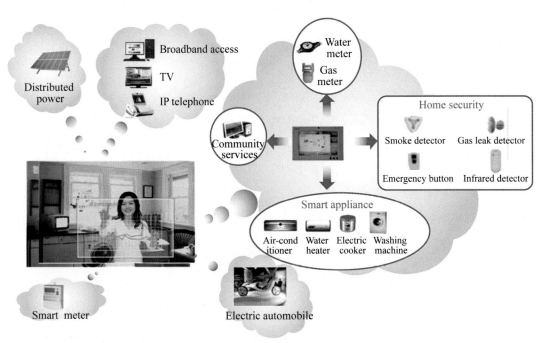

FIGURE 7.38 Major Functions of Smart Community

SGCC has basically completed the construction of a smart charging/swapping network in Beijing. As at the end of 2014, a total of 72 charging stations and 4260 charging poles had been installed, with 14 charging stations designed for buses and another 16 for taxis. Together with 1688 charging poles installed in major transport hubs, science parks, colleges, universities, and parking lots in 4S stores, a 5 km-wide "green" belt of EV charging service has taken shape at the city center. A service platform for smart charging networks has been put into operation and mobile applications for customers developed, realizing navigation, appointment, and other services to provide more convenient charging services for EV users. The network has served 9,599 electric vehicles, providing 998,000 charging times with a total power output of 31.8 GWh. The total driving mileage of the serviced vehicles combined has amounted to 75.84 million km, effectively contributing to electricity substitution and CO_2 emission reduction.

In Shanghai, SGCC has made successful exploratory efforts in introducing the power fiber-to-the-home service and establishing smart communities. The power fiber-to-the-home project covers 200,000 households and, on this basis, a smart community built on an open public information platform has been developed to capitalize on a rich variety of interactive features. First, these include more reliable, faster and more convenient information network services for residents, including IPTV high definition TV programs and various interactive media services like live broadcasts, on-demand HD programs, news and information, stock information, and games. Second, smart grid features based on fiber optics are available to collect information on electricity consumption and provide mid and low voltage line loss comparison. Information on household power consumption is uploaded through a smart meter to a concentrator before being directly transferred through an optical fiber to power suppliers. The success

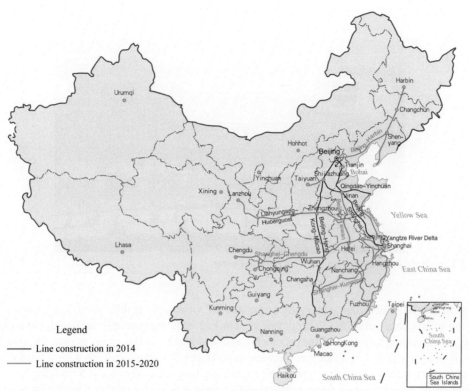

FIGURE 7.39 Quick Charging Networks for Four Horizontally Aligned and Four Vertically Aligned Expressways in 2020

rate and accuracy of data collection is up to 100% and real-time acquisition is achieved. Third, smart electricity consumption is promoted. Through smart electricity consumption, consumers can check and analyze information on their electricity consumption without leaving home, develop plans to be energy smart, promote energy conservation and emission reduction, and realize energy savings and rational use of power. The successful commercial operation of the power fiber-to-the-home project and the smart community in Shanghai has laid the foundation for the construction of a "smart city."

1.4.2.6 Power Dispatch

The key point of intelligent dispatch is to conduct active and intelligent grid monitoring, analysis, early warning, decision-making support, and self-recovery control, which is crucial for ensuring efficient utilization of clean energy and safe and economic operation of power grids. In recent years, State Grid Corporation of China (SGCC) has developed a new generation of smart grid scheduling technology support systems on the basis of unified grid-wide scheduling, gradually realizing panorama operational information presentation, network-based data transfer, online safety assessment, sophisticated dispatch decision-making, operational control automation, and optimized coordination between power generation and grid transmission. A panorama presentation of grid dispatch operations can gather and monitor information on grid operation, meteorology, social security, mass media, environmental protection

FIGURE 7.40 State Grid Corporation Dispatch Control Center

monitoring, wind forecasting, geography and industrial videos. As at the end of 2014, large-scale application of the smart grid dispatch control system had been successfully achieved, covering all of SGCC's main dispatch systems at the provincial level and higher, 7,011 access points, 890,000 data acquisition points, and 2,451 phasor measurement units in generating plants/substations, with smart grid scheduling technology support system for 33 provincial dispatch systems, and 5 grid dispatch systems completed and commissioned, forming the world's largest scheduling system with the highest controllability. It has realized real-time information sharing among all power grids at over 220 kV, three-level coordinated operation control and defense in depth. "Horizontal integration and vertical connection" of dispatch operations has also been realized, enhancing control over large grids and giving strong support to the integrated operation of the North China, Central China, and East China Power Grids and the dispatch of grid integration of clean energy. State Grid Corporation's dispatch control center is shown in Fig. 7.40.

1.4.2.7 Integrated Demonstration

In order to demonstrate and test smart grid technology in general and assess the integrated efficiency of the smart grid, SGCC has developed a number of smart grid integrated demonstration projects. As at the end of 2014, nine such projects had been set up, including the Shanghai World Expo Site, Sino-Singapore Tianjin Ecocity, Yangzhou Development Zone, Jiangxi Gongqingcheng City, Shaoxing New Area in Zhejiang Province, and Zhengzhou New Area in Henan Province, with the development of regional integrated demonstration projects in 16 other regions, including Beijing and Shandong.

As shown in Fig. 7.41, the smart grid integrated demonstration project on the Sino-Singapore Tianjin Ecocity is an iconic project on China's smart grid development. The adoption of new technologies, such as distribution automation, equipment on-line monitoring, smart scheduling, and smart substations, has improved the power quality as well as supply reliability and safety of this project. The supply reliability and voltage eligibility rate of the ecocity will reach 99.999 and 100%, respectively, while integrated line loss will decrease 1.18%, resulting in higher supply reliability. The ecocity has

FIGURE 7.41 Smart Grid Integrated Demonstration Project of Sino-Singapore Tianjin Ecocity

achieved the integration of wind and solar photovoltaic energy generation and storage at 36 kW with a microgrid, with the construction of a microgrid energy management system based on a smart dispatch support platform to fully leverage the distributed generation sources. An integrated data management platform for smart services is also established to carry out flexible and diverse exchanges with customers through the internet, SMS, phone, mail, fax, and other means of communications and to facilitate interactions in terms of onsite or remote management. Depending on their needs, customers can make inquiries about power supply and consumption, tariffs, energy efficiency analysis, and other information, while enjoying remote control and management of different types of smart home equipment. The Ecocity smart grid demonstration project will reduce oil consumption by 1,074 tons and save the equivalent of 5,930 tons of standard coal per year, with annual carbon dioxide emissions to go down 18,488 tons, producing significant results in energy conservation and emission reduction.

1.4.3 Development of Clean Energy

In recent years, China's clean energy development has shown phenomenal growth, building up an installed capacity of clean energy of 444 GW by the end of 2014, with a generation capacity of 1370 TWh. By installed capacity and generation capacity, hydropower accounts for 302 GW and 1066.1 TWh, respectively, compared with wind energy (95.81 GW and 156.3 TWh, respectively) and solar power (26.52 GW and 23.1 TWh, respectively). China's wind power grids rank first in the world in grid capacity and growth.

1.4.3.1 Hydropower Projects

As at the end of 2014, there were 54 hydropower stations with an installed capacity of more than 1 GW in China. The Three Gorges, Xiluodu, and Xiangjiaba Hydropower Station rank among the top three in installed capacity in China. See Table 7.6 for an overview of China's large hydropower stations.

Table 7.6 Overview of Large Hydropower Stations in China

S/N	Hydropower Station	Production Capacity (MW)	Total Capacity (MW)	Unit Capacity (unit × MW)
1	Three Gorges Hydropower Station	22,500	22,500	2 × 50 + 32 × 700
2	Xiluodu Hydropower Station	13,860	13,860	18 × 770
3	Xiangjiaba Hydropower Station	6,400	6,400	8 × 800
4	Nuozhadu Hydropower Station	5,850	5,850	9 × 650
5	Longtan Hydropower Station	4,900	6,300	9 × 700
6	Pubugou Hydropower Station	3,600	3,600	6 × 600
7	Xiaowan Hydropower Station	4,200	4,200	6 × 700
8	Jinping Grade II Hydropower Station	4,800	4,800	8 × 600
9	Jinping Grade I Hydropower Station	3,600	3,600	6 × 600
10	Laxiwa Hydropower Station	3,500	4,200	6 × 700

Source: Ref. [14].

Three Gorges Hydropower Station: As shown in Fig. 7.42, it is the biggest hydropower station in the world. The first unit was connected to the grid to generate power in July 2003, and all the units were put into operation by July 2012. The Three Gorges Hydropower Station is equipped with 34 generating units with a total installed capacity of 22.5 GW. The project is composed of three sections. Power stations on the left and on the right bank feature 26 Francis water-turbine generating units with a unit capacity of 700,000 kW (14 units are on the left and 12 units on the right bank), amounting to a total installed capacity of 18.2 GW. An underground hydropower station is located in a mountain on the right

FIGURE 7.42 Three Gorges Hydropower Station

bank, featuring $6 \times 700,000$ kW units, with a total installed capacity of 4.2 GW. The power source station is situated in a mountain on the left bank, featuring $2 \times 50,000$ kW units with a total installed capacity of 100,000 kW. The average annual generation of the Three Gorges Hydropower Station is 88.2 TWh, equivalent to approximately 50 million tons of standard coal. The project transmits hydropower to Central China, East China, South China, and Chongqing (Sichuan), benefiting more than half the mainland population.

Xiluodu Hydropower Station: Commissioned since June 2014 with a total installed capacity of 13.86 GW, the project is the second largest hydropower station in China. On either bank of the river is located an underground station, featuring nine 770,000 kW water–turbine generating units. Annual average generation is expected to reach 57.1 TWh. A backbone project under China's program to carry electricity from west to east, the Xiluodu Hydropower Station mainly supplies power to East China and Central China while also catering to electricity requirements in Sichuan and Yunnan. Located in the Jinsha River, it is a backbone project closest to the "west–east power transmission" route.

Xiangjiaba Hydropower Station: The project has been put into operation since July 2014 as the backbone power source in the middle of China's "west-east power transmission" route. It boasts of eight 800,000 kW water turbines, with an installed capacity of 6.4 GW and annual average generation of 30.7 TWh. Power from this hydropower station is delivered to Central China and East China through UHV DC.

1.4.3.2 Wind Power Bases

China has achieved remarkable results in the development of large-scale wind power bases with plans for nine large wind power bases (Fig. 7.43). With a planned capacity of 11 GW, the wind power base in Hebei is already in operation with a grid-connected capacity of 9.131 GW. The wind power base in Eastern Inner Mongolia, with a planned capacity of 8 GW, was commissioned with a grid-connected capacity of 7.825 GW. With a planned capacity of 13 GW, the wind power base in western Inner Mongolia has started producing electricity with a grid-connected capacity of 12.381 GW. The comparative figures of planned capacity/grid-connected capacity for other wind power bases in operation are: Jilin Wind Power Base (6 GW/4.08 GW), Gansu Wind Power Base (11 GW/10.076 GW) (Fig. 7.44), Xinjiang Wind Power Base (10 GW/8.039 GW), Jiangsu Wind Power Base (6 GW/3.023 GW), Shandong Wind Power Base (8 GW/6.224 GW), and Heilongjiang Wind Power Base (6 GW/4.537 GW).

1.4.3.3 Photovoltaic Power Generation

China's installed capacity of photovoltaic power generation had increased from less than 300,000 kW in 2009 to 26.52 GW at the end of 2014. With the continued development of photovoltaic generation technology, the efficiency of equipment utilization continues to improve, with a steady decline in system costs. Investment in photovoltaic power systems had decreased from RMB 25,000 per kWh in 2010 to RMB 9,000 per kWh in 2013. Qinghai is the largest solar photovoltaic power generation base in China. As at the end of 2014, 153 grid-connected photovoltaic power plants in Qinghai were in operation (Fig. 7.45), with a total installed capacity of 4.23 GW.

1.4.3.4 Distributed Power Supply

Development of distributed clean energy has also grown relatively rapidly, supported, and promoted by the government and power companies. On the whole, China's distributed clean energy is at its

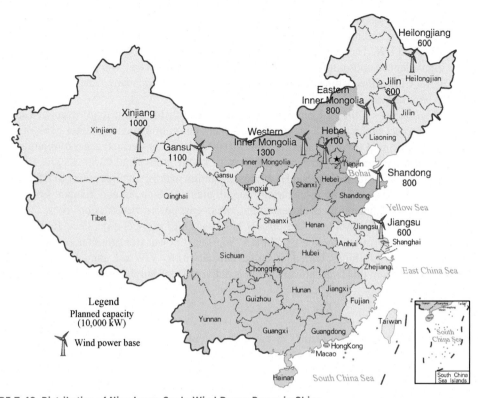

FIGURE 7.43 Distribution of Nine Large-Scale Wind Power Bases in China

FIGURE 7.44 Wind Power Base in Jiuquan, Gansu Province

FIGURE 7.45 Photovoltaic Power Base in Golmud, Qinghai Province

early stage of development, currently with limited capacity, although its long-term growth potential is enormous. As at the end of 2013, in the regions, in which SGCC operates, there were 1677 distributed generation projects (Fig. 7.46) at the predevelopment stage, under development or with access to grids at 10 kV and less. Of these projects with a total installed capacity of 4.313 GW, 688 were commissioned with a total installed capacity of 1.504 GW. The installed capacity of distributed photovoltaic power generation connected to the grid amounts to 3.1 GW, or 16% of the installed capacity of all photovoltaic power generation.

1.4.4 Nationwide Network Connection and Transnational Power Transmission

The global energy network is built on a network of robust grids around the world. Given China's vast territory and reverse distribution of resources and demands, optimized allocation of energy resources over large areas is required, making the enhancement of nationwide grid connection and transnational transmission an inevitable trend. The practice gained from China's development of nationwide network interconnection, and transnational transmission will set an important example for the development of a global energy network.

FIGURE 7.46 Distributed Photovoltaic Power Base in Haining, Zhejiang Province

Network connection on a nationwide basis in China dates back more than half a century. Before 1949, China's electric power industry developed slowly and power supplies were only available in a few large to medium-sized cities, with small power grids at various voltage grades. China unified its voltage grades after 1949 and a voltage grade sequence fell into place. The 110 kV transmission grid covering Beijing–Tianjin–Tangshan took shape after 1952. The 220 kV backbone of north-eastern China's power networks quickly came into being after 1954. The 330 kV backbone supporting China's north-western power network was gradually developed after 1972. After 1981, the construction of a series of 500 kV transmission lines gave shape to the Central China Power Grid's 500 kV backbone framework as it stands today. Successively, 500 kV backbone frameworks were also developed for the power grids in North China, East China, north-eastern China, and South China. The completion of the ±500 kV DC line between Gezhouba and Shanghai in 1989 paved the way for DC grid interconnections between Central China and East China. With the construction of a number of 750 kV transmission and transformation projects, a 750 kV backbone framework for northwest China gradually took shape starting from 2005. In 2009, the commissioning of an experimental and demonstration project on UHV AC transmission made a cross-regional grid interconnection between North China and Central China possible. In 2010, Xinjiang was connected to the main power network in northwest China through 750 kV lines. At the end of 2011, a±400 kV DC interconnection between Qinghai and Tibet was completed and commissioned, marking the end of Tibet's reliance on a single power network.

Six synchronous power grids have so far been developed in China, covering North–Central China, East China, north-eastern China, north-western China, South China, and Tibet to realize nationwide interconnections, with the exception of Taiwan Province. North China and Central China are interconnected synchronously through 1000 kV UHV AC, while asynchronous DC interconnections have been achieved between north-eastern China and North China, between north-western China and Central China, between north-western China and North China, between Central China and East China, and between Central China and South China. As at the end of 2013, China's installed generation capacity amounted to 1.25 TW, serving a population of more than 1.3 billion.

By building on UHV grids, China will accelerate the exploitation and outward transmission of hydropower in Sichuan Province and Tibet. It will also expedite the construction of a power grid in the southwest, to be connected back-to-back with Central China by DC. The country's grid interconnections may hopefully give rise to three locations at the sending end, in south-western China, north-western China and north-eastern China, and one region (North/Central/East China) at the receiving end, together with five synchronous grids under China Southern Power Grid. The result will be a more logically structured system of grid interconnections, with more well-defined functions, stronger allocation capability and higher safety levels. Capabilities will also be enhanced in terms of transmission of power from west to east and from north to south, as well as mutual support and backup between the hydropower and thermal generation sectors and between the wind and photovoltaic power generation sectors. The future of China's grid interconnections is shown in Fig. 7.47.

To build transnational connections, China has actively pursued transmission projects at the transnational level. Currently, electric power is transmitted from Russia to China through three AC lines (two 220 and one 110 kV lines) and one DC line (500 kV), from China to Mongolia through four AC lines (one 35 kV line and three 10 kV lines), from Kyrgyzstan to China through two 12 kV AC lines. China's power grids also transmit power to Vietnam through three 220 and four 110 kV lines, to northern Laos through one 115 kV line, and to Myanmar through one 35 and one 10 kV line. As at October 2014, China imported 16.5 TWh of electricity from Russia and 900,000 kWh of electricity from Kyrgyzstan,

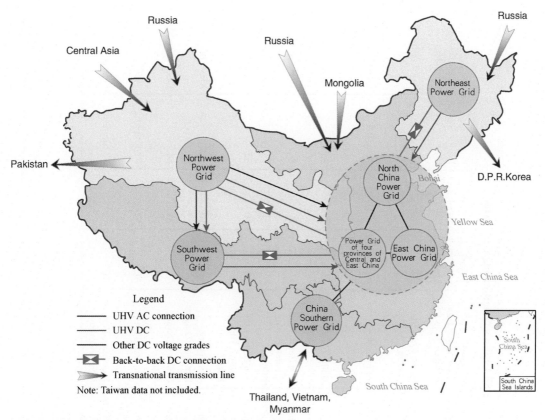

FIGURE 7.47 The Future of China's Grid Interconnections

while exporting 600 GWh of electricity to Mongolia. The existing grid interconnections between State Grid Corporation and China's neighboring countries are shown in Table 7.7.

Looking ahead, China will continue to strengthen power interconnections with neighboring countries: *Interconnections with Russia* will be achieved through UHV DC to deliver electricity from large power bases in the Russian far east and Siberia to China, with the focus on promoting the supply of power from the coal base in Yerkovtsy in the Russian Far East to Hebei Province through ±800 kV HV transmission and from the Kuzbass coal base in western Siberia to Henan Province through ± 1100 kV UHV transmission. For *interconnections with Mongolia*, China will look at the possibility of working with Mongolia to build coal-mine pit-head power plants in Mongolia to export electricity to China, with the focus on promoting the supply of electricity from Sibo Obo in Mongolia to Tianjin through ±660 kV DC transmission, and from Busi Obo to Shandong through ±800 kV UHV transmission. For *interconnections with Kazakhstan*, the focus is on promoting the supply of electricity from Ekibastuz through ±1100 kV UHV power transmission to Nanyang City, Henan Province. Through *interconnections with Pakistan*, China will transmit power to mitigate power shortages in Pakistan, with the focus on power delivery through ±660 kV DC transmission from Ili to Islamabad.

Table 7.7 Existing Interconnections between State Grid and China's Neighboring Countries

S/N	Countries	Routes	Offshore Interconnect Ion Points	Onshore Interconnection Points	Transmission Modes	Voltage Grades (kV)	Line Lengths (km)	Commissioning Dates
1	Russia	Amur substation – Heihe Converter Station	Blagoveshchensk	Heihe Converter Station	DC	500	160	April 2010
2 3		Blagoveshchensk – Aihui Circuit A and Circuit B		Aihui Station	AC	220	2 × 10.9	December 2006
4		Blagoveshchensk – Heihe Circuit		Heihe City	AC	110	8.24	July 1992
5	Mongolia	Small port power transmission lines	Habbie Riga Port	Xin Barag Right Banner, Hulun Buir	AC	10	8.2	December 2001
6		Small port power transmission lines	Baiyin Hushuo Port	Xin Barag Left Banner, Hulun Buir	AC	10	82	January 2008
7		Small port power transmission lines	Songbeier Port	Arxan City, Hinggan League	AC	10	1.7	May 2008
8		Yarant – Mongolia Circuit	Khovd Province	Qinghe County, Altay	AC	35	64	December 2009
9	Kyrghyzstan	Small port power transmission lines	Tulu Geerte Port	Torugart Port	AC	12	5 (domestic)	1997
10		Small port power transmission lines	Sary–Tash of Osh Province	Irkeshtam Port, Wuqia County	AC	12	12 (domestic)	2001

Based on the successful execution of its UHV and interconnection projects, China will strengthen exchanges and cooperation with all countries and alliances for promoting grid interconnections. Technical know-how and experience will be shared, with cooperative opportunities pursued in fundamental research, technological breakthrough, equipment research and manufacture, and project construction in respect of energy projects at the transnational and transcontinental levels. China will also be involved in research on energy development in key regions of the Arctic and the Equator and on major intercontinental transmission channels between Asia and Europe, in a bid to promote optimized allocation of global resources in concert with other countries around the world.

2 INTERNATIONAL PRACTICE

Driven by a new energy revolution, major countries around the world are actively developing the smart grid and clean energy, while promoting grid interconnections. Some countries have also started research and development work on UHV technology. These practices constitute an important foundation for building a globally interconnected energy network.

2.1 ULTRA HIGH VOLTAGE

Since the 1960s, the world's major nations in electric power have conducted a series of R&D projects on key UHV transmission technology and equipment manufacture. The former Soviet Union, Japan, the United States, and Italy proposed the development of UHV transmission technology, with work on UHV transmission planning, design, test, and equipment development achieving good results. However, UHV transmission projects in countries like the former Soviet Union and Japan were later suspended, deferred or operated at a lower voltage in response to lower demand for high capacity and/or long-distance transmission due to sluggish load growth. Out of consideration for building technological capability, the United States and Italy have conducted relevant research whereas India and Brazil have carried out engineering work on project construction based on the requirement for developing and transmitting renewable energy.

The former Soviet Union: As one of the world's first countries to pioneer research on UHV transmission technology, the former Soviet Union had set up several research institutes, such as a technology bureau under the Ministry of Power and Electrification, since 1960 to conduct fundamental research on UHV transmission. The country started building a 1.17 km three-phase UHV test line at the Brix Paster Substation in 1973, and started work on a 270km industrial test line between Yitate and Novokuznetsk in 1978. It owned a UHV test base featuring 3×1200 kV, 10–12 A cascaded test transformers and a 1000 kV impulse generator. In 1981, work started on a five-section UHV line with a total length of 2344 km. The 1150 kV Ekibastuz–Kirk Chitav line was completed and commissioned in August 1985. It has since been operating at a lower voltage because of the collapse of the Soviet Union and lower demand. Moreover, the country initiated work on a ±750 kV, 6 GW DC transmission project from Ekibastuz to Tambov, but the project was suspended and left uncompleted.

The United States: The country started work on UHV transmission technology in the latter half of the 1960s, demonstrating the feasibility of this technology through a series of research projects and tests. In 1974, American Electric Power Company and the General Electric Company carried out actual measurements of audible noise, radio interference, corona losses, and other environmental effects in the UHV transmission technology research and test station in Pittsfield. In 1974, the American Electric Power Research Institute built a three-phase 1000–1500 kV test line, and gathered experience with the electromagnetic environment, iron tower installation and transformer design through trial operations. In 1976, the United States Bonneville Power Administration started research on UHV line mechanical structure, corona, ecological environment, operation and lightning strike insulation on the Lyons testing ground and the Molo test line. To meet the needs of economic and energy development, the United States has paid more attention to the upgrading of power grid infrastructures, putting forward a vision for nationwide interconnections in Grid 2030.

Japan: Japan started R&D work on UHV transmission technology in 1972, involving Central Research Institute of Electric Power Industry, Tokyo Electric Power Company, and NGK Insulators. Under the leadership of CRIEPI, UHV test and research bases in Akagi and Shiobara were built, with field studies of electric corona noise, radio interference, wind noise, electric corona noise, and the impact on the ecological environment conducted at the Akagi base. In addition, an experimental study of air clearance of poles and towers as well as insulator strings was carried out at the Shiobara test field. In 1998, Japan started building a 426 km 1000 kV UHV transmission line designed to supply power to Tokyo. The line has been operating at a step-down voltage.

Italy: After establishing a research program on 1000 kV transmission, ENEL had been conducting UHV technology research and technological development at its test stations and laboratories since 1971. Among the 1000 kV test facilities at the Sava Reto test field are electric corona and electromagnetic environment testing devices comprised of a 1 km-long test line and a 40 m-long test cage. Experimental studies covering operation and lightning over-voltage tests and insulation characteristics had been carried out. In 1984, Italy started building a 3 km UHV overhead test line at the Sava Reto test field. The project was completed in October 1995. Some operation experience was acquired through a trial run of the project at full voltage for 2 years.

India: India embarked on a study of 1200 kV UHV AC technology in 2007. Work started in September 2011 on an outdoor test station and test line on 1200 kV UHV AC technology, with plans for building a six-circuit UHV AC line and achieving synchronous interconnections nationwide in 2020 based on 1200 kV UHV AC transmission. India also incorporated plans for developing two ±800 kV UHV DC transmission projects into its "Twelfth 5-year Plan for Grid Development" to deliver hydropower from the northeast to the west and thermal power from the eastern and central regions to the north. The two transmission lines measure 1728 km and 3700 km long. Work on one of the projects, the Biswanath Chariali–Agra ±800 kV UHV DC transmission line, commenced in March 2011, as shown in Fig. 7.48. Expected to be completed in 2016, this project has a rated power of 6 GW and a length of 1728 km. Having initiated research on 1200 kV UHV DC transmission, India started to operate the Bina test station in December 2012 and built a 2 km test line. The further development and outward transmission of thermal power from the eastern and central regions as well as hydroelectric power from the north will call for more UHV projects to be undertaken.

Brazil: The development of UHV technology in Brazil is driven mainly by the need to develop hydropower resources concentrated along the northern Amazon River and its tributaries. As the country's load centers are located in the south-eastern area 1000–2500 km away from its hydroelectric bases, UHV technology is best able to support the high capacity transmission of electricity over this long distance. In February 2014, a joint venture between SGCC and ELETROBRAS won the bid for the Belo Monte ±800 kV UHV DC transmission project in Brazil, as shown in Fig. 7.49. As the first ±800 kV UHV DC transmission line of the Americas, the 2092 km project runs from the Xingu River

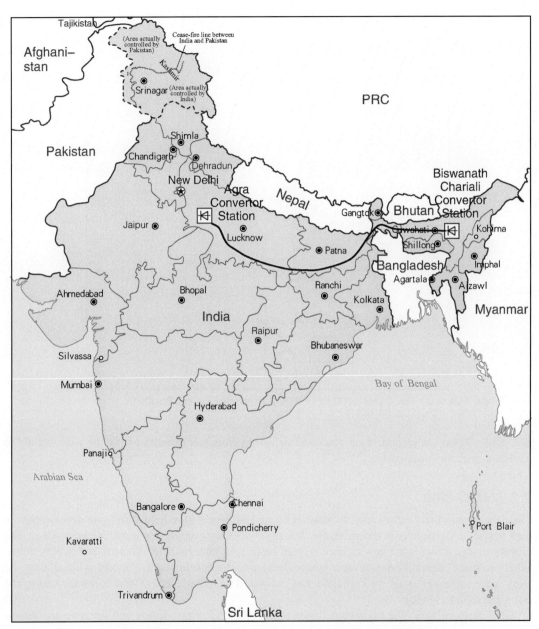

FIGURE 7.48 Map of ±800 kV UHV DC Transmission Project in India (Under Construction)

FIGURE 7.49 Map of the Belo Monte ±800 kV UHV DC Transmission Project, Brazil

to Estreito. Upon completion, the project will deliver hydropower resources from the west directly to load centers in the south-east.

2.2 SMART GRID

Since the twenty-first century, major countries around the world have held smart grid development in high regard, in the hope of overcoming the challenges of energy supply, environmental protection, and climate change through this new technology and ensuring a safe, reliable, efficient, and quality power supply to meet interactive and diverse power demands. The development of smart grids all over the world covers four key areas: grid infrastructure, advanced metering, electric vehicle infrastructure, and energy storage technology.

Grid infrastructure: The main objective is to upgrade grid infrastructure and improve grid operational safety and reliability, by adopting advanced monitoring and control technologies. All countries around the world are very conscious about the role of smart grids in further ensuring the safe and stable operation of their power networks, while contributing to a reliable and quality power supply, and improved efficiency of utilizing grid resources. In its outlook on grid development, the United States has seen the low efficiency of domestic power grids, transmission congestion, and a lack of supply reliability and power quality as having a serious impact on energy reform and innovation. As a result, the United States has been actively promoting the modernization of power grids in the past decade by promulgating a flurry of policy documents, including the *Energy Independence and Security Act, the Recovery and Reinvestment Act,* and *Strategic Plan – 2010*, providing funds to enhance grid infrastructure, build an early warning system on grid stability, provide real-time monitoring of system disturbance, and prevent major power outages. *Europe* also faces the problem of an ageing power infrastructure. The European Union looks to make grid infrastructure development the core element of its *Europe 2020 Strategy*, by committing € 200 billion to transform and upgrade power and natural gas transmission networks and accelerate the construction progress of transnational energy networks. Under the plan, the investment in power grids is estimated at € 142 billion, accounting for 70% of the total investment. The grid infrastructure projects given priority for development are: (1) an offshore power grid to connect the wind power-rich North Sea to northern and central Europe, covering over 90% of the European Union's offshore wind farms; (2) strengthening network connections in southwest Europe and the region's interconnections with Central Europe and building an undersea interconnection between North Africa and southern Europe to export renewable energy generated in North Africa; (3) connecting the power grids in Central and Eastern Europe with those in south-eastern Europe; and (4) undertaking a grid connection project in support of a uniform Baltic Sea market. *Japan* has imposed higher requirements on the automation, safety, and reliability levels of power grids. The Japanese Government considers it necessary to focus more on the operational stability of power grids against a background of large-scale clean energy development. In 2010, the *New Roadmap to International Smart Grid Standards* was released by the International Standardization Institute for New Energy under the Ministry of Economy, Trade, and Industry of Japan, stating in unequivocal terms the need to build a network of robust grids that can withstand the impact of disasters. The key technologies required to be developed include wide area monitoring and control systems for power transmission, storage batteries for power system, distribution grid management, demand side response, and advanced metering devices, which can help improve grid operational capability.

Advanced metering: The main objective is to carry out demand side management, deploy advanced metering systems, and improve the interactivity between electricity users and the power grid. *The United States* has conducted work on research and practice on smart power services to improve grid operational efficiency and the quality of power services. The US Department of Energy has been mandated to launch an R&D project on grid digital information technology in support of relevant technology evaluation and research on smart meters, demand response, distributed generation, energy storage, wide area measurement, information and communication networks, and electric vehicles. All state governments are also required to provide electricity consumers with information on energy consumption, such as real-time tariffs. In terms of project construction, America's first smart grid city has been built in Boulder, Colorado, to provide every household with a smart meter. With an intuitive understanding of real-time tariffs through the meter, customers can shift power consumption to time periods when tariffs are lower. As the largest demonstration project of its kind in the United States, the Pacific Northwest Smart Grid Demonstration Project involves 11 power utilities and 5 technology partners from the states of Idaho, Montana, Oregon, Washington, and Wyoming, covering 60,000 electricity users. The project seeks to verify the feasibility of two-way interactive communication among distributed generation, power storage, and grid infrastructure while ensuring grid safety. In *Europe*, two-way information flows between electricity users and power utilities are achieved through public utility and data networks based basically on smart meters, intelligent power terminals, and intelligent home appliances. *Italy* has installed and upgraded 30 million smart meters as part of an intelligent metering network. *Germany* has mandated the installation of smart meters for every new housing project and existing residential properties having undergone a major renovation. *France* is working to legislate a mandatory requirement that smart meters be adopted for all electricity consumers. The *UK* Government has passed into law a white paper on smart meters and electricity bills, requiring public utilities to adopt a market-oriented approach to operating all smart metering equipment to bring about concerted efforts among smart meter manufacturers, suppliers, operators, and data acquirers to promote smart metering.

Electric vehicle infrastructure: The main objective is to accelerate the construction of electric vehicle charging/swapping infrastructure, promote the technology and industry development of electric vehicles, and realize electric energy substitution. As an important service component of smart grid, electric vehicle charging and swapping represents an emerging segment of the electric power industry. Power utilities around the world are striving to develop an electric vehicle infrastructure to promote electric cars. The *United States*, mindful of the importance of a sound policy system, is the first country to introduce competition into the electric vehicle market and enhance infrastructure development, including charging points and home smart rechargers, in the hope of playing a leading and enabling role in the electric car industry. In 2012, the US Department of Energy allocated a US$120 million budget for building 14,000 charging piles. In California where electric vehicles have received one of the strongest promotional boosts ever, the Air Quality Committee earmarked US$ 27 million to develop both pure and hybrid electric automobiles. Among all European nations, Germany's electric vehicle industry has seen faster growth. As a major car manufacturer, *Germany* is oriented toward R&D on key technology in order to build core competencies, with the focus on promoting electric automobiles from the perspective of energy system optimization. The German Government's investment is mainly in R&D on related core technologies with the objective of improving basic industrial capacity. Great importance is attached to the role of electric vehicles and related charging infrastructure in improving the utilization of renewable energy and the overall efficiency of power grids. And integration of electric automobile and smart grid technologies is seen as an integral part of an action plan for electric automobiles. *Japan* is pursuing a diversification strategy covering both pure and hybrid electric cars and oriented toward high performance battery technology. The focus is on the setting of international standards and uniting the country's industry alliances to promote the development of the electric vehicle industry. Advanced battery technology, together with the solid foundation that Japanese car manufacturers have laid for gasoline engines and hybrid power technology, has given the Japanese electric car industry an important advantage. Car producers, battery manufacturers, and power utilities have come together to build a pillar of strength in the electric vehicle market.

Energy storage technology: The main objective is to develop energy storage technology and demonstrative application so as to build technological capabilities for large-scale development of clean energy. In terms of project numbers and installed capacity, the United States and Japan are two major countries in the demonstrative application of energy storage. The United States is one of the forerunners in stored energy development, accounting for half of the world's energy storage demonstration projects with successful examples of commercial application. The US Government has provided full and continuous policy and financial support for development and application of energy storage technology, especially in lithium ion battery manufacture, and system integration. *Japan* is a world leader in energy storage technology, including sodium-sulfur batteries, flow cells and lead acid batteries. Since the Fukushima nuclear incident in 2011, Japan has been promoting home energy storage as a key area of industry support, with the release in April 2012 of a subsidy policy on home energy storage systems. The *European Union* has supported national R&D and demonstration projects on energy storage in 14 countries in Europe. Through greater efforts in promoting the energy storage industry, *Germany* has effectively supported the development of the home energy storage market with a total investment of 50 million Euros in 2013 and 2014, to directly subsidize new purchases of energy storage systems.

2.3 CLEAN ENERGY

Major countries around the world value the exploitation and application of clean energy, with targets set for clean energy development. Clean energy is growing rapidly in the world thanks to policy, industry, and funding support. As at the end of 2013, the installed generating capacity of clean energy amounted to approximately 1940 GW, or 33.8% of the world's total installed capacity. Clean energy generation was around 4420 TWh, or 19.6% of the world's total generation.

Playing an important role in the global energy market, hydropower is the world's most widely developed form of clean energy. As at the end of 2013, the installed hydropower capacity globally stood at 1012 GW, generating 3190 TWh or approximately 14.2% of the global power supply. The major hydropower stations in operation worldwide continue to provide abundant clean energy. See Table 7.8 for the important large-scale hydropower stations in overseas countries.

Table 7.8 Important Large-Scale Hydropower Stations Overseas

Countries	Hydropower Stations	Rivers	Maximum Heads (m)	Installed Capacities (MW)	Annual Generations (GWh)	Operation Year
Brazil and Paraguay	Itaipu Hydropower Station	Parana	126	14,000	90,000	1984
Venezuela	Guri Hydropower Station	Caroni	146	10,300	51,000	1968
US	Grand Coulee Hydropower Station	Columbia	108	6,490	24,800	1941
Brazil	Tucurui Hydropower Station	Tocantins	68	8,120	32,400	1984
Russia	Sayano-Shushenskaya Hydropower Station	Yenisei	220	6,400	23,500	1978
Russia	Krasnoyarsk Hydropower Station	Yenisei	100.5	6,000	20,400	1968
Canada	La Grande II Hydropower Station	La Grande	142	5,330	35,800	1979
Canada	Churchill Falls Hydropower Station	Churchill	322	5,230	34,500	1971

Itaipu Hydropower Station: Jointly developed by Brazil and Paraguay in October 1975, the facility is the world's second largest hydropower station over the Parana River that divides the two countries. In 1984, the first generation set was put into production. In 1992, 18 generation sets were used to produce electricity for Phase I of the project. Two sets had been added between 2006 and 2007, this number was increased to 20, totaling 700 MW and 14 GW installed capacity. The annual energy output of the station is 90 TWh, accounting for more than 15% of the power supply in Brazil and over 70% in Paraguay.

Guri Hydropower Station: The world's fourth largest hydropower station over the Caroni River, the station was constructed in two phases from 1963 to 1968, with a total installed capacity of 10.3 GW. The No. 1 plant under Phase I boasts three 180 MW and seven 303 MW generation units, with a total installed capacity of 2.66 GW. The No. 2 plant as an addition to Phase II features 10×730 MW generation units, with a total installed capacity of 7.3 GW. Following a dam heightening project, the installed capacity of the No. 1 plant was increased from 2.66 GW to 3 GW. The station currently generates 51 TWh of electricity per year. Phase I started producing electricity 5 years after work commencement. While maintaining uninterrupted operation, the project underwent expansion to add capacity to meet load growth. The low cost hydropower generated produces marked economic benefits by substantially cutting down on Venezuela's oil consumption.

Grand Coulee Hydropower Station: The world's sixth largest hydropower station, the facility was built on the Columbia River in 1933, with a total installed capacity of 6.49 GW and room reserved for installing four 2.4 GW generating units. The first generating unit was commissioned in 1941 and work on the No. 3 plant was completed in 1978. The No. 1 and No. 2 plants were built during the preliminary stage of development, each equipped with nine 108 MW turbine generators. The No. 1 plant features three 10 MW auxiliary generating units. The No. 3 plant was built under an expansion program, with three 600 MW and three 805 MW generating units to provide a total installed capacity of 3.9 GW. After the generating units built in the early years were recoiled, output was raised to 125 MW. The three plants boast a total installed capacity of 6.49 GW with annual generation of 24.8 TWh, enabling the station to play a pivotal role in the Columbian power system.

Wind power is a more widely developed form of new energy. By the end of 2013, the installed capacity of global wind power amounted to 318 GW. The world's top 10 countries in terms of total installed wind power capacity are shown in Fig. 7.50. Overseas, the largest wind farm is the United States-based Alta Wind Energy Center, located in Bakersfield, California with 342 wind turbine generators each of 1.02 GW. The center is being expanded and expected to provide an installed capacity of 1.55 GW. The project can reduce over 5.2 million tons of carbon dioxide emissions each year, equivalent to removing 446,000 gasoline-powered cars from the roads. The world's largest offshore wind farm is the London Array project composed of 175 wind turbine generators, with a total installed capacity of 630,000 kW. It can cut down on 925,000 tons of carbon emissions each year and provide electricity for half a million households following the completion of Phase I.

Solar power holds great promise. By the end of 2013, the installed capacity of global solar power was approximately 140 GW. The world's top five nations in installed capacity of photovoltaic generation in 2013 are shown in Fig. 7.51. With abundant solar energy resources, Africa is expanding generation capacity. South Africa launched a project to encourage independent power producers to purchase renewable energy, with plans to build 47 power stations, including photovoltaic, wind, and small hydro projects. Among these facilities are 27 photovoltaic power stations with a total installed capacity of 1.048 GW. In September 2013, the photovoltaic power station under the project was put into operation, as Africa's largest facility of its kind with an installed capacity of 75,000 kW. In the same year,

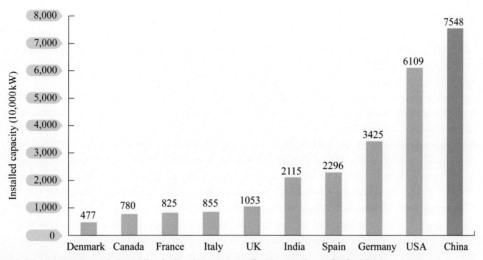

FIGURE 7.50 World's Top 10 Countries in Total Installed Wind Power Capacity in 2013

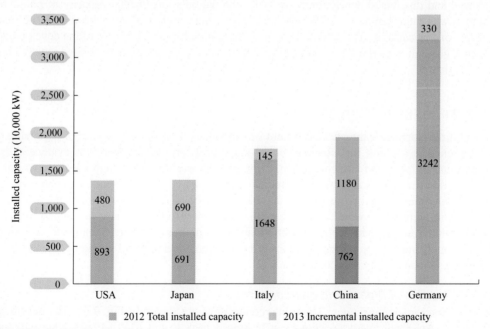

FIGURE 7.51 World's Top Five Nations in Total Installed Capacity of Photovoltaic Energy in 2013

the Solar I photothermal power station in the United Arab Emirates, the largest facility of its kind in the Middle East, was commissioned. Covering an area of 2.5 km² in a desert on the outskirts of Zayed City, the project is the Middle East's largest and most technologically advanced centralized solar power project, with an installed capacity of 100,000 kW. It supplies enough electricity for 20,000 households, reducing 175,000 tons of carbon emissions every year.

In addition to wind and solar power, exploratory efforts are seen in some countries in the utilization of other forms of new energy, such as tidal power. South Korea has built the Sihwa Lake Tidal Power Plant, the world's largest plant of its kind. Construction began in 2004 and the project became operational in August 2011. The station is equipped with 10 generating units, with a total installed capacity of 254,000 kW and an annual energy output of 550,000 MWh.

Distributed generation is developing rapidly around the world. By the end of 2011, the installed capacity of global distributed generation was 81.87 GW. Europe is one of the world's largest distributed generation markets, best represented by Germany with 83% of its photovoltaic energy projects being of a distributional nature. By the end of 2012, Germany's installed capacity of photovoltaic generation amounted to 32.28 GW, with 18% of projects being mounted on detached residential roofs (1–10 kW), 59% being installed on the roofs of small and medium-sized apartment or commercial buildings (10–100 kW), 6% being set up on the roofs of large-sized commercial buildings (greater than 100 kW), and only 17% being ground-mounted. Countries in the Asia–Pacific region have been introducing incentive policies to support the development of distributed generation. Japan's distributed generation comes mainly from heat and power cogeneration and photovoltaic energy generation, collectively accounting for 13% of the national installed capacity. North America is also an important market for distributed photovoltaic energy generation. In the United States, photovoltaic energy features a mixed mode of centralized and distributed development. In 2012, the nation's incremental capacity of photovoltaic generation was estimated at 3.31 GW, with residential, industrial, and commercial users accounting for 46.2% of this capacity. The Middle East and Africa are also pushing forward the development of distributed generation markets. Recently, off-grid distributed generation has developed rapidly in some parts of Africa.

2.4 INTERCONNECTED GRIDS

Power grid development generates the benefit of economies of scale. At the core of a global energy interconnection is a vision for interconnecting power grids around the world, as evidenced by the growing trend of large-grid interconnection. Currently, the world is seeing a quickening pace of grid interconnection resulting in increasing interconnection capacity. The individual interconnected power grids in North America, Europe and the Russia–Baltic Sea region represent important practices for grid interconnection around the world. Grid interconnections are also taking shape in southern Africa, the Gulf, and South America. More interconnections among countries on different continents are being developed, demonstrating a prominent trend toward grid interconnection and providing a practical basis for building a global energy interconnection.

2.4.1 Status Quo of Global Grid Interconnections

North American interconnections: These interconnections are shown in Fig. 7.52. The development of the North American interconnections can be traced back to the 1930s–1950s, when large-scale hydropower development triggered the first major growth of the continent's power grids. In the 1950s–1980s, voltage grades were raised in line with the rapidly expanding power demand, forming a system of interconnected grids in North America. Four synchronous power grids asynchronously interconnected are now in operation across the United States, Canada, and Mexico, including the eastern and western interconnections in North America, together with the Texas and Quebec power grids. Among the four synchronized power grids, the Eastern Interconnection is linked to the Western Interconnection through six DC tie lines, to the Texas power grid through two DC tie lines, to the Quebec power

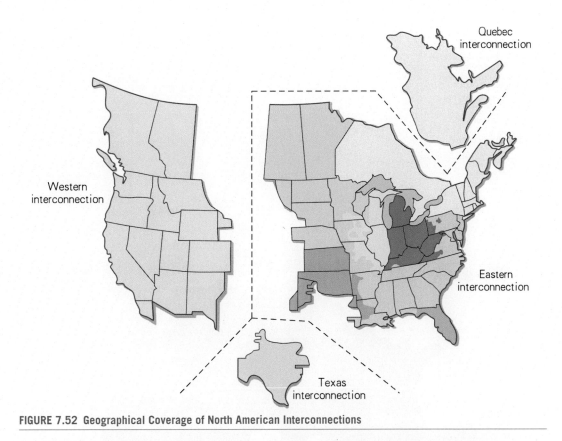

FIGURE 7.52 Geographical Coverage of North American Interconnections

grid through four DC tie lines, and one variable frequency transformer. There are more than 100 transmission lines (at 500, 230, 115 kV, etc.) in seven provincial power grids between the United States and Canada, together with a multiterminal EHV DC transmission line and a number of back-to-back DC projects. The power exchange capacity between the two countries is estimated at approximately 20 GW. Between the United States and Mexico, there are 27 mostly AC transmission lines. In North America, power is exchanged among different grid systems based on unilateral or multilateral agreements, with a mechanism for joint management, coordination, and a pool dispatch system to improve system operational reliability.

European interconnections: These interconnections are shown in Fig. 7.53. The development of Europe's grid interconnections dates back to 1958. The Western European interconnection slowly came into shape and was synchronously interconnected with the power grids in Central Europe in 1996. In July 2009, the European Network of Transmission System Operators for Electricity (ENTSO-E) was founded, bringing together 34 European countries and 42 transmission system operators. Today, the European grid system mainly comprises of five transnationally interconnected synchronous grids in continental Europe, Northern Europe, the Baltic Sea, the United Kingdom, and Ireland, as well as two independent power systems in Iceland and Cyprus. As at the end of 2013, 340 transmission lines were in operation covering the member states of the European power grid, including 318 AC and 22 DC

FIGURE 7.53 Geographic Coverage of European Interconnections

tie lines interconnected mainly through AC lines at 220/285, 330, 380, 400 kV, etc. With the continued progress of interconnection work in Europe, the capability of optimizing resource allocations has continued to grow, leading to higher levels of power exchange among the member states. In 2013, a total of 387.3 TWh was exchanged, representing 12% of total power consumption. ENTSO-E is looking to further expand the scope of Europe's interconnections to include Russia, Ukraine, Belarus, and Moldova, among other countries.

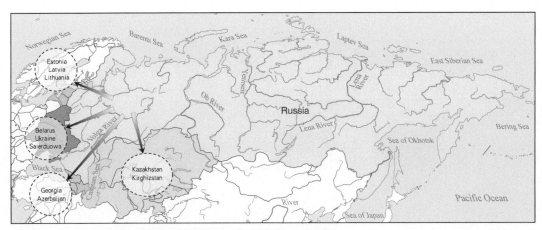

FIGURE 7.54 Geographical Coverage of Russia–Baltic Interconnection

Russia–Baltic interconnection: This interconnected grid is shown in Fig. 7.54. As the world's largest synchronous grid in terms of geographical coverage, the project spans eight time zones, linking up the grids in Russia, Azerbaijan, Belarus, Georgia, Kazakhstan, Moldova, Mongolia, Ukraine, Latvia, Lithuania, Estonia, and Kyrgyzstan, among other countries. The Finnish power grid is asynchronously connected with the Russian power grid through back-to-back HVDC transmission lines with a capacity of 1.42 GW.

Southern African interconnection: The project is shown in Fig. 7.55. Since its inception in 1995, the Southern African Power Pool (SAPP) has been actively promoting transnational grid interconnections. The alliance comprises of 12 members, including Botswana, Mozambique, Malawi, Angola, South Africa, Lesotho, Namibia, the Democratic Republic of Congo, Swaziland, Tanzania, Zambia, and Zimbabwe. With the exception of Malawi, Angola, and Tanzania, the other nine countries have developed grid interconnections, currently operated at 400, 275, 220, and 132 kV. The key interconnection projects planned by SAPP fall into two main streams. One involves the development of interconnections among Malawi, Angola, Tanzania, and other member countries. The other involves building a central African transmission channel, including a Zimbabwe–Zambia–Botswana–Namibia interconnection, a transmission corridor project in central Zimbabwe, and transmission works in Zambia. Among all countries in southern Africa, South Africa boasts the largest installed capacity, accounting for 82% of SAPP's total installed capacity.

GCC interconnection grid: Shown in Fig. 7.56, the GCC interconnection project was spearheaded by the Gulf Cooperation Council Interconnection Authority, founded in 2001 by six Gulf States to interconnect the power grids of Saudi Arabia, Kuwait, Qatar, the United Arab Emirates, Bahrain, and Oman, while taking responsibility to operate and manage the transnational grids and electricity trade among these countries. The interconnection was divided into three phases. Commissioned in December 2009, Phase I involved the development of interconnections among the power grids of the six member countries through an 800 km transmission network. Phase II, completed in 2006, saw the interconnection of power grids between the United Arab Emirates and Oman. Built in 2011, Phase III involved interconnecting the power grids developed under the first two phases. The interconnection project now provides a platform for electricity trade within the Arabian Gulf and with other regions. It has supplied

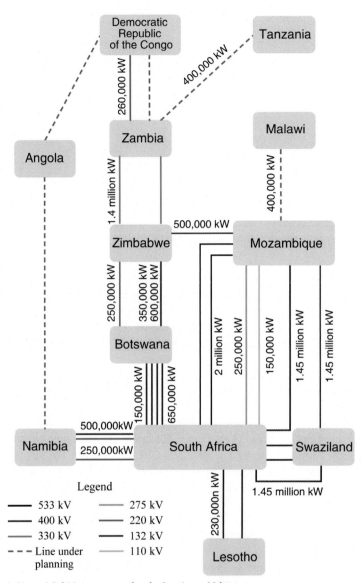

FIGURE 7.55 Sketch Map of Grid Interconnection in Southern Africa

Source: Ref. [86].

an additional 1.2 GW of power to each of Kuwait and Saudi Arabia, and additional 900, 750, 600, and 450 MW to the United Arab Emirates, Qatar, Bahrain, and Oman, respectively.

South American interconnection grid: As shown in Fig. 7.57, the South American interconnection grid covers two main sections. The northern section includes Columbia, Ecuador, and Venezuela and the southern section, Brazil, Paraguay, Argentina, and Uruguay. The interconnection among

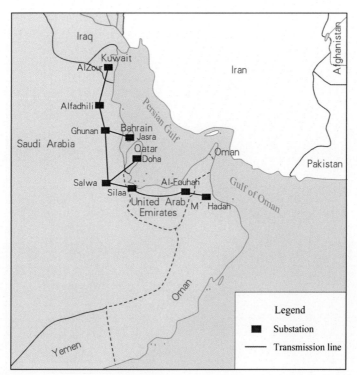

FIGURE 7.56 Sketch Map of GCC Interconnection Grid

power grids in the south is strongest. In April 2011, the Andean Electrical Interconnection System was established by the Andean Community members Peru, Columbia, Ecuador, Bolivia, and Chile, with plans to construct an Andean Power Corridor to eventually develop an interconnection grid at the regional level. In 2013, an agreement on electricity infrastructure development and promotion was signed, with plans to complete the interconnection of power grids among Peru, Ecuador, and Columbia in 2017. The interconnection will then be extended southward to reach Chile and Bolivia, as part of a grander program to integrate the electric systems of the five Andean Community nations by 2020.

Central American interconnection grid: As shown in Fig. 7.58, the Central American interconnection links up power grids among Panama, Costa Rica, Honduras, Salvador, Guatemala, and Nicaragua, including 15 substations and 230 kV transnational transmission lines measuring 1800 km. In 2014, the interconnected grid carried a total output of 40.6 TWh, more than 65%, of which was generated from renewable energy, and 20.9 TWh from hydropower.

2.4.2 Research Programs for Ultra-Large Grid Interconnections

Countries in Europe, Africa, Asia, and the Americas are actively conducting research and planning for transcontinental interconnections, with the objective of achieving optimized allocation of resources over larger areas. Typical examples of efforts in this direction include the European Super Grid, the DESERTEC initiative, the Asian Super Grid, and United States-proposed Grid 2030.

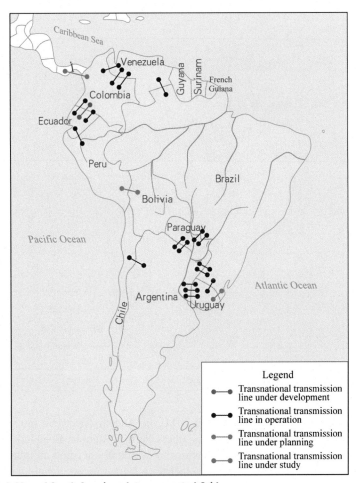

FIGURE 7.57 Sketch Map of South American Interconnected Grid

The European Super Grid: In 2010, a plan to develop an electricity supergrid was jointly announced by nine North Sea countries, including Germany, France, Belgium, Holland, Luxembourg, Denmark, Sweden, Ireland, and the United Kingdom. The planned supergrid will link clean energy bases in the coastal regions of the North Sea, integrating the wind, solar, hydropower and other energy resources in these countries, as shown in Fig. 7.59. This project will cover all of Europe, linking offshore wind power as well as pumped-storage power stations in the north and solar power stations in the south to load centers in Europe, including the United Kingdom, Germany, and France. In the future, an interconnected network of energy transmissions will evolve to connect to solar power stations on African deserts. Under Phase I (2010–2015) of the three-phase plan, old thermal plants and nuclear power stations will be replaced by new energy power plants, grid access provided for large-scale wind farms, and plans formulated to strengthen and expand current transmission systems. Phase II (2015–2020) envisages the construction of more large-scale offshore wind farms and fewer thermal and nuclear

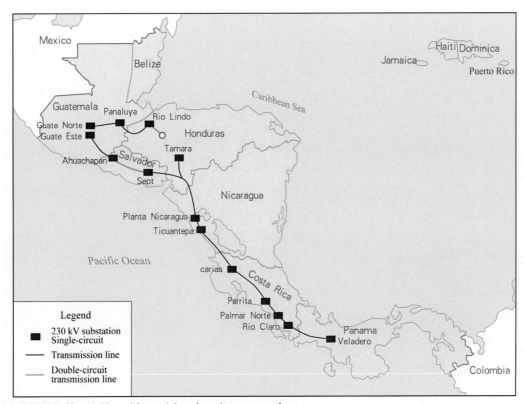

FIGURE 7.58 Sketch Map of Central American Interconnection

power plants, to realize a balance of power supply and demand across Europe, and build interconnections among offshore wind farms and across different countries. Under Phase III (2020–2050), integration of grid systems throughout Europe will continue in an effort to further establish a pan-European grid system. Load centers in continental Europe will be connected to wind farms and pumped storage hydropower stations in Northern Europe and large-scale photovoltaic power plants in southern Europe, and also to photovoltaic solar farms in Africa.

DESERTEC solar power project: In 2009, a consortium of companies and institutions in Europe and Africa agreed on a plan to construct the world's largest solar power project in the Sahara Desert in North Africa, as shown in Fig. 7.60. The €400 billion project is planned to be completed by 2050 to provide 15% of Europe's annual electricity requirements through transmission lines across deserts and the Mediterranean region. The DESERTEC Industrial Initiative (DII) was also set up to solicit wider participation in the solar power plan. The initiative was initially joined by German energy giant E.ON AG, RWE Group, Deutsche Bank, Munich Re, ABB, Abengoa Solar of Spain, Cevital from Algeria, the DESERTEC Foundation, HSH Nordbank, MAN Solar Millennium, M+W Spanish Zander, Schott Solar, and SIEMENS, among others. SGCC has participated in this program, despite the withdrawal of some companies and organizations for various reasons. Under the plan, solar power stations will mainly be located in Morocco, Tunisia, and Algeria. Morocco is

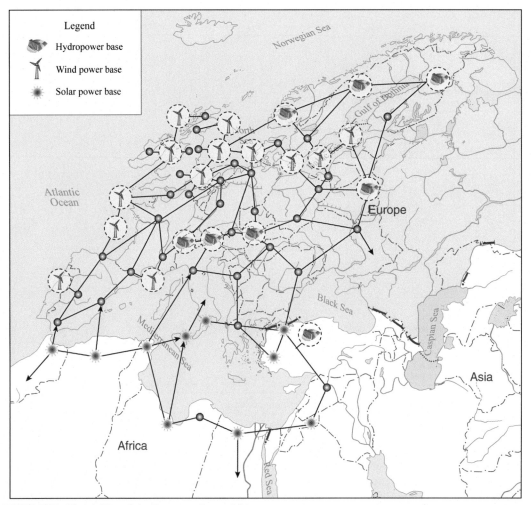

FIGURE 7.59 Sketch Map of the European Super Grid

Source: http://www.friendsoftheSuperGrid.eu.

planning to develop a solar generation capacity of 2 GW by 2020, compared with Tunisia's plans for building 4.7 GW of solar generation capacity by 2030. Algeria also plans to export 10 GW of clean electricity by 2030.

The Asian Super Grid: Originally named "Asian Super Circle," the transnational grid project was first proposed by Russia in 1998. During 1999–2000, Russia completed a feasibility study on large-scale export of electricity from Sakhalin to Japan through a network of underground cables. Under the Asian Super Grid plan, wind and solar power will be developed in the Gobi Desert, Mongolia, hydro and thermal power in the Russian far east, wind and solar power in China, as well as solar photovoltaic and wind power in South Korea and Japan, with plans to build a 36,000 km pan-Asian grid system that

FIGURE 7.60 Sketch Map of DESERTEC Solar Power Plan

Source: DESERTEC Foundation, Clean Power From Deserts–The DESERTEC Concept for Energy, Water and Climate Security, 2007.

connects Russia, China, Mongolia, South Korea, and Japan, as shown in Fig. 7.61. In recent years, some progress has been made on the project. In December 2012, Mongolia's Ministry of Energy held an international conference on Renewable Energy Cooperation and Grid Integration of north-eastern Asia. In March 2013, a memorandum of understanding on a joint feasibility study of the plan was signed among Mongolia's Ministry of Energy, Russia's Energy Research Institute, South Korea's Energy Economics Institute, Japan's Renewable Energy Sources Foundation, and the Energy Charter Secretariat. The proposal of the Asian Super Grid has met with a generally enthusiastic response from the international community. It will lay an important foundation for building a globally interconnected energy network in the future.

United States-proposed Grid 2030: The program originated from a report released in June 2003 by the US Department of Energy's Office of Electric Transmission and Distribution. The document "Grid 2030: A National Vision for Electricity's Second 100 Years" presents a vision of America's future electric system (Fig. 7.62), highlighting the importance of building interconnections on a nationwide basis and with Canada and Mexico in order to realize optimized allocation of electric power over larger areas. The plan consists of three elements. The first involves building a national electricity backbone, with the development of high capacity transmission corridors that link the east and west coasts, as well as northern Canada and southern Mexico, to fully leverage the integrated efficiency of the nationwide interconnections and improve overall grid efficiency and service quality. The second element involves building regional interconnections, with which the national backbone is connected. The connection between the regional networks is strengthened using AC or DC transmission links, whereas high capacity DC interties are employed to link adjacent, asynchronous regions. The third element involves setting up local, mini, and microgrids. The nation's local distribution systems are connected to the regional

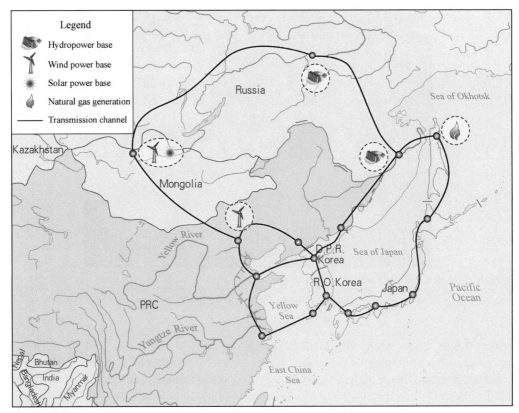

FIGURE 7.61 Sketch Map of the European Super Grid

networks, and through that to the national electric backbone. Real-time monitoring and information exchange enables markets to process transactions instantaneously and on a national basis. Customers have the ability to tailor electricity supplies to suit their individual needs. With the progress of technology, electric vehicles and fuel cell-powered vehicles may have a role in the small-scale application of distributed generation facilities. After the release of Grid 2030, the US Department of Energy funded a study on *The Future Grid to Enable Sustainable Energy Systems*, with 2050 as the target year. The research focus is on control and protection, technologies, and market mechanisms related to renewable energy development, power education and training, simulated analysis of system operations, as well as information and physical security, to effectively cope with the challenges of large-scale renewable energy development.

The research programs on grid interconnections in the aforementioned countries and regions represent valuable exploratory and application efforts, that help build an important foundation for the development of a global energy network.

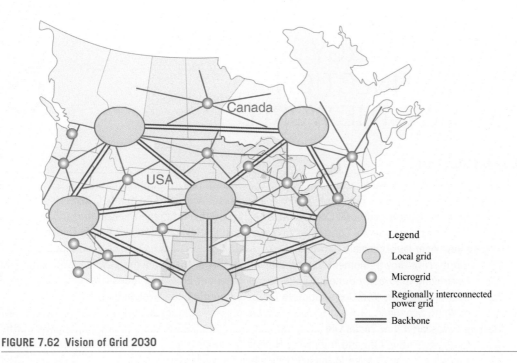

FIGURE 7.62 Vision of Grid 2030

SUMMARY

1. Countries around the world, including China, have been conducting technological research, standard setting, project construction, and planning development on UHV, smart grid, clean energy, and large grids, laying a foundation for technology research, and practice on a global energy interconnection.

2. Large-grid interconnections on each continent are beginning to take shape. This growing development demonstrates the certainty of a move toward building a global energy interconnection. The maturity and engineering application of UHV technology, provides a realistic and workable solution and technologically ensures the success of long-distance, large-scale allocation of clean energy for the global energy interconnection.

3. The development and application of smart grid technology worldwide has created a platform for the development of clean energy, of an intermittent and random nature, like wind and solar power, which will markedly improve the large-scale development and utilization of clean energy globally.

4. The development of a global energy interconnection based on UHV technology and clean energy technology will help vigorously drive the "two-replacement" policy and achieve global energy sustainability.

GLOBAL ENERGY INTERCONNECTION CHANGES THE WORLD

8

CHAPTER OUTLINE

1 CREATING A NEW ENERGY SCENARIO

Looking forward to 2050, a globally interconnected energy network supported by UHV backbones and oriented toward clean energy will basically be completed, while the "two-replacement" strategy will be pursued in earnest. Global energy development will enter a new stage focusing on electricity and clean energy, with optimized allocation of resources worldwide. Energy supply will be safer, more efficient and friendly, providing an energy security that can strongly support global economic development and the progress of human civilization.

1.1 BREAKING RESOURCE CONSTRAINTS TO PROVIDE UNIVERSAL ACCESS TO ADEQUATE CLEAN ENERGY

Clean energy such as hydro, wind, and solar energy around the globe is inexhaustible. An adequate supply of clean energy ensures the fulfillment of human demand for energy use, offering a continued strong momentum for economic and social development.

Global Energy Interconnection. http://dx.doi.org/10.1016/B978-0-12-804405-6.00008-7

More abundant energy supply. Backed by the global energy network, natural energy resources in nature such as running water, rushing wind, strong sunshine, and surging tide will be converted into electric energy by a chain of hydropower generators, wind turbines, solar photovoltaic power plants, wave generators, and other equipment for the benefit of mankind. By 2050, 66,000 TWh of clean electric energy will be generated every year on a global basis, nearly 10 times higher than in 2010. Abundant electric energy will light up every corner of the world. The world's population will grow by more than 2 billion over 2013 to reach 9.55 billion. Where there is human existence, there will be a sufficient, uninterrupted supply of electricity services, with energy access for everyone at acceptable costs. The long-standing problem of power shortage in Africa will be gone, with a large population previously denied access to electricity eventually being able to use clean electricity; oil, coal, natural gas, and other fossil energy will be mainly used as industrial raw materials to continue to serve mankind. As coal mines, oil wells, and gas fields will be developed in a more orderly manner, fossil energy will no longer be scarce – the bottleneck arising from depletion of fossil energy that inhibits our development will finally be removed. Backed by sufficient energy supplies, mankind's enormous material needs will be fully satisfied. For example, rainwater collected in large quantity, sewage discharged and salty seawater will be converted into clean, low-cost fresh water to meet the increasing demand for water and to promote food production.

Greater energy security. The energy issue is a typical global issue, long beset by resources, capital, cost, and other constraints. Relying on the global energy network, with technological progress and more stringent environmental requirements, the costs of solar and wind power generation will be significantly lower than those of fossil fuel-fired generation. Energy in the future will come at a low cost and in adequate quantities. Intelligent energy utilization systems will automatically adjust energy demand; an energy market featuring global allocations will establish an orderly mechanism for regulating energy supply and demand. Energy development will be less affected by financial manipulation, commercial speculation, monopolies, geopolitics, natural disasters, and other factors, such that wild price volatility is avoided and economic development better supported. Moreover, due to the lower demand for fossil energy and the ample supply of clean energy, together with a shift of the focal point of energy diplomacy and international cooperation to energy technology rather than energy resources themselves, fossil energy resources in the Middle East, Africa, and elsewhere will no longer be a hotbed for international disputes or potential conflicts. Energy will no longer be used as a bargaining chip to gain leverage among nations fighting for their own interests.

Stronger energy platform. UHV and smart grid technology enables energy interconnections to cover every corner of the world, eliminating dead zones or vacuums for energy supply. Through terrestrial power grids and submarine cables, a power network that covers the world will, based on its strong capability in resource allocation, will ensure large-scale access to hydropower, wind, solar, and other centralized and distributed power supply and achieve a flexible switch between the roles in power supply and demand. Backed by large-grid control technology, information and communication technology, and advanced power system simulation technology, the static state of a large system with hundreds of thousands of nodes can be simulated on a real-time basis, power loads forecasted accurately and dynamic adjustments effected to the power system structure to ensure grid operational safety and stability at the transnational and transcontinental levels. A global energy network with the characteristics of strong and smart grids can stand up well to risks by automatically prejudging and identifying most sources of failure and risk. It is also equipped with self-recovery capabilities to cope more effectively with typhoons, ice storms, earthquakes, and other calamities and external sources of damage, ensuring that incidents like the major blackouts in the United States and Canada will not be repeated.

1.2 BREAKING SPATIO–TEMPORAL CONSTRAINTS TO ACHIEVE EFFICIENT CLEAN ENERGY UTILIZATION

A global energy network can fully leverage the operational scale and economics of a grid system. Relying on a strong grid structure as well as modern information and communication technologies, the network can achieve overall optimization and real-time readjustment of the clean energy production, transmission, and distribution systems, resulting in substantially improved efficiency of clean energy development and utilization.

More efficient development. Global renewable resources will be fully utilized after UHV channels are built between large-scare energy bases and load centers. By that time, UHV transmission stretches distance over 5000 km, hence large power systems can have an efficient and coordinated operation for various energy. By fully leveraging the loads and renewable of various regions, ultimately have win–win cooperation and realize a maximization of comprehensive benefits for various energy development. High mountains, straits, deserts will not be the obstacles anymore to global energy interconnections, hydro power in Southwest China, wind power from Bering Strait, solar energy of the Sahara and marine energy from the Caribbean will be developed substantially and transmitted to power network. By then, clean energies will be fully exploited as well, including those enriched in the Arctic and equatorial areas.

More efficient allocation. Based on the global energy network, electric power in large quantities can be transmitted and allocated efficiently at the speed of light around the world. The scope of power transmission will no longer be confined to a country or a continent. Global installed capacity can be significantly cut back, given the potential to capitalize on the time difference between the Eastern and Western Hemispheres and the seasonal difference between the Northern and Southern Hemispheres. By 2050, electricity outflows from the Arctic and equatorial regions are expected to reach 12,000 TWh, accounting for 16% of global electricity demand. In the future, nighttime gale will no longer be wasted as strong wind-generated power will find its way through interconnected channels around the Northern Hemisphere to provide a reliable supply of electricity to areas where people are at work during the day. Summer's scorching sun will no longer be dreaded as it will become a source of uninterrupted power supply, delivered through intercontinental channels across the Southern and Northern Hemispheres, to warm up people in the icy cold of winter at the other end of the world. Through the tight connections in the global energy network, the globe will become an "energy village," allowing for more convenient and efficient energy allocations.

More efficient consumption. In the future, transcontinental interconnections will become possible through the global energy network. Asia, Europe and North America will be the major electricity importers, while Africa and Oceania will become electricity exporters. Large power grids will break through the problems of limited consumption in clean energy-rich regions by expanding consumption to the whole world and basically remove the compelling circumstances that force solar, wind and hydroelectric power to go wasted. At the same time, interactions with electricity users on the demand side will guide the orderly use of power and promote energy conservation, lower consumption and peak load shifting to result in improved energy system operations and energy utilization efficiency. Relying on big data, the global energy network can collect, collate, and analyze information on energy consumption from each terminal. Demand is also prejudged and the optimized allocation of various energy resources realized to improve energy efficiency by avoiding, to the greatest possible extent, waste and low efficiency of utilization.

1.3 BREAKING ENVIRONMENTAL CONSTRAINTS TO MAKE CLEAN ENERGY THE DOMINANT ENERGY OPTION

Relying on the global energy network, the "two-replacement" strategy will be fully implemented, with major development of clean energy and full application of clean electricity, making green, low-carbon energy a new fad. By 2050, clean energy is expected to represent up to 80% of primary energy consumption as the world's dominant energy.

Substitution of clean energy through energy development. To cope with climate change, fossil energy development will be subject to stringent control. Oil, coal, and natural gas production will peak successively by 2030 before being gradually replaced by clean energy. In the future, conventional coal-fired power plants will be decommissioned, while centralized and distributed clean energy generation will be carried out on a large scale. A cluster of hydro, wind and solar power stations with a single-unit capacity of 10 GW will be built where deep canyons, high mountains and barren deserts are located. Offshore wind farms with a single-unit capacity of 1000 MW will also be developed. A variety of distributed generation sources will be widely scattered in the urban and rural areas, indicating the emergence of a revolution around the world in energy production characterized by clean energy.

Substitution of electric energy through energy consumption. By 2050, electricity will gain popularity around the globe, accessible by all people. The population of once impoverished regions like Africa will bid farewell to the era of firewood as a source of energy by moving straight into a new world of electricity. Electric boilers, electric heaters, electric chillers, electric cookers, and electrified transport will be used extensively. Electricity will provide a solution to the energy requirements of water heating, cooking, heating, air conditioning, lighting, irrigation, and other basic necessities of everyday life. As clean electricity can basically meet most energy needs, energy consumption in the transport, construction, industrial, and other key sectors will fall significantly. In the industrial sector, a shift from coal/oil-fired furnaces to electric furnaces and the use of electric heating equipment will be widely promoted to put all chimneys to an end. Power-driven high-speed trains will replace short- and medium-haul aircraft, sharply reducing the carbon emissions from the aviation industry. Over 90% of cars on the roads will be powered by electricity. By 2050, electric energy is expected to account for up to 52.2% of end-use energy demand. Mankind will enter a new era of electrification where "electricity is everywhere."

Two-way interactivity through energy production and consumption. The conventional energy production and consumption process is characterized by a closed, one-way approach, leaving users with limited options. Against a backdrop of the global energy network, the Internet, the Internet of Things, mobile devices, cloud computing, big data, and many other advanced information and communication technologies will be closely integrated with energy and power technologies to make power grids more intelligent, which will drive a change in energy consumption from a one-way, passive approach to a two-way, interactive, flexible, and intelligent approach. Based on information acquired in real time on electricity usage by each user and each piece of electric equipment as well as the way electricity is utilized, power companies can provide flexible, sophisticated, and efficient demand-side response in accordance with the characteristics of users, guide users to change their power consumption behavior, and provide higher-quality electricity services. Users will be able to know in real time the electricity utilization status, electricity tariffs and other information, and take an active approach to participating in power management at the city and community levels. In addition, they can design a strategy for operating electric equipment and carry out lean management of electricity utilization. Tens of thousands of buildings, cars, and factories, which are energy consumers in the traditional sense of the word, will also become energy producers and traders on the global energy network.

2 INFUSING NEW VIGOR INTO ECONOMIC GROWTH

The global energy interconnection will create enormous productivity to unleash the elements of energy for the global economy, while driving technological reform and renewal as well as industry upgrades around the world. It will become a powerful engine for sustainable economic growth to make the world more prosperous and stronger.

2.1 REINFORCING GROWTH MOMENTUM AND RAISING QUALITY OF THE ECONOMY

The global energy interconnection will reshape the world's energy system, promote energy transition, initiate a changing mode of economic development, and drive sound and rapid economic growth.

Promoting economic globalization. The global energy interconnection will carry not just energy and electricity. It will also allow the aggregation and allocation of all the fundamental elements of economic development on a global basis to play a significant role in promoting economic globalization. In particular, the global allocation of clean energy on an unprecedented scale through the global energy network will not only replace conventional fossil energy, but also infuse new vigor into the global economy to break all the long-standing trade barriers that hinder economic integration and to advance the globalization of investment, finance, production, technology, information, commodities, services, and human capital. A transformation and upgrade of the world's economic development model will also be promoted to enable countries around the world, especially the developing nations, to benefit extensively from the globalization process.

Reducing social costs. The global energy network will significantly shorten the energy supply chain and streamline the organizational management process in the energy industry. Through the global energy network, large-scale production and network-based allocation of clean energy will be made possible to effectively reduce the costs of energy development and improve the overall level of energy utilization at the community level. The costs of renewable energy generation in sparsely populated regions such as equatorial deserts and the Arctic after 2035 are expected to increasingly become economically competitive for electricity importing regions. With the fast falling supply costs, solar and wind energy will rise to a dominant position in the generation mix. By then, countries around the world will significantly cut back on investments on the political, military, diplomatic, and economic fronts required to safeguard the security of the fossil fuel energy system. Various problems such as oil and gas storage and transportation, pollution control at thermal power plants, nuclear waste disposal, and carbon sequestration will be basically resolved. The carrying capacity of roads, railways and ships for coal, oil, and other resources will be fully released. The long-term constraints on China's economic growth of tight coal, electricity, and transport capacity, will no long be an issue and the social costs of energy consumption will fall significantly.

Optimizing economic structure. Electricity, a form of high-quality energy, has been used more extensively than ever in the world to replace fossil energy. It improves energy utilization efficiency, promotes technological progress, industry optimization and upgrading, and drives changes in production modes. It has brought about a transition from a mode of extensive economic development relying excessively on fossil energy that has led to the problems of "high consumption, high emission, high pollution, and low efficiency." As suggested in a World Bank report, low-carbon transport and higher energy efficiency achieved by factories, buildings, and home appliances could add US$1.8 trillion or 1.5% to annual GDP growth by 2030. The progress in the global energy network will create more jobs and a new division of labor, such that hundreds of millions of people who are directly or indirectly involved in the clean energy industry will profit from the global energy network. The development of the global energy network will significantly reshape economic factors, encourage economic growth through increased reliance on new technologies, new knowledge and new ideas, and accelerate the development of the green and low-carbon industries.

2.2 LEVERAGING INNOVATION BENEFIT TO FUEL DEVELOPMENT OF EMERGING INDUSTRIES

As a primary factor of production, science and technology is an inexhaustible source of strength to promote social progress. Driven by the global energy interconnection, new energy, new materials, intelligent equipment, electric vehicles, and the new-generation information industry have been revitalizing the economy, bringing profound changes to production modes and organizational structures and catalyzing a new industrial revolution, with the basis for a country's competitiveness and the competitive landscape of global industries thoroughly reshaped.

Catalyzing a new industrial revolution. As a new, cross-industry technological revolution, the global energy interconnection will open up new markets and fully promote the development of strategic emerging industries. In the previous two industrial revolutions in human history, technological innovations like steam engines, internal combustion engines, and electric motors played a leading role. The third industrial revolution, one that is induced by the global energy interconnection, will progress in the direction of new energy development and the breakthrough integration and convergence of transmission, new materials, communication, artificial intelligence, and other new technologies. The breakthroughs made in energy storage technology will enable a single battery charge for electric cars to sustain a driving range far longer than that of vehicles powered by internal combustion engines, with the substitution of battery charging stations for gas stations on the highways. Through the grid-integrated operation of hundreds of weather satellites around the world, the light intensity and wind speed at a given time and place can be accurately determined to maximize access to energy from nature. The advancement of technologies, including plasma and nanomaterials, will thoroughly reshape the energy industry by enabling the manufacture of panels more efficient in photoelectric conversion, more powerful wind turbines, lighter electric vehicles, and transmission lines with lower line loss. Accompanied by the sustainable development of the global energy interconnection and with the application of green energy, intelligent network, energy conservation, environmental, and other technologies, a wealth of innovative achievements in energy, information, materials, manufacture, biology, environmental protection, and other areas of science and technology, will emerge to precipitate the birth and growth of the third industrial revolution.

Driving innovation in business models. The global energy interconnection is a carrier of not just electric power and energy, but also information, services, technology, and civilization, with the potential to create a new economic model. The global energy interconnection will make possible an extensive body of business practices surrounding energy and electric power, giving rise to an abundance of new approaches to value creation. Massive electricity and information flows will run on a global scale, while billions of people and countless mobile phones, electric cars, and buildings will go online simultaneously on the same platform, and hundreds of millions of smart devices will be "plug and play" ready. Then on the base of big data, a full range of integrated services can be provided for a variety of energy-consuming equipment. At times of low tariffs, some electrical devices will automatically turn on and even enter the energy storage mode. Various energy projects will be available for investment to focus dispersed funds on the large-scale development of new energy. Energy and electric power will become a business enterprise of public interest. Constantly fluctuating electricity tariffs will become a medium for exchanging wealth and trading resources. Grid operators will not only sell electricity to users, but also help users save energy, and even generate electricity to obtain more economic benefits, which will bring about changes in the provision and marketing models of grid-based services.

Driving institutional and system innovations. In the next decades, a new technological revolution and a new round of industry reform will cross paths with the progressing human society in a historic encounter. The global energy interconnection, as an emerging energy infrastructure, will catalyze a new range of emerging industries and bring about transitions in the way of allocating resources, energy market operation mechanisms and regulatory system models, which will precipitate a quick formulation of different laws and regulations, industry rules, and governance systems. These new institutions and systems will break through the institutional barriers hitherto constraining the development of productivity to reshape the system for energy and economic operations, promote the mobility and convergence of key elements of advanced production, and provide an institutional assurance for the formation and development of advanced productivity. Like a transport network, a communication system or the Internet, the global energy interconnection builds a platform for innovation in social organization and management to promote the coordinated development of all industries.

2.3 ENCOURAGING WIN–WIN COOPERATION TO PROMOTE COORDINATED DEVELOPMENT OF WORLD ECONOMY

The global energy interconnection will free all elements of economic development in terms of industry structure and population distribution from energy constraints to help establish an energy-oriented, win–win global value chain and greatly enhance social productivity and labor productivity.

Promoting shared development of all economies. The global energy interconnection will spur economic change and energy transition. Different from the previous industrial revolutions, the latest transition is not initially of a local nature, but is driven by collaborative development on a global scale. The large-scale development of renewable energy around the world will turn the African and Latin American countries into energy providers in the world. This will not only solve the problem of local energy supply, but also transform the advantages of hydro, wind, and solar energy resources into an economic competitive edge. This will drive local socioeconomic development as a new source of growth in line with the world's economic expansion. The construction of the global energy interconnection will also stimulate investment, boost infrastructure development in all countries, create hundreds of millions of jobs, promote all-round development of the world economy, and significantly narrow the gap between the north and south and between the east and west for a more coordinated and balanced global economic structure.

Promoting integrated urban and rural development. By optimizing the global allocation of electric power resources, the global energy interconnection will promote integrated urban and rural development on a greater scale and substantially improve the capability for universal supply of electricity in India, Latin America, Africa and elsewhere, in fulfillment of the United Nations-advocated target of "Sustainable Energy for All." An abundant, reliable supply of electricity will ensure the normal operation of the infrastructure, transport, production, and other facilities that billions of people rely on for their daily living. The development of rural housing, education, healthcare, and other services will also be improved for the greater benefit of local residents.

Promoting integrated development of industries. The global energy interconnection will be deeply ingrained into the social fabric, complementing and positively reinforcing the operations of the Internet and the Internet of Things. It will penetrate the fields of energy, construction, transportation, manufacturing, agriculture, and education, blurring industry boundaries and making cross-industry collaboration a rising trend. In the consequent industry restructuring process, many traditional industries will, by availing themselves of the opportunities created by the emerging global energy network as an energy information system, complete their own transformation, upgrade and reinvention to generate new added value, infuse new vigor and recreate new wealth. On the one hand, an integration of different industries will be promoted and the coordinated progress of new manufacturing models, such as intelligent manufacturing and innovative design, and new business models such as service outsourcing, e-commerce and online shopping will significantly accelerate the development of the manufacturing and service industries. On the other hand, it will spur the integration of upstream and downstream operations in the same industry chain. In the future, intelligent production lines, green home appliances, and electric vehicles will be linked by energy and information technologies to achieve the restructuring and integration of upstream and downstream operations in a value chain. In a sense, there will be no place for bystanders in the development of the global energy network as governments, enterprises, social organizations and even every individual will be both participants and practitioners in a process that will benefit people from all walks of life and all economic sectors.

3 CREATING A WONDERFUL NEW SOCIAL LIFE

The global energy interconnection represents a high-level integration of "energy flows, information flows and business flows," as an intelligent, automated, and network-based system for ensuring energy security. The development of the network will produce a profound impact on life and work as well as nature, making many dreams come true for the benefit of all mankind.

3.1 CHANGING PUBLIC LIFE AND IMPROVING HUMAN DEVELOPMENT

As an energy infrastructure that mankind relies on for survival and development, the global energy interconnection creates a more dynamic economic system to realize better development in the new energy network-based economy.

Let everyone enjoy smart life. The global energy interconnection will leave an indelible mark on all aspects of human life and work. On the production front, the better integration of information technology with the clean energy system will help achieve intelligentization of the entire energy system, so much so that sensors will be everywhere and instant transfer of information realized. The information technology systems will be integrated with the energy system to drive the operation of countless mechanical equipment and automatic production lines, forming the base for joint readjustment of the linked operations between buildings, cars, factories and power sources. The attributes of energy-consuming equipment will be changed from being a single, unifunctional unit to become part of a multifunctional interconnected grid system to eventually optimize the efficiency of the energy system. To improve human life, the global energy interconnection will provide unprecedented convenience for mankind. Intelligent operations will be made possible for all purposes, ranging from car driving, train operation and uninterrupted communication to air conditioners, microwave ovens, refrigerators, washing machines and other household appliances. Through remote control, we can turn on our air conditioners at home, prepare a delicious dinner. We can automatically control washing machines and other energy-consuming equipment to maximize user convenience and save energy costs. We can automatically turn on home energy storage devices, distributed power sources, electric vehicle charging, and other facilities to become one of the hundreds of millions of electricity suppliers. Through a common platform built on the global energy interconnection, ordinary families can perform energy management, purchase electricity on a mobile device, conduct multiutility meter reading, access integrated information services, and remotely control home appliances to fully enhance the intelligence level of daily life.

Let personalized needs be fully satisfied. In terms of energy usage, everyone can produce, share, control and customize energy through the global energy interconnection, making them energy users as well as energy producers. They can even have the freedom to choose between wind power from the Arctic and solar energy from the equatorial regions or between different electricity packages at different prices during different time periods. In the future, the demand for energy, commodities and employment will see a shift from an emphasis on economies of scale to a more diversified base. Through the global energy interconnection, tens of thousands of jobs can be switched from factories and offices to homes, changing the way of life and work and solving traffic congestion and other problems of an industrialized society.

Let people fully decide on what is healthy development for them. The new technological revolution triggered by the global energy interconnection and the application of the attainments achieved will accelerate the development of productivity, significantly enhance the human capability to understand and transform the world, and initiate changes in the ways of production, lifestyle and socialization as well as in social relationships. In the future, the automation and intelligence levels of social production will rise noticeably. Boring, repetitive jobs will be performed mostly by machines and intelligent networks; factories will be sparsely manned; automated factories, production lines, and driving will gradually become a reality. Workers will relieve themselves from the direct and repetitive labor process common with the manufacturing, agriculture, distribution, social service, and domestic worker sectors as well as from some dangerous occupations in coal mining, oil and gas transportation. As a result, they will have more time and energy to engage in their favorite creative pursuits, thereby creating conditions conducive to man's full and unrestricted development. They will lay a material basis and ensure more time to enjoy more choices in life.

3.2 INITIATING SOCIAL CHANGE FOR A HIGH-EFFICIENCY SOCIAL MODEL

The global energy interconnection is positioned as a hub of energy supply and demand and exchange, supported by a strong network of equipment and exponentially growing big data. It is an integrated platform to guide industry distribution, integrate different resources, and initiate social change.

More collaborative ways of social production. As the fundamental driving force behind a new industrial revolution, the energy revolution will have a decisive impact on social production. The first and second industrial revolutions were marked by the emergence of mechanized and automated manufacturing, respectively. The global energy interconnection is the very fundamental platform to carry a new industrial revolution, with an enabling role in promoting it. With an interactive and collaborative nature, the global energy interconnection is highly integrated with the digital and intelligent manufacturing sectors, playing an increasing role in promoting "distributed" production, which in a sense is precisely the continuation of the global energy interconnection in its highest development.

Enterprises are a form of social organization that consumes the lion's share of social resources and energy. In the future, interconnections among enterprises will create a bigger, higher-level network of smart production. Manufacturing modes will undergo profound changes, with optimized energy efficiency achieved in multitype, small-batch production.

More efficient form of social organization. The concept of energy transformation is also extended to the renewal and optimization of the entire social system, including economic structure and the way of life. Against a backdrop of the global energy interconnection, a flat organizational model that is more efficient is well-received by society as a whole, while an emerging network-based organization with mobility features will gradually replace the existing rigid hierarchical organization. The global energy interconnection will create a world in which there is greater equality between people, with division of labor more extensively implemented on a global scale. The time required for the transition of new sciences, new technologies and new ideas to new products, new services and new applications will be shortened, with lower conversion costs. For individuals, the way of life is distinctly characterized by high "virtual" centralization but a dispersed geographical presence in reality. People will see themselves in a more relaxed working environment and a more efficient social organization.

Smarter social operating system. By 2050, smart grids will be deeply integrated with the Internet of Things and the Internet, providing more diversified services in energy supply, industrial monitoring, information and communication, home healthcare, logistics and transportation, distance learning, e-commerce, and other areas. This will facilitate resources sharing and multi-industry collaboration to open up a broad vista of applications. In the area of urban management, urban dwellers who account for two-thirds of the world population will enjoy unrivalled convenience, thanks to ubiquitous grids and communication systems that provide information covering almost everything, ranging from important matters like meteorological monitoring, urban energy use and fluctuating economic growth, to trivialities like whether a streetside rubbish bin is filled to capacity. The information will allow urban administrators to perform real-time analysis, processing and decision-making. On the healthcare front, human sensors will remain in contact on a real-time basis with network doctors for all-weather protection of people's health, and even perform emergency surgery by operating a mechanical scalpel through remote control. To facilitate everyday life, intelligent eyewears, watches, and other wearable devices will have powerful functions. People can activate a program using language, action and even mind power to browse through news, start vehicles, order meals, and perform other tasks. Computerized electric machines can be used to automatically perform farming, irrigation, seed sowing, fertilization, lighting, and other tasks to allow producers to control crop growth precisely with a minimum consumption of energy and water. In business operations, marketing, and logistics systems will be highly integrated while unmanned transportation will perform quick delivery and automatic battery charging. In the field of transportation, autopiloted cars will always be connected to the global energy interconnection and the global positioning system; hidden chips on the roads for real-time traffic monitoring will assure traffic safety so that traffic accidents will be substantially reduced. Sensors located everywhere and integrated with smart grids will promptly capture locations, speed, altitude, vibration, temperature, humidity, pressure, air quality, and other information, so that the whole society is in a new-found physical state that is measurable, knowable, and controllable.

3.3 IMPROVING THE NATURAL ENVIRONMENT FOR ECOLOGICAL SUSTAINABILITY

The global energy interconnection will dramatically increase the share of clean energy in global energy consumption, with significantly lower emissions of different greenhouse gases and pollutants to resolve environmental issues. The world will be a better place with green mountains, clear water, and blue skies; human beings will be in a "new normal" stage where they can enjoy the fruits of a green culture.

Climate change under control. The global energy interconnection will fundamentally resolve the problem of global climate. By 2050, the rapid development of renewable energy will bring new carbon dioxide emissions around the world down to approximately 1 trillion tons, or just half the level recorded in 1990, and the increase in global temperatures will be limited to 2°C. Threats to food production, water resources, ecology, urbanization, and people's lives and property will be reduced significantly; the risks of declining yields of rice, wheat, corns, and other major crops in some areas can also be eliminated. Accelerated melting can be prevented on Greenland, the Antarctic Ice Sheet, and plateau glaciers; sea level rise will be halted; small island states and low-lying coastal areas will not be flooded; and huge investment in dam construction can be spared in hundreds of the world's densely populated coastal cities so that hundreds of millions of people need not leave their homes to escape from seawater intrusion. Melted snow water on the Kunlun Mountains will continue to nourish the desert oasis in Xinjiang; the subsistence of the glaciers on the Himalayas will assure that drinking and irrigation water will continue to be available to hundreds of millions of people on both sides of the Ganges. Extreme weather events such as droughts, floods, cyclones, storms, heat waves, or cold waves will be reduced; and the severity of disasters such as mudslides and landslides will also be significantly reduced. As estimated by the World Health Organization, millions of lives can be saved every year as a result of lower greenhouse gas emissions from 2030 onward.

Ecological environment can be restored. Numerous energy and ecological concerns that have plagued humanity can be solved simply by the global energy interconnection. The scale of production, transmission, and consumption of traditional fossil energy will be reduced. Problems such as the surface subsidence, mining disasters, water percolation, explosions, and smoke emission caused by the mining, processing, transportation, storage, and burning of coal will be mitigated. The impact of oil and gas exploitation, transportation, and utilization on physical geology and terrestrial and marine ecosystems will be increasingly reduced. Forests and wetlands will continue to expand, and biodiversity will be protected and restored. Consumers living in densely populated or industrial cities can breathe quality air, drink quality water and eat quality food everywhere. Consumers will no longer have to pay for the costs of the prevention and control of pollution caused by fossil energy, much of the spending on medical treatment can be diverted to health care and fitness training, and quality of life will be improved and the average life expectancy will be extended. The negative impact of environmental pollution on public health will be mitigated at source.

Resource consumption significantly reduced. The global energy interconnection will steer the world away from the exploitation of fossil energy to the production of clean energy that will not result in problems of resource depletion, lowering quality or increasing prices. This is yet another prominent feature of the new renewable energy revolution that makes it different from the previous two industrial revolutions. Under the new industrial system, clean, efficient and environment-friendly energy will become the key focus of pursuit among all nations, which will contribute to a change in the human concept of production. According to the 2012 report of the World Wide Fund for Nature, if nothing is done to prevent the Earth's natural resources from continued depletion, mankind will need to consume 2.9 times the Earth's resources by 2050. Through the development of the global energy interconnection to replace fossil energy with renewable energy, the long-standing high-consumption, high-pollution mode of production, as well as high-spending, high-waste lifestyles will change. The current situation of resource consumption-led economic growth will be radically altered. The major economies around the world will be able to support sustainable economic prosperity in a world of bluer sky, more greenery, and clearer water, simply by replying on just a quarter of the level of raw materials consumed before the development of the global energy interconnection.

4 TURNING A NEW CHAPTER OF CIVILIZATION

As a major innovation with a profound impact on life and work in the twenty-first century, the global energy interconnection will change not only the way of energy development, but also the world's geopolitical situation. It will foster the concept of green culture, enhance rational knowledge of, and broaden the mind on environmental issues, and promote the continued progress of human civilization.

4.1 PROMOTING POLITICAL HARMONY AND WORLD PEACE

Energy is a base for modernization. The mechanism of energy shapes the essence of civilization. It determines the way a civilization is structured, business and trade results are distributed, and political forces work. It also guides the creation and development of social relationships.

Global energy interconnection becomes the bond that unites the world. It transforms the long-standing basis for international energy order, changes the world's geopolitical situation, and promotes the development of a multipolarized world. Different nations around the world are bounded by a common interest in new energy development and utilization, demonstrating ever-higher levels of interdependence and mutual cooperation and reinforcement among themselves. Political, military, and diplomatic contradictions and the potential of conflict arising from the competition for energy resources will be effectively controlled; geopolitical tensions in the Middle East, North Africa, and other regions will be eased. Oil and natural gas reserves in the Arctic region will no longer be the focus of international disputes, with Arctic wind to become the shared treasure of all mankind. The global shift from individual security to collective security, together with the expanding international collaborative efforts in other areas based on cooperation in resource, will effectively promote world peace.

Global energy interconnection becomes the cornerstone of social stability. The pursuit of a happy life is the most enduring source of strength for promoting the progress of human civilization. The abundant energy resources brought about by the global energy interconnection will improve productivity and support higher incomes, greater social security, better medical and public health services, more habitable living conditions, and a greener natural environment. Many factors of social instability under the traditional fossil energy system will be largely removed. As warned by the Asian Development Bank in its Report on Energy Outlook for Asia and the Pacific published in October 2013, unless the current overdependence on oil imports is reduced, the efficiency of electricity usage improved and greener energy adopted, the gap in energy terms between the Asia–Pacific region's rich and poor will widen. The global energy interconnection will induce significant structural change at all levels of the community, contributing to the easing of social conflicts. The global energy interconnection lays the groundwork for shared development by engendering international efforts to achieve balanced distribution and exchange of the resources required for human survival so as to ensure human sustainability.

Global energy interconnection serves as a platform to bring together common interests of mankind. On the global energy interconnection, the mechanism to transmit in a split second energy and information from afar will play an important role. As matters of common interest among all mankind, the world is deeply concerned about how to address global warming, seek ecological improvements, improve social well-being, and achieve peaceful development. These common interests have transcended all social systems, development levels as well as ethnic, religious and ideological differences. Mankind is now bounded by a common destiny, with equal rights for all to survival and development. Linkages underlined by common interests will be developed as an integral part of an interdependent relationship. The depth, breadth and speed of this change will be unprecedented, so much so that any nation developing and moving forward must consciously try to integrate itself into the bigger global picture where development and progress by one's own efforts will no longer be feasible. The global energy interconnection will eliminate the energy gap, laying a dynamic foundation for a world without waste, poverty, and wars. This new platform will accord various international organizations an increasingly important role, while human society will see the establishment of a better system for sharing interests.

4.2 PROMOTING ENVIRONMENTAL HARMONY AND GREEN CULTURE

Through the global energy interconnection, people will be deeply involved in the production and consumption of green energy to become the fundamental driving force behind a greener world, with a greater awareness of the importance of living a greener lifestyle and adhering to the green concept.

Green values established. For nearly 300 years, the industrialization process in the world has enabled developed countries, with a total population of about 1 billion, to enjoy a higher level of modernization. But a heavy price has been paid in terms of global resources and ecology. In the future, developing countries, including China, cannot follow this traditional economic growth model in the pursuit of modernization. The global ecosystem is a systematic, complete, and organically connected whole. Transcending industrial civilization, green culture represents a more advanced form of human civilization. The twenty-first century marks the transition of human society from industrial civilization to a new green civilization. The global energy network encourages mankind to look at nature from a systematic and holistic perspective. It promotes a sense of responsibility toward and an awareness of cherishing and protecting the environment. It puts emphasis on upholding the values of the global ecosystem for people, nature, and society and on the realization of synergistic economic, political, cultural, and ecological development.

Energy conservation and emission reduction will become a common path to take. Far exceeding the reasonable needs of humanity, the level of material consumption during the industrial age went beyond the support capacity of nature to the extent of destroying ecological balance. The development of the global energy interconnection symbolizes a complete change in the mode of production and the way of life by ushering in a green, low-carbon society. While continuing to improve quality of life, people will abandon the relentless pursuit of material desires and gradually develop a concept of moderate consumption and resource conservation. A new concept of consumption that will win more appreciation is one that is built on a healthy respect for nature and the idea of living with nature and protecting the ecosystem and lives, and one that does not exceed the world's resource and environmental carrying capacity. Intensive efforts will be devoted to change high-consumption, high-emission ways of production and life not compatible with the low-carbon economy. A scientific and healthy culture and mode of consumption will also be fostered and developed. In the future, whenever we buy a product, enjoy a service, or complete a trip, we will have access to accurate information and be able to control energy consumption on a timely basis. Everyone can manage their energy bills with precision and sees this as a virtue.

Sustainability is a deeply rooted concept. Sustainable development is a strategy of coordinated development between people, between people and society as well as between people and nature. What is achieved now will benefit many generations to come. The global energy interconnection is guided by a global view of energy, with greater emphasis on the utilization of renewable energy and a commitment to the sustainability concept. It will pave the way for a harmonious relationship between people and nature and show the way forward for the development of human society. The global energy interconnection is shaping a sustainability model that is focused on resource conservation, quality and benefits, pioneering technology and environment-friendliness. It promotes green, low-carbon, energy-efficient, and environment-friendly industries to become a new driver of sustainable economic growth, ultimately encouraging mankind to break down the boundaries of races and nations and transcend narrow personal and group interests to share responsibilities and uphold the ideals of equal cooperation – all toward the goal of protecting the earth and building a better world we can call home.

4.3 PROMOTING HUMAN HARMONY AND CULTURAL UPGRADE

The development of the global energy interconnection promotes not only greater improvements to the material wealth and welfare of mankind, but also a more open and equal relationship between people for creating greater harmony on an interpersonal level and between people and nature, in order to achieve the transformation and upgrade of human civilization.

Enhancing cognitive thinking. The global energy interconnection promotes an upgrade of the way of thinking and conceptualization. The first is the ability to think holistically. As we are moving into a new era of global energy interconnections, people are reconciling, with an unprecedented rigor, energy and information, time and space, virtuality and reality, production and consumption, equality and efficiency, competition and cooperation, software and hardware, and any other two elements that are mutually incompatible or are binary in nature. In this way, human civilization is constantly developing and in greater depth. The second is the development of open, innovative thinking. The global energy interconnection is a highly open system and a strong driving force behind industry development to give rise to new technologies as well as new business types and models. The continued emergence of different convergent and breakthrough innovations will create a huge impact on industry economics, social politics and the world situation. People will be driven to think in tune with the times while constantly breaking the rules to become more open and inclusive.

Restructuring knowledge system. Knowledge is power. At a time of global energy interconnection, man's boundaries of knowledge will constantly be pushed forward, with an ever-growing depth of knowledge. In a sense, the basis for promoting the world economy will switch from one oriented toward material resources to one more focused on knowledge resources. In the future world of energy, the continued emergence of multidisciplinary, cross-disciplinary, and interdisciplinary technologies based on UHV transmission, new energy generation, information network, and intelligent control technologies, will bring about an upgrade of mankind's existing system of energy knowledge. This, together with an enhanced capability to change the world based on natural sciences and social sciences, will produce a major impact on all aspects of socioeconomic development. In the future, knowledge and technology will account for a higher weighting in the development of the national economy, with greater emphasis among all countries being placed on education and technology research, cultivation of all-round talents, enhanced human creativity, and intellectual property rights, with the objective of gaining a first mover advantage.

Enhance spiritual civilization. Each substantial productivity leap in human history implies a leap forward in spiritual civilization. With a highly developed material civilization, human society will reap the fruits of success and achievement to better satisfy the spiritual needs of people and open up broader vistas for new thinking. *Human intelligence will continue to improve.* The global energy interconnection will facilitate international cooperation and sharing between people, enhance the motivation to participate in cooperative opportunities in energy and information, arouse awareness of a common destiny, and strengthen social cohesion. *Ethical standards will further rise.* The concept of global energy cooperation that transcends time and space will be widely accepted; the spirit of contract and a sense of responsibility will impel people to commit to the principle of shared responsibilities and shared interests during the development and utilization of resources, with a new atmosphere of cooperation and integrity fostered. *Civilization will be further diversified.* Against a background of globally concerted development, the civilizations of different peoples, countries and regions will be respected and their characteristics accepted as they are, with sharing and exchange activities conducted to lift civilization to a higher level.

SUMMARY

1. The global energy interconnection is driving a change in the mode of energy development. By freeing energy development from the resource, time, space and environmental constraints, efficient exploitation, and utilization of clean energy will be realized and clean energy promoted as the dominant energy option, with access for all to an adequate supply of energy.
2. The global energy interconnection is driving a change in the mode of economic development. It promotes a shift to a growth model focused on innovation, full coordination, and quality enhancement so as to inject new vigor and propel the world economy into a new era of prosperity.
3. The development of the global energy interconnection has a profound impact on the way of production and life. Significant improvements in development quality, the efficiency of social management and environmental well-being will help create a new, better life for mankind.
4. With the help of the global energy interconnection, civilization will reach a higher level, marked by global scenes of political, ecological, and human harmony and turning a new chapter of world civilization.
5. Looking to 2050, the global energy interconnection will basically be completed, a development that will not only provide a strong impetus to the rejuvenation of the Chinese nation, but also make an outstanding contribution to the sustainable development of the world economy.

References

[1] Jinping X. The governance of China. Beijing: Foreign Languages Press; 2014.

[2] Zemin J. Research on China's energy problem. Shanghai: Shanghai Jiao Tong University Press; 2008.

[3] Peng L. Electric power must come first, Peng Li's power diary. Beijing: China Electric Power Press; 2005.

[4] Guobao Z. Report on China's energy development 2010. Beijing: Economic Science Press; 2010.

[5] Zhenya L. Electric power and energy in China. Beijing: China Electric Power Press; 2012.

[6] Zhenya L. Ultra-high voltage AC/DC grids. Beijing: China Electric Power Press; 2013.

[7] Zhenya L. Smart grid technology. Beijing: China Electric Power Press; 2010.

[8] Zhenya L. A serial publication on UHV DC technology. Beijing: China Electric Power Press; 2009.

[9] Zhenya L. UHV questions and answers. Beijing: China Electric Power Press; 2006.

[10] Zhenya L. Innovation of UHV AC transmission technology in China. Power System Technology; 2013;37(3): 566–574.

[11] Zhenya L. Smart grid and the Third Industrial Revolution. Science Daily; December 5, 2013.

[12] Zhenya L, Qiping Z, Cun D, et al. Efficient and security transmission of wind, photovoltaic and thermal power of large-scale energy resource bases through UHVDC projects. Proc. CSEE 2014;16.

[13] Zhenya L. Global energy Internet. IEEE Spectr. 2014;10:54–56.

[14] China Electric Power Encyclopedia Editorial Board. China electric power encyclopedia. 3rd ed. Beijing: China Electric Power Press; 2014.

[15] National Bureau of Statistics of China. China statistical yearbook 2013. Beijing: China Statistics Press; 2013.

[16] Department of Energy Statistics, National Bureau of Statistics of China. China energy statistical yearbook 2013. Beijing: China Statistics Press; 2013.

[17] National energy strategy towards 2030 following the scientific development approach. Policy Planning Department of National Energy Administration; 2012.

[18] 2013 State of environment report review. Ministry of Environmental Protection of the People's Republic of China; 2014.

[19] Development Research Center of the State Council, Shell International Limited. Research on China's medium and long term energy development strategy. Beijing: China Development Press; 2013.

[20] China Electricity Council, Environmental Defense Fund. The current status of air pollutant control for coal-fired power plants in China 2009. Beijing: China Electric Power Press; 2009.

[21] Compilation of statistical materials of electric power industry 2013. China Electricity Council; 2014.

[22] Global new energy report 2014. ACFIC New Energy Chamber of Commerce; 2014.

[23] Hefner III RA. The grand energy transition [Yuanchun M, Boshu L, Trans.]. Beijing: China Citic Press, Citic Publishing House; 2013.

[24] Yergin D. The quest: energy, security, and the remaking of the modern world [Yuben Z, Zhimin Y, Trans.]. Beijing: Petroleum Industry Press; 2012.

[25] Rifkin J. The Third Industrial Revolution: how lateral power is transforming energy, the economy, and the world [Tiwei Z, Yuning S, Trans.]. Beijing: China Citic Press, Citic Publishing House; 2012.

[26] Montgomery SL. The powers that be: global energy for the twenty-first century and beyond [Yang S, Wenbo J, Trans.]. Beijing: China Machine Press; 2012.

[27] Mackay DJC. Sustainable energy – without the hot air [Ju Z, Meng D, et al., Trans.]. Beijing: Science Press; 2013.

[28] Botkin DB, Perez D. Powering the future [Mu C, Trans.]. Beijing: Publishing House of Electronics Industry; 2012.

[29] Scheer H. The energy imperative: 100 percent renewable now [Qiankun W, Trans.]. Beijing: Posts & Telecom Press; 2013.

[30] Lovins A, Rocky Mountain Institute. Reinventing fire: bold business solutions for the New Energy Era [Haiyan Q, China General Certification Center, Trans.]. Changsha: Hunan Science & Technology Press; 2014.

Global Energy Interconnection. http://dx.doi.org/10.1016/B978-0-12-804405-6.00009-9

[31] McElroy MB. Energy: perspectives, problems and prospects [Yuxuan W, Jiming H, Xi L, Trans.]. Beijing: Science Press; 2011.

[32] Shenghai Y. Energy war. Beijing: Peking University Press; 2012.

[33] Hongyu Z, Li X. The Third Industrial Revolution and modern China. Wuhan: Hubei Education Press; 2013.

[34] Chongshan D. Dilemma and breakthrough: mankind's energy crisis and the way out. Beijing: People's Publishing House; 2006.

[35] Chuan-kun W, Wei L. Analysis methods and reserves evaluation of ocean energy resources. Beijing: China Ocean Press; 2009.

[36] Asarin AE, Qiuyun Z. Water Resources of Russia and their Use. Express Water Resour. Hydropower Inform. 2008;29(5).

[37] Xinzhong F. Fact-finding Report on Exploitation and Utilization of Water Resources in the US and Canada. China Hydropower Electrification 2008;(4):23–28.

[38] Haiying W, Xiaomin B. Assessment on generation adequacy and capacity credit for large-scale wind farm. Power Syst. Technol. 2012;36(6):200–206.

[39] Xiaosheng Y. Wind power technology and wind farm project. Beijing: Chemical Industry Press; 2012.

[40] Mathew S. Wind energy fundamentals, resource analysis and economics [Fengfei X, Trans.]. Beijing: China Machine Press; 2011.

[41] Xing Z, Renxian C, et al. Taiyangneng Guangfu Bingwang Fadian Jiqi Nibian Kongzhi. Beijing: China Machine Press; 2011.

[42] Daoqing H, Tao H, Honglin D. Taiyangneng Guangfu Fadian Xitong Yuanli Yu Yingyong Jishu. Beijing: Chemical Industry Press; 2012.

[43] Suyi H, Shuhong H, et al. The principle and technology of solar thermal generation. Beijing: China Electric Power Press; 2012.

[44] Xiang H. Solar thermal generation technology. Beijing: China Electric Power Press; 2013.

[45] Guangfu T, Xiang L, Xiaoguang W. Multi-terminal HVDC and DC-grid technology. Proc. CSEE 2013;33(10):8–17.

[46] Yushuang W. Review on submarine cable projects for power transmission worldwide. Southern Power Syst. Technol. 2012;2:26–30.

[47] Luguang Y, Liye X, Liangzhen L, et al. A proposal for development of high-voltage, long-distance, high-capacity high temperature superconducting power transmission. Adv. Technol. Electric. Engineer. Energy 2012;31(1):1–7.

[48] Worzyk T. Submarine power cables: design, installation, repair, environmental aspects [Qiliang Y, Xiaofeng X, Jiansheng S, Trans.]. Beijing: China Machine Press; 2011.

[49] Zheng X. HVDC system based on voltage source converter. Beijing: China Machine Press; 2013.

[50] Minxiao H, Jun W, Yonghai X. Principle and applications of HVDC transmission system. 2nd ed. Beijing: China Machine Press; 2013.

[51] Yinshun W. Basics of superconducting power technology. Beijing: Science Press; 2011.

[52] Jianhua Z, Wei H. Weidianwang Yunxing Kongzhi Yu Baohu Jishu. Beijing: China Electric Power Press; 2010.

[53] Qian A, Zhiyu Z. Distributed generation and smart grids. Shanghai: Shanghai Jiao Tong University Press; 2013.

[54] Barnes FS, Levine JG, et al. Large energy storage systems handbook [Xi X, Zanxiang N, et al., Trans.]. Beijing: China Machine Press; 2013.

[55] Jialiang W, Gansheng C, Tiehui Y, et al. Fengguang Hubu Yu Chuneng Xitong. Beijing: Chemical Industry Press; 2012.

[56] Xiu Y, Hongzhong L, Jingjing Z. Basics of distributed generation and energy storage technologies. Beijing; 2012. Available from: www.waterpub.com.cn.

[57] Yuejin T, Jing S, Li R. Superconducting magnetic energy storage system (SMES) and its application in electric power systems. Beijing: China Electric Power Press; 2009.

[58] Hongzhong L, Jianmin D, Chengmin W. Zhineng Dianwang Zhong Xudianchi Chuneng Jishu Jiqi Jiazhi Pinggu. Beijing: China Machine Press; 2012.

[59] Wei S, Shinian L, Guobin Z, et al. Chemical energy storage technology and its application in electric power systems. Beijing: Science Press; 2013.

[60] Zipei T. The Big Data Revolution. Guilin: Guangxi Normal University Press; 2013.

[61] Zhenghong Y. Big Data, the Internet of Things and Cloud Computing. Beijing: Tsinghua University Press; 2014.

[62] Mayer-Schönberger V, Cukier K. Big Data [Tao Z, et al., Trans.]. Hangzhou: Zhejiang People's Publishing House; 2013.

[63] Keiser G. Optical fiber communications, 4th ed. [Tao P, Junhua X, Yang S, Trans.]. Beijing: Publishing House of Electronics Industry; 2012.

[64] Zhenya L. Smart grid in China: development and practice. Paper C1-203, CIGRE Session 2014. Paris, France; August 2014.

[65] BP statistical review of world energy 2014; 2014.

[66] UN world energy statistics yearbook 2013; 2014.

[67] IEA energy balances of non-OECD countries 2014; 2014.

[68] IEA energy balances of OECD countries 2014; 2014.

[69] IEA electricity information 2014; 2014.

[70] IEA world energy outlook 2014; 2014.

[71] IEA CO_2 emissions from fuel combustion 2013; 2013.

[72] IEA CO_2 emissions from fuel combustion highlights 2014; 2014.

[73] IEA energy prices and taxes Q2 2014; 2014.

[74] IEA energy supply security: emergency response of IEA countries 2014; 2014.

[75] IMF world economic outlook 2014; 2014.

[76] WEC world energy scenarios: composing energy futures to 2050; 2013.

[77] WEC world energy resources: 2013 survey; 2013.

[78] WEC world energy perspective smart grids: best practice fundamentals for a modern energy system; 2012.

[79] WNA world nuclear power reactors & uranium requirements; 2013.

[80] IAEA uranium 2014 resources, production and demand; 2014.

[81] GWEC global wind report 2013; 2014.

[82] IPCC climate change 2014 impacts, adaptation, and vulnerability: summary for policymakers; 2014.

[83] IPCC renewable energy and climate change; 2011.

[84] IPCC special report on renewable energy resources and climate change mitigation; 2011.

[85] IRENA estimating the renewable energy potential in Africa; 2014.

[86] IRENA southern African power pool: planning and prospects for renewable energy; 2014.

[87] REN21 renewables 2014 global status report; 2014.

[88] NREL eastern wind integration and transmission study; 2010.

[89] NREL US renewable energy technical potentials: a GIS-based analysis; 2012.

[90] EPIA global market outlook for photovoltaics 2014–2018; 2014.

[91] BTM consult world market update 2013; 2014.

[92] US energy information administration international energy outlook 2014; 2014.

[93] EU energy road map 2050; 2012.

[94] EUenergy, transport and GHG emissions trends to 2050 reference scenario 2013; 2014.

[95] EU renewable energies in Africa final report 2011; 2011.

[96] European environmental agency Europe's onshore and offshore wind energy potential: an assessment of environmental and economic constraints; 2009.

[97] ENTSO-E statistical factsheet; 2013.

[98] IEEE power system operations committee organization and procedures; 2009.

[99] Vasil'ev VA, Saparov MI, Tarnizhevskii BV. Possibilities of use and promising layouts of power plants employing renewable energy sources in Arctic regions of Russia. Power Technol Engineer; 2005;39(5):308–311.

[100] Boute A, Willems P. RUSTEC: greening Europe's energy supply by developing Russia's renewable energy potential energy policy; 2012.

[101] Elliott D. Emergence of European super grids: essay on strategy issues energy strategy reviews; 2013.

[102] Conner AM, Francfort JE, Rinehart BN. DOE US hydropower resource assessment final report; 1998.

[103] Chen H, Cong TN, Yang W, Tan C, et al. Progress in electrical energy storage system: a critical review. Prog. Nat. Sci. 2009;19:291–312.

[104] de Wild-Scholten MJ. Energy payback time and carbon footprint of commercial photovoltaic systems. Sol. Energ. Mat. Sol. C. 2013;119:296–305.

[105] Hydropower of World's Major Rivers; Hydroelectric Generation, January 1957.

[106] Investigation Results for Hydropower in the People's Republic of China; National Leading Group for Nationwide Hydropower Review, 2005.

[107] Asselin, Russian Water Resource and Utilization; Hydropower Engineering Construction, June 2007.

[108] GWEC, Annual Market Update 2013.

[109] Japan Electric Power Information Center, Statistics of Overseas Electric Industries 2013.

[110] IEA, 2014 Key World Energy Statistics.

Postscript

The discovery and application of electricity is one of the greatest achievements in human history, accelerating the progress of global civilization at an unprecedented rate and giving economic and social development a powerful thrust. I started my career in China's electric power industry in the 1970s, and from 2000 onward, I worked in State Power Corporation and State Grid Corporation of China. I have been the President of China Electricity Council since late 2009. My 40 years of experience in electricity research, construction, and management have infused me with a passion for work related to energy and electric power. I continue to study, explore, and think about its future growth, and my knowledge is constantly deepened and perfected.

In both the past and present, energy and electricity are important factors that affect long-term economic and social development. By the early years of twenty-first century, the continuous growth in global energy production and consumption, and the massive development and utilization of fossil fuels have resulted in global problems such as resources depletion, environmental pollution, and climate change, problems that pose a serious threat to human survival and development. Faced with such tough challenges, my basic argument is this: the mode of energy development based on traditional fossil fuels is no longer sustainable, and the major trend is for renewable energy to totally replace fossil fuels. This trend should be on everybody's radar screen and actively promoted.

As the world's largest developing country, as well as the world's biggest energy producer and consumer, China's energy and environmental problems are quite acute. Problems such as smog, coal transportation, and oil and gas supply have become increasingly serious in recent years. Firmly rooted in China's national circumstances and energy resources, State Grid Corporation of China proposed the "One Ultra and Four Large" strategy, and the electricity replacement strategy of "replacing coal with electricity, replacing oil with electricity, and delivering clean electricity from afar." The idea is to expedite the construction progress of a robust smart grid system supported by backbone UHV grids and well-coordinated development of grids at different levels to promote optimized allocation of energy resources on a nation-wide basis. Stellar results have been achieved and the successful implementation of the strategies confirms the importance and feasibility of developing UHV grids and clean energy.

My research and practice in the energy and electricity industry in China have inspired deeper thoughts into the global energy problem. I believe that climate change is a long-term problem that affects global energy development and human survival and growth both in the present and future, and the emission of greenhouse gases through the use of fossil fuels is a problem that cannot be ignored. The only way out of the climate change conundrum is the development of green, low-carbon energy; the fundamental path is substitution of clean energy and electric power; the key approach is one of reliance on UHV grids to create a global energy interconnection to leverage its important role in optimizing energy resources on a large scale. This will bring about the massive development and use of clean energy on a global scale, ensuring that everyone will enjoy a sufficient supply of clean energy. This is the only way forward to attain sustainable development of the world's energy resources.

The technological and engineering foundations are already in place for the creation of a global energy interconnection. Developments in UHV grids and clean energy technology, in particular, have created excellent conditions for the interconnection. In recent years, SGCC has completed three UHV AC and four UHV DC projects. The operational safety and stability of these projects over a long period of time ensures the development of large-scale energy bases, and the long-distance, large-scale,

Global Energy Interconnection. http://dx.doi.org/10.1016/B978-0-12-804405-6.00010-5

and high-efficiency transmission of electricity in China. The technology for the UHV, smart grids, and clean energy mentioned in this book is based on current levels of science, and it is real and feasible. It is also the basis for developing a global network of energy interconnections.

The international community has been very enthusiastic about the global energy interconnection concept. SGCC's research findings were very well received after the company presented them in 2014 in various platforms such as the Institute of Electrical and Electronics Engineers Power and Energy Society (IEEE PES) General Meeting, United Nations Climate Summit, Accenture Global Energy Board meeting, the first CEO Council Meeting for Sustainable Urbanization, IEEE Popular Science Forum, and Forbes magazine. From developing large-scale energy bases in the Arctic and equatorial regions and elsewhere to constructing a global UHV grid system, from promoting the wide application of smart grids to inspiring innovation in energy and power technology, both developed and developing countries have common talking points, common visions, and common interests. Most people believe that the construction of a global energy network and the realization of the "two-replacement" policy are the fundamental solutions to the problems of global energy resources and climate change.

I am also aware that energy issue is global and extensive; it is the focus around which international politics, economics, diplomacy, national defense, and the climate change game are played out. No single country can solve the problem on its own. On the one hand, the creation of a global energy interconnection is a historic opportunity for the accelerated development of renewable energy. On the other hand, there are huge challenges in geopolitics, economic interests, social environment, energy policies, market creation, and technological innovation. The process will not be smooth sailing all the way, nor will it be quick. It will be a difficult and arduous journey. Still, the first step must be taken. As long as we work together with a common purpose, the global energy interconnection will be a great success. I have confidence and I expect that this will come to pass. This is also my reason and motivation for writing this book.

Given the limitations of my knowledge about energy, this book contains many imperfections. Its purpose is to inspire and invite feedback from experts, academics, and colleagues, as well as to gather different strands of wisdom and form a consensus, so that more interested parties will dedicate themselves to the great enterprise of the global energy interconnection. Many leaders, experts, and colleagues rendered their assistance during the writing of this book, for which I wish to express my heart-felt gratitude.

Zhenya Liu
January 2015

Index

365

Printed in the United States
By Bookmasters